WHO FOOD ADDITIVES SERIES: 67

Safety evaluation of certain food additives

Prepared by the
Seventy-sixth meeting of the Joint FAO/WHO
Expert Committee on Food Additives
(JECFA)

World Health Organization, Geneva, 2012

WHO Library Cataloguing-in-Publication Data

Safety evaluation of certain food additives / prepared by the Seventy-sixth meeting of the Joint FAO/WHO Expert Committee on Food Additives (JECFA).

(WHO food additives series ; 67)

1.Food additives - toxicity. 2.Flavoring agents - analysis. 3.Flavoring agents - toxicity. 4.Risk assessment. I.Joint FAO/WHO Expert Committee on Food Additives. Meeting (76th : 2012 : Geneva, Switzerland). II.World Health Organization. III.Series.

ISBN 978 92 4 166067 9   (NLM classification: WA 712)
ISSN 0300-0923

© **World Health Organization 2012**

All rights reserved. Publications of the World Health Organization are available on the WHO web site (www.who.int) or can be purchased from WHO Press, World Health Organization, 20 Avenue Appia, 1211 Geneva 27, Switzerland (tel.: +41 22 791 3264; fax: +41 22 791 4857; e-mail: bookorders@who.int).

Requests for permission to reproduce or translate WHO publications—whether for sale or for non-commercial distribution—should be addressed to WHO Press through the WHO web site (www.who.int/about/licensing/copyright_form/en/index.html).

The designations employed and the presentation of the material in this publication do not imply the expression of any opinion whatsoever on the part of the World Health Organization concerning the legal status of any country, territory, city or area or of its authorities, or concerning the delimitation of its frontiers or boundaries. Dotted lines on maps represent approximate border lines for which there may not yet be full agreement.

The mention of specific companies or of certain manufacturers' products does not imply that they are endorsed or recommended by the World Health Organization in preference to others of a similar nature that are not mentioned. Errors and omissions excepted, the names of proprietary products are distinguished by initial capital letters.

All reasonable precautions have been taken by the World Health Organization to verify the information contained in this publication. However, the published material is being distributed without warranty of any kind, either expressed or implied. The responsibility for the interpretation and use of the material lies with the reader. In no event shall the World Health Organization be liable for damages arising from its use.

This publication contains the collective views of an international group of experts and does not necessarily represent the decisions or the policies of the World Health Organization.

Typeset in India

Printed in Malta

CONTENTS

Preface v

Specific food additives (other than flavouring agents) 1
    Magnesium dihydrogen diphosphate 3
    3-Phytase from *Aspergillus niger* expressed in *Aspergillus niger* 19
    Serine protease (chymotrypsin) from *Nocardiopsis prasina* expressed in
        Bacillus licheniformis 39
    Serine protease (trypsin) from *Fusarium oxysporum* expressed in
        Fusarium venenatum 51

Safety evaluations of groups of related flavouring agents 63
    Introduction 65
    Aliphatic and aromatic amines and amides (addendum) 69
    Aliphatic and aromatic ethers (addendum) 95
    Aliphatic linear α,β-unsaturated aldehydes, acids and related alcohols,
        acetals and esters (addendum) 117
    Amino acids and related substances (addendum) 133
    Epoxides (addendum) 147
    Furfuryl alcohol and related substances (addendum) 161
    Linear and branched-chain aliphatic, unsaturated, unconjugated alcohols,
        aldehydes, acids and related esters (addendum) 171
    Miscellaneous nitrogen-containing substances (addendum) 185
    Phenol and phenol derivatives (addendum) 201
    Pyrazine derivatives (addendum) 223
    Pyridine, pyrrole and quinoline derivatives (addendum) 245
    Sulfur-containing heterocyclic compounds (addendum) 277

Annexes
    Annex 1 Reports and other documents resulting from previous meetings
                      of the Joint FAO/WHO Expert Committee on Food Additives 303
    Annex 2 Abbreviations used in the monographs 315
    Annex 3 Participants in the seventy-sixth meeting of the Joint FAO/WHO
                      Expert Committee on Food Additives 317
    Annex 4 Toxicological information and information on specifications 319
    Annex 5 Summary of the safety evaluation of secondary components for
                      flavouring agents with minimum assay values of less than 95% 333

PREFACE

The monographs contained in this volume were prepared at the seventy-sixth meeting of the Joint Food and Agriculture Organization of the United Nations (FAO)/World Health Organization (WHO) Expert Committee on Food Additives (JECFA), which met at WHO headquarters in Geneva, Switzerland, on 5–14 June 2012. These monographs summarize the data on selected food additives, including flavouring agents, reviewed by the Committee.

The seventy-sixth report of JECFA has been published by the World Health Organization as WHO Technical Report No. 974. Reports and other documents resulting from previous meetings of JECFA are listed in Annex 1. The participants in the meeting are listed in Annex 3 of the present publication.

JECFA serves as a scientific advisory body to FAO, WHO, their Member States and the Codex Alimentarius Commission, primarily through the Codex Committee on Food Additives, the Codex Committee on Contaminants in Food and the Codex Committee on Residues of Veterinary Drugs in Foods, regarding the safety of food additives, residues of veterinary drugs, naturally occurring toxicants and contaminants in food. Committees accomplish this task by preparing reports of their meetings and publishing specifications or residue monographs and dietary exposure and toxicological monographs, such as those contained in this volume, on substances that they have considered.

The monographs contained in this volume are based on working papers that were prepared by WHO experts. A special acknowledgement is given at the beginning of each monograph to those who prepared these working papers. The monographs were edited by M. Sheffer, Ottawa, Canada.

The designations employed and the presentation of the material in this publication do not imply the expression of any opinion whatsoever on the part of the organizations participating in WHO concerning the legal status of any country, territory, city or area or its authorities, or concerning the delimitation of its frontiers or boundaries. The mention of specific companies or of certain manufacturers' products does not imply that they are endorsed or recommended by the organizations in preference to others of a similar nature that are not mentioned.

Any comments or new information on the biological or toxicological properties of the compounds evaluated in this publication should be addressed to: WHO Joint Secretary of the Joint FAO/WHO Expert Committee on Food Additives, Department of Food Safety and Zoonoses, World Health Organization, 20 Avenue Appia, 1211 Geneva 27, Switzerland.

SPECIFIC FOOD ADDITIVES
(OTHER THAN FLAVOURING AGENTS)

MAGNESIUM DIHYDROGEN DIPHOSPHATE

First draft prepared by

D.J. Benford[1], M. DiNovi[2], Y. Kawamura[3], J.-C. Leblanc[4], Madduri Veerabhadra Rao[5] and J.R. Schlatter[6]

[1] Food Standards Agency, London, England
[2] Center for Food Safety and Applied Nutrition, Food and Drug Administration, College Park, Maryland, United States of America (USA)
[3] Division of Food Additives, National Institute of Health Sciences, Tokyo, Japan
[4] Agence nationale de sécurité sanitaire de l'alimentation, de l'environnement et du travail (ANSES), Maisons-Alfort, France
[5] Department of the President's Affairs, Al Ain, United Arab Emirates
[6] Nutritional and Toxicological Risks Section, Swiss Federal Office of Public Health, Zurich, Switzerland

1. Explanation	3
1.1 Chemical and technical considerations	5
2. Biological data	5
2.1 Biochemical aspects	5
2.2 Toxicological studies	6
2.2.1 Acute toxicity	6
2.2.2 Short-term studies of toxicity	6
2.2.3 Long-term studies of toxicity and carcinogenicity	7
2.2.4 Genotoxicity	7
2.3 Observations in humans	7
3. Dietary exposure assessment	9
3.1 Functional use and proposed use levels in foods	9
3.2 Assessment of dietary exposure	9
3.2.1 Screening by the budget method	9
3.2.2 Assessment based on individual dietary records	11
3.3 General conclusions	12
4. Comments	14
4.1 Toxicological data	14
4.2 Assessment of dietary exposure	14
5. Evaluation	15
5.1 Recommendations	15
6. References	16

1. EXPLANATION

At the present meeting, the Committee evaluated magnesium dihydrogen diphosphate for use as an acidifier, stabilizer and raising agent. It is proposed for use as an alternative to sodium-based acidifiers and raising agents, primarily in self-raising flour, noodles (oriental style), batters and processed cereals.

Magnesium dihydrogen diphosphate has not been evaluated previously by the Committee. Phosphates, diphosphates and polyphosphates were evaluated by the Committee at its sixth, seventh, eighth, ninth, thirteenth, fourteenth, seventeenth, twenty-sixth and fifty-seventh meetings (Annex 1, references *6–8*, *11*, *19*, *22*, *32*, *59* and *154*). A maximum tolerable daily intake (MTDI) of 70 mg/kg of body weight (bw) was established at the twenty-sixth meeting on the basis of the lowest dietary concentration of phosphorus (1% in the diet) that caused nephrocalcinosis in rats. It was considered inappropriate to establish an acceptable daily intake (ADI), because phosphorus (primarily as phosphate) is an essential nutrient and an unavoidable constituent of food. The MTDI is expressed as phosphorus and applies to the sum of phosphates naturally present in food and the phosphates derived from use of these food additives. At its seventy-first meeting, the Committee evaluated ferrous ammonium phosphate and concluded that consideration of the toxicity of phosphate did not indicate a need to revise the Committee's previous evaluation of this ion (Annex 1, reference *191*).

The MTDI was considered to cover a number of phosphate salts, according to the principle established by the Committee at its ninth, twenty-third and twenty-ninth meetings (Annex 1, references *11*, *50* and *70*) that the ADI (or MTDI) established for ionizable salts should be based on previously accepted recommendations for the constituent cations and anions. Magnesium-based salts previously discussed by the Committee and covered by the MTDI for phosphates included magnesium phosphate (monobasic, dibasic and tribasic) and monomagnesium phosphate. However, certain specific phosphate salts were not included, because specifications were lacking and because information was not available to indicate whether they were being used as food-grade materials.

The Committee has previously evaluated other magnesium salts, allocating ADIs "not limited"[1] or "not specified" to magnesium carbonate, magnesium hydroxide, magnesium chloride, magnesium DL-lactate, magnesium hydrogen carbonate, magnesium gluconate, magnesium di-L-glutamate and magnesium sulfate (Annex 1, references *11*, *50*, *70*, *77*, *137* and *187*). At its twenty-ninth meeting (Annex 1, reference *70*), the Committee highlighted that the use of magnesium salts as food additives was acceptable, provided that the following were taken into consideration:

- The minimum laxative effective dose is approximately 1000 mg of magnesium moiety from a magnesium salt (observed only when the magnesium salt is administered as a single dose).
- Infants are particularly sensitive to the sedative effects of magnesium salts.
- Individuals with chronic renal impairment retain 15–30% of administered magnesium.

At its present meeting, the Committee was asked to conduct a safety assessment and set specifications for magnesium dihydrogen diphosphate by

---

[1] At its eighteenth meeting (Annex 1, reference *35*), the Committee replaced the term ADI "not limited" with ADI "not specified".

the Forty-third Session of the Codex Committee on Food Additives (FAO/WHO, 2011). The Committee received a submission that included tests for acute toxicity, skin and eye irritation and genotoxicity of magnesium dihydrogen diphosphate and considered other information available in the literature of relevance to the magnesium and phosphate ions.

1.1 Chemical and technical considerations

Magnesium dihydrogen diphosphate (chemical formula: $MgH_2P_2O_7$; Chemical Abstracts Service registry number: 20768-12-1) is the acidic magnesium salt of diphosphoric acid. It is manufactured by adding an aqueous dispersion of magnesium hydroxide slowly to phosphoric acid until a magnesium to phosphorus ratio of about 1:2 is reached. The temperature is held under 60 °C during the reaction. About 0.1% hydrogen peroxide is added to the reaction mixture, and the slurry is then dried and milled.

2. BIOLOGICAL DATA

2.1 Biochemical aspects

Magnesium dihydrogen diphosphate has a solubility of 1 g/l. It dissolves quickly under stomach conditions into its component ions: magnesium, hydrogen and diphosphate. The diphosphate ion hydrolyses to orthophosphate under acidic conditions (Bioserv, 2011).

Magnesium is an essential mineral, acting as a cofactor for many enzyme systems. It is involved in energy metabolism, the synthesis of proteins and nucleotides, and the metabolism and action of vitamin D and parathyroid hormone. It is a normal constituent of the human body and is ubiquitous in foods, where it is commonly bound to phosphates. The absorption of magnesium from different foods ranges from 40% to 60% (Schwartz, 1984). Magnesium is present throughout the body, with highest amounts in bone. It is excreted primarily in urine (EVM, 2003). Recommended Dietary Allowances, which are considered to meet the nutrient needs of 97–98% of individuals in a population, have been set at 80–420 mg of magnesium per day for different age groups by the United States Institute of Medicine (Institute of Medicine, 1997).

Phosphorus is also an essential mineral, with an important role in carbohydrate, fat and protein metabolism and in bone. Phosphorus is widely present in food, largely as the phosphate ion. Absorption of phosphate has been reported to range from 55% to 70% in adults and from 65% to 90% in infants and children. Approximately 80% of the body's phosphorus is present in the skeleton, with the remainder distributed in soft tissues and extracellular fluid. Excretion is primarily in the urine. The balance of phosphorus and calcium is regulated by parathyroid hormone, which increases urinary excretion of phosphate under conditions of high phosphate and low calcium intake (EVM, 2003). Recommended Dietary Allowances have been set at 460–1250 mg of phosphorus per day for different age groups by the United States Institute of Medicine (Institute of Medicine, 1997).

2.2 Toxicological studies

2.2.1 Acute toxicity

An acute oral toxicity test has been performed with magnesium dihydrogen diphosphate (purity 97.5%), conducted in accordance with good laboratory practice (GLP) according to Organisation for Economic Co-operation and Development (OECD) Test Guideline 423. Three female Balb/c mice were dosed by gavage at the limit dose of 2000 mg/kg bw. There were no deaths and no abnormal clinical signs. Therefore, the oral median lethal dose (LD_{50}) exceeds 2000 mg/kg bw (Bioserv, 2011).

A skin irritation test was performed with magnesium dihydrogen diphosphate (purity 97.5%), conducted in accordance with GLP according to OECD Test Guideline 404. The test material was moistened with the smallest amount of water necessary to ensure good skin contact, and 0.5 ml was applied to the clipped, intact skin of three female rabbits (not less than 2 kg bw). Physiological saline was applied as a control substance. The application sites were occluded for 4 hours, and then the test material was washed off. The application sites were then observed for 72 hours. There were no signs of skin irritation (Bioserv, 2011).

An eye irritation test was performed with magnesium dihydrogen diphosphate (purity 97.5%), conducted in accordance with GLP according to OECD Test Guideline 405. The test material was ground to a fine dust and moistened, and 100 mg was placed in the lower conjunctival sacs of the right eyes of three female rabbits (not less than 2 kg bw). The left eyes were untreated and served as controls. After 1 hour, the test material was rinsed out with distilled water. The eyes were observed with a slit lamp at 1, 24, 48 and 72 hours after application. Hyperaemic blood vessels were observed in all animals, but no other ocular lesions were seen. The material met the requirements for classification as non-irritating to the eye (Bioserv, 2011).

2.2.2 Short-term studies of toxicity

Groups of 10 male and 10 female F344 rats were fed magnesium chloride hexahydrate at dietary concentrations of 0%, 0.1%, 0.5% or 2.5% for 90 days. These dietary concentrations were equivalent to 0, 100, 500 and 2500 mg/kg bw per day, corresponding to 0, 12, 60 and 300 mg/kg bw per day expressed as magnesium. Transient soft faeces and a sustained increase in water consumption were observed in both the males and females of the 2.5% group, and a slight reduction in body weight gain was noted in the high-dose males. There were no changes in feed consumption, organ weights, haematology, biochemistry or histopathology. The no-observed-adverse-effect level (NOAEL) was reported to be 0.5% magnesium chloride hexahydrate in the diet (Takizawa et al., 2000), which is equivalent to 60 mg/kg bw per day expressed as magnesium.

Beta-calcium pyrophosphate was administered orally to groups of 10 male and 10 female 5-week-old Sprague-Dawley rats at 0 or 30 mg/kg bw per day for 90 days. The route of administration was not specified, but as saline was used as control, the administration was presumably by gavage. Body weights, feed and water consumption, clinical signs and urine analysis were recorded regularly throughout the study. Ophthalmological examinations were conducted on all animals at baseline and towards the end of the treatment period. At autopsy, blood was sampled for haematology and serum biochemistry, organ weights were recorded and a wide range of tissues were sampled for histopathological examination. No treatment-related changes were observed (Lee et al., 2009).

2.2.3 Long-term studies of toxicity and carcinogenicity

Groups of 50 male and 50 female B6C3F1 mice were given magnesium chloride hexahydrate ($MgCl_2 \cdot 6H_2O$) at dose levels of 0% (control), 0.5% or 2% in the diet for 96 weeks, after which all animals received the control diet for 8 weeks and were then necropsied. The dietary concentrations were equivalent to 0, 750 and 3000 mg/kg bw per day, corresponding to 0, 90 and 360 mg/kg bw per day expressed as magnesium. In females of the high-dose group, a decrease in body weight was observed. Survival rates did not differ between the treatment and control groups for males or females. Clinical signs and urinary, haematological and serum clinical chemistry parameters showed no treatment-related effects. With the exception of a significant decrease in the incidence of liver tumours among males of the high-dose group, no differences were noted in tumour incidence between the treated and control animals (Kurata et al., 1989). Based on the decreased body weight gain in the high-dose females, the NOAEL was 0.5% magnesium chloride hexahydrate in the diet, equivalent to 90 mg/kg bw per day expressed as magnesium.

An analysis of the incidence of nephrocalcinosis in rats, the basis of the MTDI, has highlighted that this lesion is common in untreated rats fed purified diets. The calcium and phosphorus contents of diets, the calcium to phosphorus ratio in diets, deficiency of magnesium, deficiency of chloride and high urinary pH are considered to be associated with its development. Increasing the calcium to phosphorus molar ratio to greater than 1.0 either markedly decreased the incidence and severity of the lesion or prevented it altogether. Most of the commercially available diets commonly used for toxicology studies are either high in phosphorus or low in calcium, with a net calcium to phosphorus molar ratio of less than 1.0, which contributes to the development of this lesion (Rao, 2002). Thus, addition of phosphate salts to the diet will further increase this imbalance.

2.2.4 Genotoxicity

The Committee was provided with genotoxicity data on magnesium dihydrogen diphosphate. The material was negative for bacterial and cell mutation in the presence and absence of metabolic activation (Table 1).

2.3 Observations in humans

Excessive ingestion of magnesium compounds from non-food (medicinal) sources has been reported to cause osmotic diarrhoea. Diarrhoea and other gastrointestinal effects were reported in some patients or healthy volunteers receiving supplemental magnesium at 384 mg/day for 6 weeks (Bashir et al., 1993) or 470 mg/day for 60 days (Marken et al., 1989). In contrast, no adverse effects

Table 1. Genotoxicity of magnesium dihydrogen diphosphate

End-point	Test system	Concentration	Results	Reference
In vitro				
Reverse mutation[a]	*Salmonella typhimurium* TA97a, TA98, TA100, TA102, TA1535	0.5–50 mg/ml	Negative	Bioserv (2011)
Cell mutation[a]	Mouse lymphoma L5178Y cells, Tk$^{+/-}$ locus	5–50 µg/ml[b]	Negative	Bioserv (2011)

[a] In the presence and absence of rat liver S9 (9000 × *g* supernatant) mix.
[b] A slight decrease in cloning efficiency was reported at the highest concentration.

were reported in other studies at 393 mg/day for 4 weeks (Paolisso et al., 1992), 452 mg/day for 6 days (Altura et al., 1994), 486 mg/day for 3 months (Zemel et al., 1990), 750 mg/day for 6 months (Stendig-Lindberg, Tepper & Leichter, 1993) or 1200 mg/day for 6 weeks (Nagy et al., 1988).

The United Kingdom Expert Group on Vitamins and Minerals concluded that there were insufficient data to establish a safe upper level for magnesium, but that 400 mg of supplemental magnesium per day, in addition to that present in the diet, would not be expected to result in any significant adverse effects (EVM, 2003). The European Union (EU) Scientific Committee on Food noted that no laxative effects had been observed in adult men and women at magnesium doses up to 250 mg/day in studies in which a pharmaceutical type of dosage formulation was taken in addition to magnesium present in normal foods and beverages. This dose was considered as a NOAEL. An uncertainty factor of 1 was considered appropriate in view of the fact that data were available from many human studies involving a large number of subjects from a spectrum of life stage groups; and the fact that the NOAEL was based on a mild, transient laxative effect, without pathological sequelae, which is readily reversible and for which considerable adaptation can develop within days. The Scientific Committee on Food therefore concluded that an upper level of 250 mg/day could be established for readily dissociable magnesium salts (e.g. chloride, sulfate, aspartate, lactate) and compounds like magnesium oxide in nutritional supplements, in water or added to food and beverages. This upper level does not include magnesium normally present in foods and beverages. Dietary exposures for different EU countries and the USA were noted as being 200–250 mg/day for average consumption and 340–630 mg/day for high-level consumption (SCF, 2001). Similarly, the United States Institute of Medicine concluded that the tolerable upper level of magnesium to avoid the laxative effects of magnesium in pharmacological agents was 350 mg/day. It stressed that this does not include magnesium in food or water (Institute of Medicine, 1997).

For phosphorus, the Institute of Medicine set tolerable upper levels of 3–4 g/day for different age groups, equivalent to about 70 mg/kg bw per day in a 60 kg adult (Institute of Medicine, 1997). These were derived on the basis of the dietary exposure associated with the upper boundary of normal levels of serum phosphorus in adults. The European Food Safety Authority (EFSA) concluded that the available data were not sufficient to establish an upper level for phosphorus, but indicated that normal healthy individuals can tolerate phosphorus (phosphate) exposures up to at least 3000 mg/day without adverse systemic effects. In some individuals with supplemental exposures above 750 mg of phosphorus per day, however, mild gastrointestinal symptoms had been reported. Estimates of habitual dietary exposures in European countries were on average around 1000–1500 mg/day, ranging up to about 2600 mg/day (EFSA, 2006).

3. DIETARY EXPOSURE ASSESSMENT

Magnesium dihydrogen diphosphate is the acidic magnesium salt of the diphosphoric acid and is produced by chemical synthesis. Its International Numbering System (INS) number is 450ix. It is chemically very similar to calcium dihydrogen diphosphate (INS 450vii) and monomagnesium phosphate (INS 343i), both of which have previously been evaluated by the Committee (Annex 1, reference *154*).

Magnesium dihydrogen diphosphate is a product of the ongoing development of alternatives to sodium-based raising agents/acidifiers, primarily used for baking purposes, without affecting the taste, in order to gain a high customer acceptance. At its present meeting, the Committee was requested to evaluate the safety of the use of magnesium dihydrogen diphosphate as a raising agent. A submission was received from a sponsor (Chemische Fabrik Budenheim KG, 2011), including a chemical and technical assessment.

3.1 Functional use and proposed use levels in foods

The sponsor stated that the additive is intended to be used as a replacement or substitute for other existing phosphate salt additives for industrial application as a raising agent in baking.

The proposed use levels are in accordance with, or even below, current use levels of other phosphates and are based on the ratio of "needed leavening" versus flour content, but considering sugar-rich formulations, which need more raising action. Table 2 provides the current proposed use levels according to the Codex General Standard for Food Additives (GSFA) food categorization.

The sponsor proposed to maintain the current approach of grouping "phosphates" in accordance with toxicological considerations and the GSFA approach, where commonly phosphorus from all food sources has an MTDI of 70 mg/kg bw per day assigned (Annex 1, reference *59*).

3.2 Assessment of dietary exposure

3.2.1 Screening by the budget method

The "budget method" is used to assess theoretical maximum daily dietary exposure. The budget method has been used as a screening method in assessing food additives by JECFA (FAO/WHO, 2001) and for assessments within the EU Scientific Cooperation Task 4.2 (EC, 1998).

Table 2. Proposed use levels of magnesium dihydrogen diphosphate in individual food uses according to the GSFA food categorization

Food category	Category name (restriction)	Maximum use level	Typical use level
06.2.1	Flours (self-raising flour)	7 g/kg (as phosphorus)	6 g/kg (as phosphorus)
06.4.2 06.4.3	Dried and pre-cooked pastas and noodles and like products	1 g/kg (as phosphorus)	0.5–1 g/kg (as phosphorus)
06.6	Batters	5.6 g/kg (as phosphorus)	4–5 g/kg (as phosphorus)
06.7	Pre-cooked or processed rice products, including rice cakes (puffed products)	6 g/kg (as phosphorus)	4–5 g/kg (as phosphorus)
07.1.1 07.1.2 07.1.3 07.1.6	Bread and rolls (non-yeast-raised only products, crackers, bagels, tortillas, and premixes of those)	7 g/kg (as phosphorus)	4–5 g/kg (as phosphorus)
07.2	Fine bakery wares	7 g/kg (as phosphorus)	5.5–6.5 g/kg (as phosphorus)

The method relies on assumptions regarding 1) the level of consumption of foods and of non-milk beverages, 2) the level of presence of the substance in foods and in non-milk beverages and 3) the proportion of foods and of non-milk beverages that may contain the substance. More specifically:

- The levels of consumption of foods and beverages considered are maximum physiological levels of consumption—i.e. the daily consumption of 0.1 litre of non-milk beverages per kilogram of body weight and the daily consumption of 418.4 kJ/kg bw from foods (equivalent to 0.05 kg/kg bw based on an estimated energy density of 8.4 kJ/g). These levels correspond to the daily consumption of 6 litres of non-milk beverages and 3 kg of food in a person with a body weight of 60 kg (typical adult) and a daily consumption of 1.5 litres of non-milk beverages and 750 g of food in a person with a body weight of 15 kg (typical 3-year-old child) (FAO/WHO, 2009).
- The level present in foods is assumed to be the highest proposed use level of magnesium dihydrogen diphosphate, which corresponds, as reported in Table 2, to a maximum use level of 7 g/kg (as phosphorus) in solid food uses. Liquid food is not taken into account in this scenario, as no use levels are proposed for beverages.
- The proportion of solid foods that may contain the substance is generally set at 12.5%. However, for this proposed additive, it was decided to set the proportion at 50%, as some uses concern food such as rice or bread, which are staple foods.

Assuming the conservative scenario above, the Committee estimated that the theoretical maximum daily exposure to magnesium dihydrogen diphosphate from its proposed uses would be approximately 175 mg/kg bw per day (7000 × 0.05 × 50%), expressed as phosphorus, in the general population.

As the proposed use is a substitutional use, not an add-on use to other phosphates or other additives (it is an alternative to existing sodium-based acidifiers and raising agents), the Committee concluded that combined exposure from other phosphorus additives already regulated did not need to be assessed.

3.2.2 Assessment based on individual dietary records

An assessment of exposure from the typical proposed use levels in individual bakery products was provided by the sponsor using EFSA's Comprehensive European Food Consumption Database. The total exposure to phosphorus estimated by the sponsor using the German data as a reference was 63.7 mg/kg bw per day for the 97.5th percentile of consumers.

The Committee noted that anticipated dietary exposures provided in the dossier for other population groups among member states, including children, were not expressed as a range in the assessment. The Committee also noted that use levels used in the calculation were the typical proposed use levels rather than the maximum proposed use levels. Therefore, the Committee decided to provide its own assessment based on the whole EFSA food consumption data set and the maximum proposed use levels for the respective individual food uses. The EFSA Comprehensive European Food Consumption Database has been built from existing national information on food consumption at a detailed level. Competent authorities in European countries provide EFSA with data on the level of food consumption by the individual consumer from the most recent national dietary surveys in each country (EFSA, 2011b).

For calculation of chronic exposure, intake statistics have been calculated based on individual average consumption over the total survey period, excluding surveys with only 1 day per subject. High-level consumption was calculated only for those foods and population groups where the sample size was sufficiently large to allow calculation of the 95th percentile (EFSA, 2011b). The Committee estimated chronic exposure for the following population groups: toddlers, children, adolescents, adults and the elderly. Calculations were performed using individual body weights. Thus, for the present assessment, food consumption data were available from 26 different dietary surveys carried out in 17 different European countries, as summarized in Table 3.

Consumption records were codified according to the FoodEx classification system (EFSA, 2011a). Nomenclature from the FoodEx classification system has been linked to the Food Classification System as presented in Commission Regulation (EU) No. 1129/2011, part D, to perform exposure estimates.

Exposure to magnesium dihydrogen diphosphate from its use as a food additive has been calculated using the maximum proposed use levels as listed in Table 2 and national consumption data for the five population groups listed in Table 3.

High-level exposure (typically 95th percentile of consumers only) was calculated by adding the 95th percentile of exposure from one food group (i.e. the one having the highest value) to the mean exposure resulting from the consumption of all other food groups. This is based on the assumption that an individual might be

Table 3. Population groups considered for the exposure estimates for magnesium dihydrogen diphosphate

Population	Age range	Countries with food consumption surveys covering more than 1 day
Toddlers	From 12 months up to and including 35 months of age	Bulgaria, Finland, Germany, the Netherlands
Children[a]	From 36 months up to and including 9 years of age	Belgium, Bulgaria, Czech Republic, Denmark, Finland, France, Germany, Greece, Italy, Latvia, the Netherlands, Spain, Sweden
Adolescents	From 10 years up to and including 17 years of age	Belgium, Cyprus, Czech Republic, Denmark, France, Germany, Italy, Latvia, Spain, Sweden
Adults	From 18 years up to and including 64 years of age	Belgium, Czech Republic, Denmark, Finland, France, Germany, Hungary, Ireland, Italy, Latvia, the Netherlands, Spain, Sweden, the United Kingdom
Elderly[a]	65 years of age and older	Belgium, Denmark, Finland, France, Germany, Hungary, Italy

[a] The terms "children" and "elderly" correspond, respectively, to "other children" and the merger of "elderly" and "very elderly" in the EFSA guidance on the use of the EFSA Comprehensive European Food Consumption Database in exposure assessment (EFSA, 2011b).

a high-level consumer of one food category and an average consumer of the others. This approach, used in the work of the EFSA Panel of Food Additives and Nutrient Sources Added to Food, has been tested several times in the re-evaluation of food colours and has been shown to give reasonable correlation with high-level total exposures when using the raw food individual consumption data by computerized method. Therefore, this approach was preferred for the calculations based on the maximum proposed use levels in order to avoid excessively conservative estimates.

However, the Committee noted that its estimates should be considered as being conservative, as it is assumed that all processed foods contain the additive added at the maximum proposed use levels. Table 4 summarizes the estimated exposure to magnesium dihydrogen diphosphate from its use as a food additive for all five population groups, and Table 5 summarizes the main food categories contributing to the overall dietary exposure within population groups.

3.3 General conclusions

For the evaluation of magnesium dihydrogen diphosphate as a new food additive intended to be used as an alternative to sodium-based acidifiers and raising agents, the Committee evaluated an anticipated dietary exposure based on individual food consumption data from the EU with the maximum proposed use levels of magnesium dihydrogen diphosphate (0.1% up to 0.7% by weight in solid

Table 4. **Summary of anticipated exposure to magnesium dihydrogen diphosphate from its use as a food additive using maximum proposed use levels in five population groups**

	Estimated exposure (minimum–maximum) using maximum proposed use levels (mg/kg bw per day, expressed as phosphorus)				
	Toddlers (12–35 months)	Children (3–9 years)	Adolescents (10–17 years)	Adults (18–64 years)	Elderly (≥65 years)
Mean exposure	28.6–66.3	24.5–68.1	18.9–38.2	10.2–23.6	10.3–19.7
High-level exposure[a]	58.3–104.5	42.5–114.8	32.4–60.9	20.6–40.0	19.8–34.0

[a] Typically 95th percentile of consumers only.

Table 5. **Main food categories contributing to the total anticipated dietary exposure to magnesium dihydrogen diphosphate using maximum proposed use levels, and number of surveys in which each food category contributes over 5%**

Food category	% contribution to total exposure (number of surveys)				
	Toddlers	Children	Adolescents	Adults	Elderly
Flours and starches	8–93 (2)	7–84 (5)	6–20 (4)	5–97 (6)	9–98 (4)
Pasta	—[a]	5–6 (3)	7 (1)	—[a]	—[a]
Bread and rolls	59–76 (3)	7–84 (15)	46–79 (12)	61–84 (14)	68–86 (6)
Fine bakery wares	19–33 (3)	7–58 (14)	5–48 (12)	5–35 (14)	6–31 (6)

[a] Pasta contributes <5% in all surveys.

food, as phosphorus) in GSFA food categories such as flours, pasta, noodles and similar products, puffed products, bread and rolls and fine bakery wares.

Based on this conservative scenario, assuming that 100% of food products would be manufactured and consumed at the maximum proposed use levels, the Committee concluded that anticipated average dietary exposures to magnesium dihydrogen diphosphate would be up to approximately 20 mg of phosphorus per kilogram of body weight per day for an adult and up to 70 mg of phosphorus per kilogram of body weight per day for a child. The 95th percentiles of exposure are estimated to be up to 40 mg of phosphorus per kilogram of body weight per day for an adult and up to 115 mg of phosphorus per kilogram of body weight per day for a child. The main food groups contributing to these overall dietary exposures within all population groups were bread and rolls (7–86%), fine bakery wares (6–58%) and flours and starches (5–98%).

The dietary exposure to magnesium estimated from the anticipated use of magnesium dihydrogen diphosphate would be 39% of the estimated exposure to phosphorus, based on the contribution to molecular weight. This corresponds to an average dietary exposure of up to approximately 8 mg of magnesium per kilogram of body weight per day for an adult and up to 27 mg of magnesium per kilogram of body weight per day for a child. The 95th percentiles of exposure are estimated to be up to 16 mg of magnesium per kilogram of body weight per day for an adult and up to 45 mg of magnesium per kilogram of body weight per day for a child.

4. COMMENTS

4.1 Toxicological data

Magnesium dihydrogen diphosphate ionizes into its component ions: magnesium, hydrogen and diphosphate. Therefore, the safety assessment should be based on previously accepted recommendations for the constituent cations and anions. Magnesium and phosphorus (primarily as phosphate) are essential minerals that are naturally present in the human body and in food.

The Committee received data showing that magnesium dihydrogen diphosphate does not exert acute toxicity, skin or eye irritation or genotoxicity.

At previous meetings, the Committee noted that toxicity can arise from an imbalance of calcium, magnesium and phosphate. Excessive dietary phosphorus causes hypocalcaemia, which can result in bone loss and calcification of soft tissues. The MTDI of 70 mg/kg bw was derived from studies demonstrating nephrocalcinosis in rats at dietary concentrations of 1% phosphorus. Nephrocalcinosis has been defined as calcified deposits, mainly in the form of calcium phosphate, in tubules located predominantly at the corticomedullary junction of the kidney. The exact approach taken in deriving the MTDI from this end-point is unclear. In addition, the Committee noted that there is evidence that rats are particularly sensitive to mineralization in the kidneys resulting from an imbalance of calcium and phosphate in the diet. Therefore, the relevance of mineralization in the rat kidney for safety assessment is unclear. The available toxicological information on phosphate salts did not indicate that the MTDI is insufficiently health protective.

4.2 Assessment of dietary exposure

For the evaluation of magnesium dihydrogen diphosphate as a new food additive intended to be used as an alternative to sodium-based acidifiers and raising agents, the Committee evaluated an anticipated dietary exposure based on individual food consumption data from the EU with the maximum proposed use levels of magnesium dihydrogen diphosphate (0.1% up to 0.7% by weight in solid food, as phosphorus) in GSFA food categories such as flours, pasta, noodles and similar products, puffed products, bread and rolls and fine bakery wares.

Based on this conservative scenario, assuming that 100% of food products would be manufactured and consumed at the maximum proposed use levels, the Committee concluded that anticipated average dietary exposures to magnesium dihydrogen diphosphate would be up to approximately 20 mg of phosphorus per kilogram of body weight per day for an adult and up to 70 mg of phosphorus per kilogram of body weight per day for a child. The 95th percentiles of exposure are estimated to be up to 40 mg of phosphorus per kilogram of body weight per day for an adult and up to 115 mg of phosphorus per kilogram of body weight per day for a child. The main food groups contributing to these overall dietary exposures within all population groups were bread and rolls (7–86%), fine bakery wares (6–58%) and flours and starches (5–98%).

The dietary exposure to magnesium estimated from the anticipated use of magnesium dihydrogen diphosphate would be 39% of the estimated exposure to phosphorus, based on the contribution to molecular weight. This corresponds to an average of up to approximately 8 mg of magnesium per kilogram of body weight per day for an adult and up to 27 mg of magnesium per kilogram of body weight per day for a child. The 95th percentiles of exposure are estimated to be up to 16 mg of magnesium per kilogram of body weight per day for an adult and up to 45 mg of magnesium per kilogram of body weight per day for a child.

5. EVALUATION

Although an ADI "not specified" has been established for a number of magnesium salts used as food additives, the estimated chronic dietary exposures to magnesium (960 mg/day for a 60 kg adult at the 95th percentile) from the proposed uses of magnesium dihydrogen diphosphate are up to twice the background exposures from food previously noted by the Committee (180–480 mg/day) and in the region of the minimum laxative effective dose of approximately 1000 mg of magnesium when taken as a single dose. The estimates of dietary exposure to phosphorus from the proposed uses of magnesium dihydrogen diphosphate are in the region of, or slightly exceed, the MTDI of 70 mg/kg bw for phosphate salts, expressed as phosphorus, from this source alone. Thus, the MTDI is further exceeded when other sources of phosphate in the diet are taken into account. The Committee therefore concluded that the proposed use levels and food categories result in an estimated dietary exposure to magnesium dihydrogen diphosphate that is a potential concern.

The Committee emphasized that in evaluating individual phosphate-containing food additives, there is a need for assessment of total dietary exposure to phosphorus.

5.1 Recommendations

The Committee noted that an ADI "not specified" has been allocated individually to a number of magnesium-containing food additives and recommended that total dietary exposure to magnesium from food additives and other sources in the diet should be assessed.

The information submitted to the Committee and in the scientific literature did not indicate that the MTDI of 70 mg/kg bw for phosphate salts, expressed as phosphorus, is insufficiently health protective. On the contrary, because the basis for its derivation might not be relevant to humans, it could be overly conservative. Therefore, there is a need to review the toxicological basis of the MTDI for phosphate salts expressed as phosphorus.

6. REFERENCES

Altura BT et al. (1994). Comparative effects of magnesium enriched diets and different orally administered magnesium oxide preparations on ionized Mg, Mg metabolism and electrolytes in serum of human volunteers. *Journal of the American College of Nutrition*, 13:447–454.

Bashir Y et al. (1993). Effects of long-term oral magnesium chloride replacement in congestive heart failure secondary to coronary artery disease. *American Journal of Cardiology*, 72:1156–1162.

Bioserve (2011). Acute toxicity and genotoxicity test results—magnesium dihydrogen diphosphate. Unpublished test report for sample no. 2010080293. Rostock, Germany, Bioserv Analytik und Medizinprodukte Gmbh. Submitted to WHO by Chemische Fabrik Budenheim KG, Budenheim, Germany.

Chemische Fabrik Budenheim KG (2011). Magnesium dihydrogen diphosphate (INS 450ix). Draft chemical and technical assessment (CTA) prepared by T. Janssen. Submitted to WHO by Chemische Fabrik Budenheim KG, Budenheim, Germany, 5 December 2011.

EC (1998). *Report on methodologies for the monitoring of food additive intake across the European Union.* Final report submitted by the Task Coordinator, 16 January 1998. Brussels, Belgium, European Commission, Directorate General III Industry, Reports of a Working Group on Scientific Cooperation on Questions Relating to Food, Task 4.2 (SCOOP/INT/REPORT/2).

EFSA (2006). *Tolerable upper intake levels for vitamins and minerals.* Parma, Italy, European Food Safety Authority, Scientific Committee on Food, Scientific Panel on Dietetic Products, Nutrition and Allergies (http://www.efsa.europa.eu/en/ndatopics/docs/ndatolerableuil.pdf).

EFSA (2011a). Evaluation of the FoodEx, the food classification system applied to the development of the EFSA Comprehensive European Food Consumption Database. *The EFSA [European Food Safety Authority] Journal*, 9(3):1970 (27 pp.).

EFSA (2011b). Use of the EFSA Comprehensive European Food Consumption Database in exposure assessment. *The EFSA [European Food Safety Authority] Journal*, 9(3):2097 (34 pp.).

EVM (2003). *Safe upper levels for vitamins and minerals.* London, England, Food Standards Agency, Expert Group on Vitamins and Minerals (http://cot.food.gov.uk/cotreports/cotjointreps/evmreport/).

FAO/WHO (2001). *Guidelines for the preparation of working papers on intake of food additives for the Joint FAO/WHO Expert Committee on Food Additives.* Rome, Italy, Food and Agriculture Organization of the United Nations; Geneva, Switzerland, World Health Organization (http://www.who.int/foodsafety/chem/jecfa/en/intake_guidelines.pdf).

FAO/WHO (2009). Dietary exposure assessment of chemicals in food. In: *Principles and methods for the risk assessment of chemicals in food.* Rome, Italy, Food and Agriculture Organization of the United Nations; Geneva, Switzerland, World Health Organization (Environmental Health Criteria 240; http://whqlibdoc.who.int/ehc/WHO_EHC_240_9_eng_Chapter6.pdf).

FAO/WHO (2011). *Report of the Forty-third Session of the Codex Committee on Food Additives, Xiamen, China, 14–18 March 2011.* Rome, Italy, Food and Agriculture Organization of the United Nations, and Geneva, Switzerland, World Health Organization, Joint FAO/WHO Food Standards Programme, Codex Alimentarius Commission, Codex Committee on Food Additives (REP11/FA).

Institute of Medicine (1997). *Dietary reference intakes for calcium, phosphorus, magnesium, vitamin D and fluoride.* Prepared by the Standing Committee on the Scientific Evaluation of Dietary Reference Intakes, Food and Nutrition Board, Institute of Medicine. Washington, DC, USA, National Academy Press, pp. 190–249.

Kurata Y et al. (1989). Lack of carcinogenicity of magnesium chloride in a long-term feeding study in B6C3F1 mice. *Food and Chemical Toxicology*, 27(9):559–563.

Lee JH et al. (2009). A 90-day subchronic toxicity study of beta-calcium pyrophosphate in rat. *Drug and Chemical Toxicology*, 32(3):277–282.

Marken PA et al. (1989). Effects of magnesium oxide on the lipid profile of healthy volunteers. *Atherosclerosis*, 77:37–42.

Nagy L et al. (1988). Human tolerability and pharmacodynamic study of Tisacid tablet in duodenal ulcer patients, a prospective, randomized, self-controlled clinicopharmacological study. *Acta Medica Hungarica*, 45(2):231–247.

Paolisso G et al. (1992). Daily magnesium supplements improve glucose handling in elderly subjects. *American Journal of Nutrition*, 55:1161–1167.

Rao GN (2002). Toxicologic pathology of the kidney and urinary bladder. *Toxicologic Pathology*, 30(6):651–656.

SCF (2001). *Opinion of the Scientific Committee on Food on the tolerable upper intake level of magnesium (expressed on 26 September 2001).* Brussels, Belgium, European Commission, Scientific Committee on Food (SCF/CS/NUT/UPPLEV/54; http://ec.europa.eu/food/fs/sc/scf/out105_en.pdf).

Schwartz R (1984). Magnesium absorption in human subjects from leafy vegetables. *American Journal of Clinical Nutrition*, 39:571–576.

Stendig-Lindberg G, Tepper R, Leichter I (1993). Trabecular bone density in a two year controlled trial of personal magnesium in osteoporosis. *Magnesium Research*, 6:155–163.

Takizawa T et al. (2000). [A 90-day repeated dose oral toxicity study of magnesium chloride in F344 rats.] *Kokuritsu Iyakuhin Shokuhin Eisei Kenkyusho Hokoku*, 118:63–70 (in Japanese).

Zemel PC et al. (1990). Metabolic and hemodynamic effects of magnesium supplementation in patients with essential hypertension. *American Journal of Nutrition*, 51:665–669.

3-PHYTASE FROM ASPERGILLUS NIGER EXPRESSED IN ASPERGILLUS NIGER

First draft prepared by

S. Choudhuri[1], M. DiNovi[1], L.R.M. dos Santos[2], J.-C. Leblanc[3], I. Meyland[4] and U. Mueller[5]

[1] Center for Food Safety and Applied Nutrition, Food and Drug Administration, College Park, Maryland, United States of America (USA)
[2] Departamento de Ciência de Alimentos, Faculdade de Engenharia de Alimentos, Universidade Estadual de Campinas, São Paulo, Brazil
[3] Agence nationale de sécurité sanitaire de l'alimentation, de l'environnement et du travail (ANSES), Maisons-Alfort, France
[4] Birkerød, Denmark
[5] Food Standards Australia New Zealand, Canberra, ACT, Australia

1. Explanation	19
1.1 Genetic modification	20
1.2 Chemical and technical considerations	20
2. Biological data	21
2.1 Biochemical aspects	21
2.2 Toxicological studies	22
2.2.1 Acute toxicity	23
2.2.2 Short-term studies of toxicity	23
2.2.3 Long-term studies of toxicity and carcinogenicity	25
2.2.4 Genotoxicity	25
2.2.5 Reproductive toxicity	25
2.3 Observations in humans	25
3. Dietary exposure	27
3.1 Assessment of dietary exposure	28
3.1.1 Screening by the budget method	28
3.2 General conclusions	34
4. Comments	34
4.1 Assessment of potential allergenicity	34
4.2 Toxicological data	34
4.3 Assessment of dietary exposure	36
5. Evaluation	36
6. References	36

1. EXPLANATION

At the request of the World Food Programme and the Global Alliance for Improved Nutrition, the Committee evaluated the safety of the 3-phytase enzyme preparation (3-phytase: *myo*-inositol hexakisphosphate 3-phosphohydrolase; Enzyme Commission number 3.1.3.8), which it had not evaluated previously. 3-Phytase catalyses the sequential hydrolysis of phosphate monoesters from phytate (phytic acid), also known as *myo*-inositol (1,2,3,4,5,6) hexakisphosphate or *myo*-inositol hexakisphosphate. Hydrolysis of phytate by 3-phytase generates a series of lower (pentakis-, tetrakis-, etc.) *myo*-inositol phosphates and inorganic phosphates. The catalytic activity of 3-phytase is relatively specific; it does not have any significant levels of secondary enzyme activities. In this monograph, the expression "3-phytase" refers to the 3-phytase enzyme and its amino acid sequence, and the expression "3-phytase enzyme preparation" refers to the preparation formulated for commercial use. The 3-phytase enzyme preparation is used as a food additive, in the processing of phytate-rich food, such as cereal grains and legumes, and as a dietary supplement, for co-consumption with phytate-rich foods.

1.1 Genetic modification

The enzyme 3-phytase is produced from a genetically modified *Aspergillus niger* strain containing multiple copies of 3-phytase gene from *A. niger*. *Aspergillus niger* is a filamentous fungus that commonly occurs in the environment and is non-pathogenic. It has a long history of use as a source of citric acid and enzymes used in food processing, including enzymes from genetically engineered strains of the organism.

Prior to the introduction of the 3-phytase gene, the *A. niger* host strain ISO-500 was genetically modified by deletion of the genes encoding glucoamylase activity. The modified host strain was then transformed with an amplifiable deoxyribonucleic acid (DNA) cassette containing the phytase gene from *A. niger* and the *Aspergillus nidulans* acetamidase (*amdS*) gene, which was the selectable marker. The recombinant production strain is genetically stable and does not contain any antibiotic resistance markers or any other heterologous DNA. Batch analysis demonstrated that 3-phytase enzyme preparations from *A. niger* were free of aflatoxin B1, T2 toxin, ochratoxin A, zearalenone and sterigmatocystin.

1.2 Chemical and technical considerations

3-Phytase is produced by submerged, fed-batch, aerobic, pure culture fermentation of the genetically modified *A. niger* production strain. The enzyme is secreted into the fermentation broth and is subsequently purified and concentrated by ultrafiltration. The enzyme concentrate is formulated with glycerol (liquid form) or with maltodextrin (powder form) to achieve the desired phytase activity and stability. The 3-phytase enzyme preparation contains food-grade materials and conforms to the General Specifications and Considerations for Enzyme Preparations Used in Food Processing (Annex 1, reference *184*). Phytase activity is measured in phytase units, or FTU. One FTU is defined as the amount of enzyme that liberates 1 µmol of inorganic phosphate per minute from sodium phytate at a concentration of 5.1 mmol/l at 37 °C and pH 5.5. The mean activity of 3-phytase calculated from three different batches of the ultrafiltrate concentrate was 106 FTU per milligram of total organic solids (TOS). TOS consists of the enzyme of interest and residues of organic materials, such as proteins, peptides and carbohydrates, derived from the production organism and the manufacturing process. The 3-phytase enzyme preparation is typically used at a range of 5–70 g/kg of food product, depending on the intended application and on the phytate content. 3-Phytase is expected to be inactivated during processing or cooking.

2. BIOLOGICAL DATA

In this monograph, the expression "3-phytase" is used when referring to the pure 3-phytase enzyme and its amino acid sequence, whereas the expression "3-phytase enzyme preparation" is used when referring to the enzyme preparation used for various applications.

2.1 Biochemical aspects

3-Phytase from *A. niger* has an apparent molecular weight of 85 kDa, as determined by sodium dodecyl sulfate–polyacrylamide gel electrophoresis analysis. Based on the amino acid sequence, the molecular weight of 3-phytase from *A. niger* is 50 kDa.

3-Phytase from *A. niger* has been evaluated for potential allergenicity (Wilms, 2011) using bioinformatics criteria recommended in the report of the Joint FAO/WHO Expert Consultation on Allergenicity of Foods Derived from Biotechnology (FAO/WHO, 2001a). According to the FAO/WHO allergenicity recommendation, the possibility of cross-reactivity between a query protein and a known allergen has to be considered if there is 1) more than 35% identity[1] in the amino acid sequence of the expressed protein using a sliding window of 80 amino acids and a suitable gap penalty or 2) an identity of six contiguous amino acids. Using these two criteria, the 3-phytase sequence from *A. niger* was compared with the allergenic protein sequences in two databases: the Allermatch database and the Structural Database of Allergenic Proteins. These two databases allow for the bioinformatics analysis of protein allergenicity using the FAO/WHO recommendation, specifically the six contiguous amino acids match.

A similarity search using the Allermatch database (http://www.allermatch.org/) did not produce a match with any sequence showing greater than 35% identity over any sliding window of 80 amino acids. However, two stretches of six contiguous amino acids in the *A. niger* 3-phytase sequence were found to be identical to a six contiguous amino acid sequence in each of two allergenic proteins from the WHO-International Union of Immunological Societies (WHO-IUIS) list. One match is to Zea m 14 protein in maize, and the other match is to Der f 18 protein in house dust mite. A comparison using the Structural Database of Allergenic Proteins (http://fermi.utmb.edu/SDAP/) produced one match with Asp n 25 protein in *A. niger* that has greater than 35% identity over several windows of 80 amino acids.

The Zea m 14 protein is a nonspecific lipid transfer protein in maize and is identified as a food allergen in the WHO-IUIS list. Similar nonspecific lipid transfer proteins from many species from the Rosaceae family (e.g. peach, plum, cherry,

---

[1] *Sequence identity* is quantified as per cent identity and is based on the number of identical amino acid residues in positions in the two sequences being aligned for comparison. Historically, the expression *sequence homology* has been frequently used to mean *sequence identity*, but *sequence homology* is truly an evolutionary term. Sequences are considered homologous if they are derived from a common ancestral sequence. Therefore, sequences are either homologous or not homologous, and there is no quantification of homology.

apricot, almond), as well as outside the Rosaceae family (e.g. wheat, rice, grape, orange, peanut) (Pastorello et al., 2000, 2003; Pasquato et al., 2006), are also identified as food allergens in the WHO-IUIS list. The sequence of six contiguous amino acids of *A. niger* 3-phytase that is present in Zea m 14 is not present in any of these other nonspecific lipid transfer proteins identified as food allergens, nor is it present in any other maize proteins identified as food allergens in the WHO-IUIS list. Additionally, further search and bioinformatics analysis using the National Center for Biotechnology Information protein database revealed that the sequence of six contiguous amino acids is present in over 100 other proteins, which include a number of maize phospholipid transfer proteins not identified as food allergens in the WHO-IUIS list.

The other six contiguous amino acids identity of *A. niger* 3-phytase is with Der f 18 protein in house dust mite. Der f 18 is a chitinase of *Dermatophagoides farinae* (American house dust mite) and is identified as an allergen in the WHO-IUIS list. Chitinases from other sources (e.g. latex, banana, papaya) have also been associated with allergenicity (Breiteneder & Ebner, 2000), but none of these allergenic proteins share the sequence of six contiguous amino acids found in *A. niger* 3-phytase. Although cross-reactivity is quite frequently reported between house dust mite allergy and allergic responses to seafood (e.g. snails, molluscs, shrimp), this cross-reactivity is thought to be due to cross-sensitization to tropomyosin (Lopata, O'Hehir & Lehrer, 2010). Further search and bioinformatics analysis using all protein sequences in the National Center for Biotechnology Information protein database revealed that in addition to Der f 18, this sequence of six contiguous amino acids is present in over 100 other proteins, which include other phytases from *A. niger* as well as phytases from other species of *Aspergillus*.

The totality of evidence suggests that these two stretches of six contiguous amino acids are most likely not part of any allergenic epitopes. Therefore, these sequences in *A. niger* 3-phytase will most likely be of no concern for cross-reactivity to allergic or sensitive individuals. The Asp n 25 protein with which greater than 35% identity over several windows of 80 amino acids has been identified is one of the phytases from *A. niger*. Hence, some degree of sequence similarity is expected.

Therefore, despite some marginal similarity to a few sequences in allergen databases, 3-phytase from *A. niger* does not appear to have the characteristics of a potential food allergen, and the Committee considered that oral intake of 3-phytase is not anticipated to pose a risk of allergenicity.

2.2 Toxicological studies

Aspergillus niger is a non-pathogenic and non-toxigenic fungus that has been utilized as a source of enzymes in food for many years. The 3-phytase enzyme preparations from *A. niger* have been tested for the presence of aflatoxin B1, T2 toxin, ochratoxin A, zearalenone and sterigmatocystin, and none were found.

Several toxicological studies were conducted using the *A. niger* 3-phytase enzyme preparation, such as a 14-day oral toxicity study, a 90-day oral toxicity study and genotoxicity studies. The representative batch (FYT/9901/DPP/07) used for the toxicological studies had an activity of 15 500 FTU per gram and a TOS content of 18.5%. The enzyme activity/TOS ratio of this batch was therefore 83.8 FTU per milligram of TOS.

2.2.1 Acute toxicity

No information was available.

2.2.2 Short-term studies of toxicity

A 14-day oral range-finding study (De Hoog, 2000a) was performed in order to examine the tolerability and possible adverse effects of *A. niger* 3-phytase. The study was conducted following the Organisation for Economic Co-operation and Development (OECD) principles of good laboratory practice (GLP). There were four groups of rats: one control group and three treatment groups. Aqueous suspensions of the 3-phytase enzyme preparation (batch FYT/9901/DPP/07) were administered daily to groups of 10 Wistar rats (5 males and 5 females; approximately 6 weeks old at the start of the treatment) by oral gavage for 14 consecutive days at doses of 92.5 mg of TOS per kilogram of body weight (bw) per day (low dose), 278 mg of TOS per kilogram of body weight per day (middle dose) and 833 mg of TOS per kilogram of body weight per day (high dose). The control group received an equivalent volume of water by gavage. Animals were evaluated daily for mortality/viability and clinical signs; weekly for body weight and feed consumption; and at termination for haematology, clinical chemistry, macroscopy and organ weights. No histopathology of organs was performed.

No treatment-related adverse effects were observed with respect to mortality, clinical signs, body weight, organ weights, feed consumption, haematology, clinical chemistry or macroscopic examination. However, some statistically significant differences (at the 5% level) were observed, such as a lower per cent body weight gain (26% less gain than controls) in low- and mid-dose females on day 8 but not on day 14; a decrease (4%) in mean corpuscular haemoglobin concentration (MCHC) in mid-dose males; a decrease (5%) in mean total protein concentration in low-dose males; a decrease (10%) in mean potassium concentration in mid-dose females; and an increase (15%) in absolute heart weight in low-dose females. These effects were random, were not dose dependent, did not occur in both sexes and were not corroborated by other related parameters (e.g. there were no changes in haematocrit values and haemoglobin concentration in the case of the decrease in MCHC, and there were no changes in the heart to body weight ratio in the case of the increase in absolute heart weight). Therefore, these effects were not toxicologically significant, because they were not treatment related.

Overall, it can be concluded that the daily administration of the 3-phytase enzyme preparation at levels up to 833 mg of TOS per kilogram of body weight per day by gavage for 14 days did not result in any treatment-related adverse effects in rats. Therefore, the same dose levels (92.5, 278 and 833 mg of TOS per kilogram of body weight per day) were maintained for a 13-week subchronic study, as described below.

A subchronic oral toxicity study (De Hoog, 2000b) was performed under GLP and conducted in accordance with OECD Test Guideline 408 (Repeated

Dose 90-Day Oral Toxicity Study in Rodents); EEC Directive 87/302/EEC, Part B, Sub-chronic Oral Toxicity Test: 90 day repeated oral dose using rodent species, 1988; and United States Environmental Protection Agency Office of Prevention, Pesticides and Toxic Substances (71010 EPA 712-C-96-199) "Health Effects Test Guidelines" OPPTS 870.3100, 90-Day Oral Toxicity, Public Draft, 1996. There were four groups of rats: one control group and three treatment groups. Based on the results of the 14-day oral range-finding study, the dose levels of the 3-phytase enzyme preparation for the subchronic oral toxicity study were selected to be 92.5 mg of TOS per kilogram of body weight per day (low dose), 278 mg of TOS per kilogram of body weight per day (middle dose) and 833 mg of TOS per kilogram of body weight per day (high dose). Aqueous suspensions of the 3-phytase enzyme preparation (batch FYT/9901/DPP/07) were administered daily to groups of 20 Wistar rats (10 males and 10 females; approximately 6 weeks old at the start of the treatment) by oral gavage for 90 consecutive days. The control group received an equivalent volume of water by oral gavage. Animals were evaluated with respect to general clinical observations, functional observations, ophthalmoscopic examination, body weight, feed consumption, haematology, clinical chemistry, organ weights, macroscopic examination, histopathology of principal organs (control and high-dose groups only), histopathology of liver, kidney and lungs in the mid-dose group and histopathology of all lesions. Mortality/viability and clinical signs were evaluated daily; body weight and feed consumption were evaluated weekly; and all other parameters were evaluated at study termination. Functional observations included testing for hearing ability, pupillary reflex, static righting reflex, grip strength and motor activity. Ophthalmoscopic observations were made for the control and the high-dose treatment groups. As no treatment-related findings were noted, the low- and mid-dose groups were not examined.

In the high-dose group, one female died of pyelonephritis, which was unrelated to treatment. No toxicologically relevant, treatment-related adverse effects were observed with respect to clinical signs, body weight, organ weights, feed consumption, haematology, clinical biochemistry, macroscopic examination or histopathology. Some statistically significant increases (mostly at the 5% level; some at the 1% level) were noted in the body weights of low-dose males from day 8 through day 64 (average increase 8%; determined weekly) and of mid-dose males from day 29 through day 36 (average increase 7%; determined weekly). However, the per cent body weight gains in both males and females were not significantly different in these treatment groups compared with the respective controls. It was also noted that the average weekly feed consumption in low-dose males was slightly higher than that in control, mid-dose and high-dose males, although the values were not statistically significant. In addition, the body weight increases did not show a dose–response relationship and were observed only in low-dose males. Therefore, these body weight increases were not considered to be treatment related. Some other statistically significant differences (at the 5% level) were observed between some treatment groups and the respective controls: for example, increased (15%) absolute liver weight (but not relative liver weight) in low-dose males, decreased (13%) relative testes weight (but not absolute testes weight) in mid-dose males, increased (20%) relative spleen weight (but not absolute spleen weight) in high-dose females, increased (47%) mean neutrophilic granulocyte count in high-dose males, decreased (6%) partial thromboplastin time in low-dose males, decreased (23%) white blood cell count in mid-dose females, decreased (20%) total bilirubin level and increased (46%) triglyceride level in mid-dose males, and increased (14–16%) blood glucose levels in low- and mid-dose males. These changes were not considered to be treatment related or toxicologically relevant, because the effects were random, were not dose dependent and did not occur in both sexes. In the case of organ weights, a change in absolute weight was not corroborated by a corresponding change in relative weight, and vice versa. Additionally, there was no evidence of histopathological abnormalities. Two clinical chemistry parameters showed significant change (at the 5% level) in all treatment groups relative to the respective controls; these were decreased serum chloride level in males (3% across all groups) and increased serum sodium level in females (1% across all groups). However, these changes did not occur in both sexes and were not dose dependent either; only the values in individual treatment groups were significantly different from the respective controls. Therefore, these effects were not considered to be treatment-related adverse effects.

Overall, it can be concluded that no toxicologically relevant effects were seen in this study of general toxicity in rats when the 3-phytase enzyme preparation was administered daily by oral gavage for 90 consecutive days at doses up to 833 mg of TOS per kilogram of body weight per day. The no-observed-adverse-effect level (NOAEL) was identified as the highest dose tested, 833 mg of TOS per kilogram of body weight per day (De Hoog, 2000b).

2.2.3 Long-term studies of toxicity and carcinogenicity

No information was available.

2.2.4 Genotoxicity

The results of two genotoxicity studies with the 3-phytase enzyme preparation from *A. niger* (batch FYT/9901/DPP/07) are summarized in Table 1. The bacterial reverse mutation assay was conducted in accordance with OECD Test Guideline 471 (Bacterial Reverse Mutation Test) and EEC Directive 92/69/EEC (B13 Mutagenicity: *Escherichia coli* Reverse Mutation Assay with WP2*uvrA* only; B14 Mutagenicity: *Salmonella typhimurium* Reverse Mutation Assay). The in vitro chromosomal aberration assay was conducted in accordance with OECD Test Guideline 473 (In Vitro Mammalian Chromosome Aberration Test; 1997). Both studies were certified for compliance with GLP and quality assurance. The results of the studies showed that the 3-phytase enzyme preparation was not genotoxic.

2.2.5 Reproductive toxicity

No information was available.

2.3 Observations in humans

In a double-blind, placebo-controlled intervention study (Troesch et al., 2011) in South African schoolchildren from two primary schools in lower socioeconomic areas (n = 200 at enrolment; mean age ~8 years; mean weight 22.6 kg; 5 in the

Table 1. Genotoxicity of the 3-phytase enzyme preparation

End-point	Test system	Concentration	Result	Reference
Reverse mutation[a]	*Salmonella typhimurium* TA98, TA100, TA1535 and TA1537 and *Escherichia coli* WP2*uvr*A	100–5000 µg dry matter (in solution) per plate ± S9[b]	Negative	Verspeek-Rip (2000)
Chromosomal aberrations in vitro[c]	Human lymphocytes	1st experiment: 1000, 3330 and 5000 µg/ml ± S9 2nd experiment: 1000, 3330 and 5000 µg/ml ± S9	Negative	Bertens (2000)

[a] Positive control substances, sodium azide, 9-aminoacridine, daunomycin, methylmethane sulfonate, 4-nitroquinoline and 2-aminoanthracene, produced the expected increase in the number of revertants. The test material precipitated on the plates at the highest dose level, but the bacterial background lawn was not reduced, and no decrease in the number of revertant colonies was observed, indicating the absence of toxicity of the test material at all dose levels tested. In this assay, the test substance is considered positive (mutagenic) if it produces at least a doubling in the mean number of revertants per plate in one or more strains compared with the solvent control, with or without metabolic activation. The positive response should be reproducible in at least one independently repeated experiment.

[b] The S9 is the $9000 \times g$ supernatant fraction of liver homogenate. It contains microsomes and cytosol and is devoid of mitochondria. The livers used to obtain the S9 fraction were obtained from rats treated with Aroclor 1254, which is a broad-spectrum inducer of cytochrome P450 enzymes. Hence, the S9 fraction acts as an exogenous metabolic activation system.

[c] Positive control chemicals, mitomycin C and cyclophosphamide, both produced a statistically significant increase in the incidence of cells with chromosomal aberrations. In this test, the test substance is considered positive (clastogenic) if 1) it induced a dose-related statistically significant increase in the number of cells with chromosomal structural aberrations; and 2) there is a statistically significant increase in the proportion of cells with chromosomal structural aberrations, even in the absence of a clear dose–response relationship.

treatment group and 3 in the control group did not complete the study), a micronutrient mixture with 3-phytase was used. Ethical approval for the study was obtained from the Medical Research Council of South Africa and ETH Zurich, Switzerland. The study was planned and conducted according to the general principles of good clinical practice and supervised by an independent safety monitoring board. The baseline screening included all children attending preschool through grade 5 in these two schools whose parents had given written informed consent.

All children received a daily bowl of 250 g (wet weight) sweetened maize porridge, 5 times per week. For the treatment group, the micronutrient powder with the 3-phytase enzyme preparation was added to the porridge, whereas for the control group, an identical-appearing powder consisting of the unfortified carrier (dextrose) was added to the porridge. The amount of the 3-phytase enzyme preparation used was 380 FTU per child (190 FTU plus 100% overage to guarantee sufficient activity during storage, although there was no appreciable loss of activity on storage). This resulted in an exposure of 0.160 mg of TOS per kilogram of body weight per day. The intervention lasted for a total of 19 weeks (the intervention was targeted for 23 weeks, but was interrupted by holidays for 3 weeks in July and for 1 week in September). Anthropometric measurements (height, weight, skinfold) were performed, and haemoglobin, iron, zinc and C-reactive protein status in the blood were measured. It was observed that the micronutrient mixture decreased iron deficiency by 75% in the treatment group compared with 35% in the control group and decreased zinc deficiency by 36% in the treatment group compared with 9% in the control group. The authors concluded that adding exogenous phytase to food just before consumption could enhance the release of minerals, leading to increased bioavailability. This study also demonstrated that the 3-phytase enzyme preparation was tolerated by children when administered in the diet for a total of 19 weeks, 5 times per week.

The results of the study by Troesch et al. (2011) corroborated the findings of an earlier study by Sandberg, Hulthén & Türk (1996) that was conducted on 20 healthy adult human volunteers aged 20–52 years. In their study, Sandberg and co-workers investigated the effects of intrinsic wheat bran phytase and *A. niger* phytase on iron absorption. Iron absorption was measured from meals containing white wheat rolls supplemented with wheat bran with or without phytase activity (experiment 1) and phytase-deactivated wheat bran with or without the addition of *A. niger* phytase (experiment 2). The addition of *A. niger* phytase preparation to the phytate-containing meal just before consumption increased iron absorption almost 2-fold (from 14% to 26%). No adverse effects of phytase consumption were reported by the authors. These observations demonstrated that orally administered *A. niger* phytase preparation, in addition to improving iron absorption, was tolerated by adult humans.

The *A. niger* phytase preparation used by Sandberg, Hulthén & Türk (1996) was not the same as the 3-phytase enzyme preparation used by Troesch et al. (2011). Nevertheless, fungal phytases are 3-phytases, and they mostly show acidic pH optima (Wyss et al., 1999). Therefore, the phytases used in these two studies are very similar. Thus, the findings of these two studies are complementary, and together they indicate the safety of oral exposure to 3-phytase in humans.

3. DIETARY EXPOSURE

3-Phytase is an enzyme from *A. niger* expressed in *A. niger* used to degrade phytate found in plant-derived foods, particularly cereal grain and legumes, in order to improve mineral bioavailability. Phytate is a mixed potassium, magnesium and calcium salt of phytic acid that is present as a chelate and storage form for phosphorus in cereals, legumes and oilseeds (Rimbach et al., 2008). The enzyme preparation is intended to be used in large food categories, such as ready-to-use foods (therapeutic and supplementary foods), vitamin and mineral supplements (e.g. multiple micronutrient powders or "Sprinkles"), fortified blended foods (precooked and milled cereals, soya, beans, pulses fortified with micronutrients), fortified flour (wheat, corn and sorghum), breakfast cereals and beverages (ready-to-drink and instant hot beverages), in order to reduce the natural phytic acid/phytate content of food through hydrolysis to phosphate and inositol.

The Committee evaluated one chemical and technical assessment submission received from a manufacturer of the enzyme preparation (DSM, 2011). Depending on the application, the use of the enzyme may exert its effect in two general ways:

- Where the intended use (action) of the enzyme takes place prior to ingestion—for example, during processing of phytate-rich food, such as cereals and legumes—the phytase is not intended to be functional at the time of consumption (technological use with 50–760 FTU per 100 g of cereal or leguminous ingredients is recommended). Enzymatic dephytinization is required to degrade phytate to the recommended low levels (the molar ratio of phytate to iron should be decreased to less than 1:1 and ideally to less than 0.4:1).

- Where the intended use (action) of the enzyme takes place post-ingestion—for example, its inclusion in supplements intended for co-consumption with phytate-rich foods—the phytase is expected to remain functional during the residence time of a meal in the stomach (nutritional use with 190–380 FTU active enzyme per serving plus overage as required, depending on the stability of the enzyme in the food product, is recommended).

Tables 2 and 3 describe applications, anticipated function, intended population, use levels and recommended dosages of 3-phytase as provided by the manufacturer of the enzyme preparation.

3.1 Assessment of dietary exposure

3.1.1 Screening by the budget method

The "budget method" is used to assess theoretical maximum daily dietary exposure. The budget method has been used as a screening method in assessing food additives by JECFA (FAO/WHO, 2001b) and for assessments within the European Union Scientific Cooperation Task 4.2 (EC, 1998).

The method relies on assumptions regarding 1) the level of consumption of foods and of non-milk beverages, 2) the level of presence of the substance in foods and in non-milk beverages and 3) the proportion of foods and of non-milk beverages that may contain the substance. More specifically:

- The levels of consumption of foods and beverages considered are maximum physiological levels of consumption—i.e. the daily consumption of 0.1 litre of non-milk beverages per kilogram of body weight and the daily consumption of 418.4 kJ/kg bw from foods (equivalent to 0.05 kg/kg bw based on an estimated energy density of 8.4 kJ/g). These levels correspond to the daily consumption of 6 litres of non-milk beverages and 3 kg of food in a person with a body weight of 60 kg (typical adult) and a daily consumption of 1.5 litres of non-milk beverages and 750 g of food in a person with a body weight of 15 kg (typical 3-year-old child) (FAO/WHO, 2009).

- The level present in foods and beverages is assumed to be the highest maximum use level of the enzyme preparation, expressed in milligrams of

Table 2. Applications of 3-phytase intended for use in developing countries and/or developed countries (commercially and/or in nutritional intervention programmes)

Intended application	Description and anticipated function of 3-phytase	Intended population	Recommended use level
Vitamin and mineral food supplements	In the case of "MixMe" Multiple Micronutrient Powder, active 3-phytase will be included in the vitamin and mineral powder, to be added to food at the point of consumption ("home fortification"). The enzyme is intended to function during standing and to continue to function in the stomach following consumption.	Infants 6–59 months and pregnant and lactating women	190 FTU plus 100% overage per serving (sachet). The use level was calculated to be sufficient to degrade the phytate contained in a porridge prepared with 60 g of maize flour, and the overage was calculated to ensure sufficient remaining activity after 2 years of storage under field conditions.
Fortified blended foods	A vitamin and mineral premix including active 3-phytase will be added to the dry fortified blended food—e.g. corn soya blend (CSB+/++) or wheat soya blend (WSB+/++)—before packaging. The enzyme is intended to function during porridge preparation and will be denatured once the porridge reaches boiling temperature. In the case of instant CSB (ICSB), 3-phytase is intended to function during the porridge preparation with potable water and to continue to function in the stomach following consumption.	Infants 6–23 months, malnourished infants 6–59 months, malnourished adults, general population	Maximum 320 FTU per 100 g ingredients, depending on the phytate content of the raw materials

continued

Table 2 (continued)

Intended application	Description and anticipated function of 3-phytase	Intended population	Recommended use level
Ready-to-use foods, including ready-to-use supplementary foods, ready-to-use therapeutic foods and lipid-based nutrient supplements	Two uses of phytase in Nutriset's ready-to-use foods: • 3-Phytase is intended to be used during raw material processing when it exerts its function during a preconditioning phase and becomes denatured during extrusion cooking. • 3-Phytase will be incorporated in the packaged product as an active enzyme and is intended to function in the stomach after consumption.	Infants 6–59 months, malnourished infants 6–59 months, malnourished adults, general population	Maximum 760 FTU per 100 g product, depending on the phytate content of the raw materials 380 FTU plus 100% overage per serving
Addition to flour after milling, for use in home baking	A vitamin and mineral premix including active 3-phytase will be added to flour after milling. The enzyme is intended to function during home baking, during the dough leavening and during baking until the dough temperature exceeds 70 °C, when it becomes denatured.	General population	50–320 FTU per 100 g flour, depending on the phytate content of the local flour and the conditions for phytase action (dough pH, temperature, leavening time, kneading time) in different countries

Beverages	Three uses of 3-phytase in beverages are proposed:	General population	
	• Instant cereal-based hot beverages: 3-phytase is intended to exert its function during the preparation of the beverage ingredients and will be denatured by addition of hot water during preparation of the drink.		50–320 FTU per 100 g ingredients, depending on the phytate content of the raw materials
	• Instant cold beverages: incorporation of active 3-phytase in the powder from which the drink is prepared; the enzyme is intended to function in the stomach on co-consumption with cereal-based foods.		190 FTU plus 100% overage per serving
	• Ready-to-drink beverages: incorporation of active 3-phytase in the bottled beverage; the enzyme is intended to function in the stomach on co-consumption with cereal-based foods.		190 FTU plus 600% overage per serving
Dietary or food supplements	Active 3-phytase may be included in a vitamin and mineral or multienzyme supplement, intended to be consumed with or after a phytate-containing meal. The enzyme is intended to function on the co-consumed food in the stomach following consumption.	Adult population, especially vegetarians	190 FTU plus 100% overage per dose
Beverages	Ready-to-drink beverages: incorporation of active 3-phytase in the bottled beverage; the enzyme is intended to function in the stomach on co-consumption with cereal-based foods.	General population, especially vegetarians	190 FTU plus 100% overage per serving

continued

Table 2 (continued)

Intended application	Description and anticipated function of 3-phytase	Intended population	Recommended use level
Breakfast cereals	Breakfast cereals are typically made from cereal pastes or doughs that are cooked and shaped through a variety of methods (roasting, pressing, extrusion, extrusion cooking). 3-Phytase may be applied at the paste or dough phase. This requires a holding time of 2–15 minutes to allow the enzyme to work.	General population, especially vegetarians and children	50–320 FTU per 100 g ingredients, depending on the phytate content of the raw materials

Table 3. Applications, use levels of 3-phytase in food ingredients, amount of TOS in final food and recommended dosage per day

Application	Enzyme use level in food ingredient	Amount of TOS in final food (mg/kg) or per serving (mg)	Recommended dosage per day
Food supplements (including MixMe)	190 FTU plus 100% overage per serving	3.6 mg TOS per serving	1 serving per day (infants, 6–59 months) 2 servings per day (pregnant and lactating women)
Fortified blended foods	Maximum 320 FTU per 100 g ingredients	30 mg TOS per kilogram ingredients	Maximum 200 g fortified blended food
Ready-to-use foods	760 FTU per 100 g product	72 mg TOS per kilogram product	Maximum 200 g food
	380 FTU plus 100% overage per serving	7.2 mg TOS per serving	Maximum 2 servings per day
Addition to flour, breakfast cereals	50–320 FTU per 100 g flour or product	4.7–30 mg TOS per kilogram flour or product	Maximum 200 g food, general population
Beverages	50–320 FTU per 100 g ingredients	4.7–30 mg TOS per kilogram ingredients	Maximum 200 g food, general population
	190 FTU plus 100% overage per serving	3.6 mg TOS per serving	1–3 servings per day, general population
	190 FTU plus 600% overage per serving	12.6 mg TOS per serving	1 serving per day, general population
Dietary or food supplements	190 FTU plus 100% overage per dose	3.6 mg TOS per dose	Maximum 3 doses per day, general population

TOS per kilogram of product, reported in any representative category by the manufacturer, respectively, for foods (72 mg of TOS per kilogram of ready-to-use foods) and for beverages (30 mg of TOS per kilogram of ingredients).

- The proportion of, respectively, solid foods and beverages that may contain the substance is generally set at 12.5% and 25% (100% for children).

Assuming the conservative scenario described above, the Committee estimated that the theoretical maximum daily exposures to the 3-phytase enzyme preparation would be 1.2 mg of TOS per kilogram of body weight in adults ([30 × 0.1 × 0.25] + [72 × 0.05 × 0.125]) and 3.45 mg of TOS per kilogram of body weight in children ([30 × 0.1 × 1] + [72 × 0.05 × 0.125]).

The Committee noted that the dietary exposure to the 3-phytase enzyme preparation provided by the manufacturer based on individual dietary recommended intake would give, respectively, when upper estimates per day per population group for foods and beverages were combined, 0.4 mg of TOS per kilogram of body weight per day in adults and 4.1 mg of TOS per kilogram of body weight per day in children (Table 4).

3.2 General conclusions

3-Phytase is expected to be inactivated in processed food. An estimate of the theoretical maximum dietary exposure to the 3-phytase enzyme preparation from *A. niger* was made by the Committee using the conservative budget method approach. Based on the level of TOS of 28% in the enzyme preparation and its maximum proposed use levels in a variety of phytate-rich food applications, such as ready-to-use foods, vitamin and mineral supplements, fortified blended foods, fortified flour, breakfast cereals and beverages, the Committee estimated theoretical maximum daily exposures of 1.2 mg of TOS per kilogram of body weight in adults and 3.5 mg of TOS per kilogram of body weight in children. These estimates are conservative, as they are made assuming that 100% of food products would be manufactured using the enzyme preparation and that 100% of the enzyme preparation would remain in the final food.

4. COMMENTS

4.1 Assessment of potential allergenicity

3-Phytase was evaluated for potential allergenicity according to the bioinformatics criteria recommended by FAO and WHO (FAO/WHO, 2001a). The amino acid sequence of 3-phytase was compared with the amino acid sequences of known allergens. A similarity search using the Allermatch database did not produce a match with any sequence showing greater than 35% identity over any sliding window of 80 amino acids. However, two stretches of six contiguous amino acids in the *A. niger* 3-phytase sequence were found to be identical to a sequence of six contiguous amino acids in each of two allergenic proteins from the WHO-IUIS list. One match is to Zea m 14 protein in maize, and the other match is to Der f 18 protein in house dust mite. A comparison using the Structural Database of Allergenic Proteins produced one match with Asp n 25 protein in *A. niger* that has greater than 35% identity over several windows of 80 amino acids. Further search and bioinformatics analysis using the National Center for Biotechnology Information protein database revealed that the sequence of six contiguous amino acids of *A. niger* 3-phytase that is present in Zea m 14 is not present in other similar allergenic food proteins, but is present in many non-allergenic proteins. Similarly, the sequence of six contiguous amino acids of *A. niger* 3-phytase that is present in Der f 18 is not present in chitinases from other sources (e.g. latex, banana, papaya) that are associated with allergenicity, but is present in many other proteins, including other phytases from *A. niger* as well as phytases from other species of *Aspergillus*. The Asp n 25 protein is one of the phytases from *A. niger*. Thus, some degree of sequence similarity is expected. Therefore, the Committee considered that oral intake of 3-phytase is not anticipated to pose a risk of allergenicity.

4.2 Toxicological data

Toxicological studies were performed with the 3-phytase enzyme preparation representative of commercial material with an activity of 83.8 FTU per milligram

Table 4. Estimated daily exposure to 3-phytase from the manufacturer

Application	Amount of TOS in final food (mg/kg) or per serving (mg)	Recommended dosage per day	Estimated daily exposure to TOS (mg/kg bw)[a]
Food supplements (including MixMe)	3.6 mg TOS per serving	1 serving per day (infants, 6–59 months) 2 servings per day (pregnant and lactating women)	0.72 mg/kg bw (infants) 0.144 mg/kg bw (adults)
Fortified blended foods	30 mg TOS per kilogram ingredients	Maximum 200 g fortified blended food	1.2 mg/kg bw (infants) 0.12 mg/kg bw (adults)
Ready-to-use foods	72 mg TOS per kilogram product	Maximum 200 g food	2.9 mg/kg bw (infants) 0.29 mg/kg bw (adults)
	7.2 mg TOS per serving	Maximum 2 servings per day	2.9 mg/kg bw (infants) 0.29 mg/kg bw (adults)
Addition to flour, breakfast cereals	4.7–30 mg TOS per kilogram flour or product	Maximum 200 g food, general population	0.19–1.2 mg/kg bw (infants) 0.019–0.12 mg/kg bw (adults)
Beverages	4.7–30 mg TOS per kilogram ingredients	Maximum 200 g food, general population	0.19–1.2 mg/kg bw (infants) 0.019–0.12 mg/kg bw (adults)
	3.6 mg TOS per serving	1–3 servings per day, general population	0.72–2.1 mg/kg bw (infants) 0.072–0.21 mg/kg bw (adults)
	12.6 mg TOS per serving	1 serving per day, general population	2.5 mg/kg bw (infants) 0.25 mg/kg bw (adults)
Dietary or food supplements	3.6 mg TOS per dose	Maximum 3 doses per day, general population	0.22 mg/kg bw (adults)

[a] For infants (6–59 months), a body weight of 5 kg is assumed, and for adults, a body weight of 50 kg is assumed.

of TOS. In a 13-week study of general toxicity in rats, no treatment-related, toxicologically relevant effects were seen when the 3-phytase enzyme preparation was administered daily by gavage at doses up to 833 mg of TOS per kilogram of body weight. The NOAEL was identified as the highest dose tested (i.e. 833 mg of TOS per kilogram of body weight per day). The 3-phytase enzyme preparation was not mutagenic in a bacterial reverse mutation assay in vitro and was not clastogenic in an assay for chromosomal aberrations in human lymphocytes in vitro.

4.3 Assessment of dietary exposure

3-Phytase is expected to be inactivated in processed food. An estimate of the theoretical maximum dietary exposure to the *A. niger* 3-phytase enzyme preparation was made by the Committee using the conservative budget method approach. Based on the level of TOS of 28% in the enzyme preparation and its maximum proposed use levels in a variety of phytate-rich food applications, such as ready-to-use foods, vitamin and mineral supplements, fortified blended foods, fortified flour, breakfast cereals and beverages, the Committee estimated theoretical maximum daily exposures of 1.2 mg of TOS per kilogram of body weight in adults and 3.5 mg of TOS per kilogram of body weight in children. These estimates are conservative, as they are made assuming that 100% of food products would be manufactured using the enzyme preparation and that 100% of the enzyme preparation would remain in the final food.

5. EVALUATION

Comparing the conservative exposure estimate with the NOAEL from the 13-week study of oral toxicity in rats, the margin of exposure is approximately 250. The Committee allocated an acceptable daily intake (ADI) "not specified" for the 3-phytase enzyme preparation from *A. niger* expressed in *A. niger* used in the applications specified and in accordance with good manufacturing practice.

6. REFERENCES

Bertens AMC (2000). Evaluation of the ability of enzyme preparation from *Aspergillus niger* (NPH54) to induce chromosome aberrations in cultured peripheral human lymphocytes. Unpublished study CRO-report no. 263723 from NOTOX, 's-Hertogenbosch, the Netherlands. Submitted to WHO by DSM Nutritional Products, Basel, Switzerland.

Breiteneder H, Ebner C (2000). Molecular and biochemical classification of plant-derived food allergens. *Journal of Allergy and Clinical Immunology*, 106:27–36.

De Hoog SCM (2000a). 14-day range finding study with enzyme preparation from *Aspergillus niger* (NPH54) by daily gavage in the rat. Notox project 270574. Unpublished study CRO-report no. 270574 from NOTOX, 's-Hertogenbosch, the Netherlands. Submitted to WHO by DSM Nutritional Products, Basel, Switzerland.

De Hoog SCM (2000b). 90-day oral toxicity study with enzyme preparation from *Aspergillus niger* (NPH54) by daily gavage in the rat. Unpublished study CRO-report no. 263745 from NOTOX, 's-Hertogenbosch, the Netherlands. Submitted to WHO by DSM Nutritional Products, Basel, Switzerland.

DSM (2011). Chemical and technical assessment of 3-phytase from *Aspergillus niger* expressed in *Aspergillus niger*. Submitted to WHO by DSM Nutritional Products, Basel, Switzerland, 6 December 2011.

EC (1998). Report on methodologies for the monitoring of food additive intake across the European Union. Final report submitted by the Task Coordinator, 16 January 1998. Brussels, Belgium, European Commission, Directorate General III Industry, Reports of a Working Group on Scientific Cooperation on Questions Relating to Food, Task 4.2 (SCOOP/INT/REPORT/2).

FAO/WHO (2001a). Evaluation of allergenicity of genetically modified foods. Report of a Joint FAO/WHO Expert Consultation on Allergenicity of Foods Derived from Biotechnology, 22–25 January 2001. Rome, Italy, Food and Agriculture Organization of the United Nations (http://www.who.int/foodsafety/publications/biotech/en/ec_jan2001.pdf).

FAO/WHO (2001b). Guidelines for the preparation of working papers on intake of food additives for the Joint FAO/WHO Expert Committee on Food Additives. Rome, Italy, Food and Agriculture Organization of the United Nations; Geneva, Switzerland, World Health Organization (http://www.who.int/foodsafety/chem/jecfa/en/intake_guidelines.pdf).

FAO/WHO (2009). Dietary exposure assessment of chemicals in food. In: *Principles and methods for the risk assessment of chemicals in food*. Rome, Italy, Food and Agriculture Organization of the United Nations; Geneva, Switzerland, World Health Organization (Environmental Health Criteria 240; http://whqlibdoc.who.int/ehc/WHO_EHC_240_9_eng_Chapter6.pdf).

Lopata AL, O'Hehir RE, Lehrer SB (2010). Shellfish allergy. *Clinical and Experimental Allergy*, 40:850–858.

Pasquato N et al. (2006). Crystal structure of peach Pru p 3, the prototypic member of the family of plant non-specific lipid transfer protein pan-allergens. *Journal of Molecular Biology*, 356:684–694.

Pastorello EA et al. (2000). The maize major allergen, which is responsible for food-induced allergic reactions, is a lipid transfer protein. *Journal of Allergy and Clinical Immunology*, 106:744–751.

Pastorello EA et al. (2003). Lipid-transfer protein is the major maize allergen maintaining IgE-binding activity after cooking at 100 degrees C, as demonstrated in anaphylactic patients and patients with positive double-blind, placebo-controlled food challenge results. *Journal of Allergy and Clinical Immunology*, 112:775–783.

Rimbach G et al. (2008). Effect of dietary phytate and microbial phytase on minerals and trace element bioavailability—a literature review. *Current Topics in Nutraceutical Research*, 6(3):131–144.

Sandberg AS, Hulthén LR, Türk M (1996). Dietary *Aspergillus niger* phytase increases iron absorption in humans. *Journal of Nutrition*, 126:476–480.

Troesch B et al. (2011). A micronutrient powder with low doses of highly absorbable iron and zinc reduces iron and zinc deficiency and improves weight-for-age Z-scores in South African children. *Journal of Nutrition*, 141:237–242.

Verspeek-Rip CM (2000). Evaluation of the mutagenic activity of enzyme preparation from *Aspergillus niger* (NPH54) in the *Salmonella typhimurium* reverse mutation assay and the *Escherichia coli* reverse mutation assay (with independent repeat). Unpublished study CRO-report no. 263701 from NOTOX, 's-Hertogenbosch, the Netherlands. Submitted to WHO by DSM Nutritional Products, Basel, Switzerland.

Wilms L (2011). Phytase amino acid sequence comparison. Unpublished study report no. REG00058010 from DSM Nutritional Products, Basel, Switzerland. Submitted to WHO by DSM Nutritional Products, Basel, Switzerland.

Wyss M et al. (1999). Biochemical characterization of fungal phytases (*myo*-inositol hexakisphosphate phosphohydrolases): catalytic properties. *Applied and Environmental Microbiology*, 65:367–373.

SERINE PROTEASE (CHYMOTRYPSIN) FROM NOCARDIOPSIS PRASINA *EXPRESSED IN* BACILLUS LICHENIFORMIS

First draft prepared by

S. Choudhuri[1], M. DiNovi[1], J.-C. Leblanc[2], I. Meyland[3], U. Mueller[4] and J. Srinivasan[1]

[1] *Center for Food Safety and Applied Nutrition, Food and Drug Administration, College Park, Maryland, United States of America (USA)*
[2] *Agence nationale de sécurité sanitaire de l'alimentation, de l'environnement et du travail (ANSES), Maisons-Alfort, France*
[3] *Birkerød, Denmark*
[4] *Food Standards Australia New Zealand, Canberra, ACT, Australia*

1. Explanation 39
   1.1 Genetic modification 40
   1.2 Chemical and technical considerations 40
2. Biological data... 41
   2.1 Biochemical aspects 41
   2.2 Toxicological studies 42
      2.2.1 Acute toxicity 42
      2.2.2 Short-term studies of toxicity.. 42
      2.2.3 Long-term studies of toxicity and carcinogenicity........ 43
      2.2.4 Genotoxicity 43
      2.2.5 Reproductive toxicity 44
   2.3 Observations in humans... 44
3. Dietary exposure... 44
   3.1 Assessment of dietary exposure 45
      3.1.1 Screening by the budget method 45
   3.2 General conclusions 46
4. Comments... 46
   4.1 Assessment of potential allergenicity 46
   4.2 Toxicological data 47
   4.3 Assessment of dietary exposure 47
5. Evaluation 48
6. References.. 48

1. EXPLANATION

At the request of the Codex Committee on Food Additives at its Forty-third Session (FAO/WHO, 2011), the Committee evaluated an enzyme preparation containing a serine protease with chymotrypsin specificity (chymotrypsin: Enzyme Commission number 3.4.21.1), which it had not evaluated previously. Serine protease (chymotrypsin) catalyses the hydrolysis of peptide bonds in a protein, preferably at the carboxyl end of Tyr (Tyr-X), Phe (Phe-X) and Trp (Trp-X), where X is not proline. It also catalyses the hydrolysis of peptide bonds at the carboxyl end of other amino acids, primarily Met and Leu, albeit at a slower rate. In this monograph, the expression "serine protease (chymotrypsin)" refers to the serine protease (chymotrypsin) enzyme and its amino acid sequence, and the expression "serine protease (chymotrypsin) enzyme preparation" refers to the serine protease (chymotrypsin) enzyme preparation as formulated for commercial use. The serine protease (chymotrypsin) enzyme preparation is used as a food additive to produce partially or extensively hydrolysed proteins of vegetable and animal origin. Such protein hydrolysates may be used for various applications as ingredients in food and/or beverages.

1.1 Genetic modification

Serine protease (chymotrypsin) is produced from a genetically modified strain of *Bacillus licheniformis* containing the serine protease (chymotrypsin) gene from *Nocardiopsis prasina*. *Bacillus licheniformis* is a Gram-positive bacterium that is widely distributed in nature. It has a long history of use in the production of enzymes used in food processing, including enzymes from genetically engineered strains of the organism.

Prior to the introduction of the serine protease (chymotrypsin) gene, the *B. licheniformis* host strain was genetically modified through deletion of genes responsible for sporulation and two endoproteases. The modified host strain was then transformed with an amplifiable deoxyribonucleic acid (DNA) cassette containing the serine protease (chymotrypsin) gene from *N. prasina*. A strain containing multiple copies of serine protease (chymotrypsin) gene was selected. The recombinant production strain was free of any markers, including antibiotic resistance genes. The final production strain is genetically stable and does not contain antibiotic resistance genes or other heterologous DNA.

1.2 Chemical and technical considerations

Serine protease (chymotrypsin) is produced by submerged, fed-batch, pure culture fermentation of the genetically modified *B. licheniformis* production strain. The enzyme is secreted into the fermentation broth and is subsequently purified and concentrated. The enzyme concentrate is formulated with sodium benzoate, potassium sorbate, glycerol and sorbitol to achieve the desired activity and stability. The serine protease (chymotrypsin) enzyme preparation contains commonly used food-grade materials and conforms to the General Specifications and Considerations for Enzyme Preparations Used in Food Processing (Annex 1, reference *184*). Serine protease (chymotrypsin) activity is measured in protease units (PROT). One PROT is defined as the amount of enzyme that releases 1 µmol of *p*-nitroaniline per minute from substrate (Suc-Ala-Ala-Pro-Phe-pNA) at a concentration of 1 mmol/l at pH 9.0 and at 37 °C. The mean protease activity of three unstandardized batches of enzyme concentrate was 476.3 PROT per milligram of enzyme concentrate. Total organic solids (TOS) consist of the enzyme of interest and residues of organic materials, such as proteins, peptides and carbohydrates, derived from the production organism and the manufacturing process.

The serine protease (chymotrypsin) enzyme preparation is typically used up to a level of 20 g/kg of protein in the product. The serine protease (chymotrypsin) enzyme preparation is expected to be inactivated during processing.

2. BIOLOGICAL DATA

In this monograph, the expression "serine protease (chymotrypsin)" is used when referring to the pure serine protease (chymotrypsin) enzyme and its amino acid sequence, whereas the expression "serine protease (chymotrypsin) enzyme preparation" is used when referring to the enzyme preparation used for various applications.

2.1 Biochemical aspects

Serine protease (chymotrypsin) from *N. prasina* was evaluated for potential allergenicity (Friis, 2009) using bioinformatics criteria recommended in the report of the Joint FAO/WHO Expert Consultation on Allergenicity of Foods Derived from Biotechnology (FAO/WHO, 2001a). According to the FAO/WHO allergenicity recommendation, the possibility of cross-reactivity between a query protein and a known allergen has to be considered if there is 1) more than 35% identity[1] in the amino acid sequence of the expressed protein using a sliding window of 80 amino acids and a suitable gap penalty or 2) an identity of six contiguous amino acids. Using these two criteria, the serine protease (chymotrypsin) sequence from *N. prasina* was compared with the allergenic protein sequences in the Structural Database of Allergenic Proteins (http://fermi.utmb.edu/SDAP/), which allows for the bioinformatics analysis of protein allergenicity using the FAO/WHO recommendation, specifically the six contiguous amino acids identity match. No matches of six contiguous amino acids were found between the *N. prasina* serine protease (chymotrypsin) and any allergenic proteins in the Structural Database of Allergenic Proteins. Also, no matches of 35% amino acid identity were found between the *N. prasina* serine protease (chymotrypsin) and any allergenic proteins using a sliding window of 80 amino acids. However, a 35% amino acid identity was found with Pla a 2 (*Platanus acerifolia* [London plane tree], a large deciduous tree, thought to be a hybrid of *Platanus orientalis* [oriental plane] and *Platanus occidentalis* [American sycamore], which is not listed as a food allergen in the WHO–International Union of Immunological Societies [IUIS] list) if the window length was extended beyond 80 amino acids and gaps were introduced. Nevertheless, multiple gap openings and gap extensions needed to obtain this identity suggest that the identity is most likely not biologically meaningful.

The totality of evidence suggests that the *N. prasina* serine protease (chymotrypsin) does not have the characteristics of a potential food allergen. Therefore, the Committee considered that oral intake of serine protease (chymotrypsin) is not anticipated to pose a risk of allergenicity.

---

[1] *Sequence identity* is quantified as per cent identity and is based on the number of identical amino acid residues in positions in the two sequences being aligned for comparison. Historically, the expression *sequence homology* has been frequently used to mean *sequence identity*, but *sequence homology* is truly an evolutionary term. Sequences are considered homologous if they are derived from a common ancestral sequence. Therefore, sequences are either homologous or not homologous, and there is no quantification of homology.

2.2 Toxicological studies

Several toxicological studies were conducted using the *N. prasina* serine protease (chymotrypsin) enzyme preparation, such as a 90-day oral toxicity study and genotoxicity studies. The representative batch (PPA 26797) used for the toxicological studies had an activity of 54 600 PROT per gram and a TOS content of 9.5%. Therefore, the enzyme activity/TOS ratio of this batch was 574.7 PROT per milligram of TOS.

2.2.1 Acute toxicity

No information was available.

2.2.2 Short-term studies of toxicity

A 13-week oral toxicity study was performed under good laboratory practice (GLP) and conducted in accordance with Organisation for Economic Co-operation and Development (OECD) Test Guideline 408 (Repeated Dose 90-Day Oral Toxicity Study in Rodents) (Glerup, 2009). There were four groups of animals: one control group and three treatment groups. The dose levels of the serine protease (chymotrypsin) enzyme preparation used were 50 mg of TOS per kilogram of body weight (bw) per day (low dose), 165.1 mg of TOS per kilogram of body weight per day (middle dose) and 500.1 mg of TOS per kilogram of body weight per day (high dose). An aqueous solution of the serine protease (chymotrypsin) enzyme preparation (batch PPA 26797) was administered daily to groups of 20 Sprague-Dawley rats (10 males and 10 females; approximately 6 weeks old at the start of the treatment) by gavage for 90 consecutive days. The control group received an equivalent amount of water by gavage. Animals were evaluated with respect to general clinical observations, open field and stimuli-induced tests, body weight, feed consumption, water consumption, ophthalmoscopic examination, haematology, clinical chemistry, organ weights, macroscopic examination and microscopic examination. Major organs and tissues were collected for histopathological examination from all control and high-dose animals, as well as from any animals that died before the scheduled sacrifice.

Mortality/viability and visible signs of ill-health were evaluated daily. Body weight, feed consumption, water consumption and detailed clinical observations, such as skin/fur, eyes, mucous membranes, occurrence of secretions, excretions, autonomic activities (e.g. lacrimation, piloerection, pupil size, unusual respiratory pattern), changes in gait, posture, response to handling, clonic or tonic movements, stereotypies (e.g. excessive grooming, repetitive circling) and bizarre behaviour (e.g. self-mutilation, walking backwards), were recorded weekly. Open field and stimuli-induced tests were administered to all animals once during the last 2 weeks of the study, in order to determine grip strength, motor activity and response to auditory, visual and tactile stimuli. Ophthalmoscopic observations were made for the control and the high-dose treatment groups before the initiation and after the termination of the treatment.

No test article–related deaths occurred. However, one male from the low-dose group died on day 33, and another from the mid-dose group died on day 25. A female from the mid-dose group died on day 7 of the study, and another from the same group was sacrificed moribund on day 8. Subsequent necropsy of these animals showed that they had sustained either oesophageal injury or accidental tracheal dosing as a consequence of the gavage procedure.

No toxicologically relevant, treatment-related adverse effects were observed with respect to stimuli-induced sensory reactivity and open field behaviour, body weight, organ weights, ophthalmoscopic examination, feed consumption, water consumption, haematology, clinical chemistry, macroscopic examination and histopathology. Increased (range 19–38%, average 25%; mostly $P < 0.01$) water consumption in high-dose males was frequently noticed from days 38–42 onwards. Because of the absence of other adverse findings and the lack of increased water consumption of similar frequency in other dose groups, the increased water consumption was not considered to be a treatment-related adverse effect. Some additional statistically significant differences were observed between the control and treatment groups. For example, a decrease (20%; $P < 0.05$) in white blood cell counts was observed in mid-dose females, an increase (22%; $P < 0.01$) in alkaline phosphatase activity was observed in high-dose males, a dose-related decrease (17%, 18% and 19%; $P < 0.01$) in alanine aminotransferase activity was observed in females, a decrease (16%; $P < 0.05$) in aspartate aminotransferase activity was seen in high-dose females and a decrease (17%; $P < 0.01$) in relative liver weight was observed in mid-dose males. These differences were not treatment related and were not seen in both sexes. The decrease in relative liver weight was not accompanied by a decrease in absolute liver weight. The dose-related decrease in alanine aminotransferase activity observed in females was also not considered to be toxicologically significant, because the changes were very small and in the opposite direction (a decrease rather than an increase) to be regarded as toxicologically significant.

Overall, it can be concluded that no toxicologically relevant effects were seen in this study of general toxicity in rats when the serine protease (chymotrypsin) enzyme preparation was administered daily by gavage for 90 consecutive days at doses up to 500.1 mg of TOS per kilogram of body weight per day. The no-observed-adverse-effect level (NOAEL) was identified as the highest dose tested, 500.1 mg of TOS per kilogram of body weight per day (Glerup, 2009).

2.2.3 Long-term studies of toxicity and carcinogenicity

No information was available.

2.2.4 Genotoxicity

The results of the bacterial reverse mutation assay and the in vitro chromosomal aberration assay of genotoxicity with the serine protease (chymotrypsin) enzyme preparation (batch PPA 26797) are summarized in Table 1. The bacterial reverse mutation assay was conducted in accordance with OECD Test Guideline 471 (Bacterial Reverse Mutation Test). The in vitro chromosomal aberration assay was conducted in accordance with OECD Test Guideline 473 (In Vitro Mammalian Chromosome Aberration Test; 1997). Both studies were certified for compliance with GLP and quality assurance. The results of these studies showed that the serine protease (chymotrypsin) enzyme preparation was not genotoxic.

Table 1. Genotoxicity of the serine protease (chymotrypsin) enzyme preparation

End-point	Test system	Concentration	Result	Reference
Reverse mutation[a]	*Salmonella typhimurium* TA98, TA100, TA1535 and TA1537 and *Escherichia coli* WP2*uvr*A	156–5000 µg test substance (in solution) per millilitre[b] ± S9[c]	Negative	Pedersen (2009)
Chromosomal aberrations in vitro[d]	Human lymphocytes	1st experiment: 1582, 2813 and 5000 µg/ml ± S9 2nd experiment: 1311, 2048 and 4000 µg/ml −S9 and 2048, 3200 and 5000 µg/ml +S9	Negative	Whitwell (2009)

[a] Positive control substances, 2-nitrofluorene, 9-aminoacridine, *N*-ethyl-*N*'-nitro-nitrosoguanidine, 2-aminoanthracene and *N*-methyl-*N*'-nitrosoguanidine, produced the expected increase in the number of revertants. In this assay, the test substance is considered positive (mutagenic) if it produces at least a doubling of the mean number of revertants per plate in one or more strains compared with the solvent control.

[b] All bacterial strains were dosed per millilitre of test substance—i.e. exposed in liquid culture ("treat and plate assay").

[c] S9 is the $9000 \times g$ supernatant fraction of liver homogenate. It contains microsomes and cytosol and is devoid of mitochondria. The livers used to obtain the S9 fraction were obtained from rats treated with Aroclor 1254, which is a broad-spectrum inducer of cytochrome P450 enzymes. Hence, the S9 fraction acts as an exogenous metabolic activation system.

[d] Positive control chemicals, 4-nitroquinoline 1-oxide and cyclophosphamide, both produced a statistically significant increase in the incidence of cells with chromosomal aberrations. In this test, the test substance is considered positive (clastogenic) if a proportion of cells with chromosomal structural aberrations at one or more concentrations exceeds the historical negative control range in both replicate cultures; a statistically significant increase in the proportion of cells with chromosomal structural aberrations (excluding gaps) is observed at such concentrations; and there is a concentration-related increase in cells with chromosomal structural aberrations (excluding gaps).

2.2.5 Reproductive toxicity

No information was available.

2.3 Observations in humans

No information was available.

3. DIETARY EXPOSURE

The enzyme is a serine protease that is used to hydrolyse proteins such as casein, whey, soya isolate, soya concentrate, wheat gluten and corn gluten. This enzyme is proposed to be used as a processing aid for the production of partly or extensively hydrolysed proteins of vegetable and animal origin. Such protein hydrolysates may be used for various applications as ingredients in food and/or beverages.

The Committee evaluated one chemical and technical assessment submission received from a manufacturer of the enzyme preparation (Novozymes, 2011).

The serine protease (chymotrypsin) enzyme preparation is used at the minimum levels necessary to achieve the required hydrolysis and according to requirements for normal production following good manufacturing practice. The optimum dosage depends on the desired effect and the specific circumstances in the processing plant. For protein hydrolysis, the dosage recommended by the manufacturer would be up to a maximum of 20 g of enzyme preparation per kilogram of processed protein. The enzyme preparation will be standardized with an approximate content of 7.7% TOS (total organic solids from the fermentation, mainly protein and carbohydrate components), giving an estimate of 1.54 g of TOS per kilogram of protein hydrolysate.

3.1 Assessment of dietary exposure

Although it is expected that the enzyme would be inactivated and residues of the enzyme preparation would be removed in the production of final food products, but also because the resulting protein hydrolysates may be used for a variety of applications as ingredients in food and/or beverages, an upper-bound exposure assessment approach was used to estimate dietary exposure. This approach employed the maximizing assumptions that 100% of the food products would be manufactured using the enzyme preparation and that 100% of the enzyme preparation would remain in the final food products.

3.1.1 Screening by the budget method

The "budget method" is used to assess theoretical maximum daily dietary exposure. The budget method has been used as a screening method in assessing food additives by JECFA (FAO/WHO, 2001b) and for assessments within the European Union Scientific Cooperation Task 4.2 (EC, 1998).

The method relies on assumptions regarding 1) the level of consumption of foods and of non-milk beverages, 2) the level of presence of the substance in foods and in non-milk beverages and 3) the proportion of foods and of non-milk beverages that may contain the substance. More specifically:

- The levels of consumption of foods and beverages considered are maximum physiological levels of consumption—i.e. the daily consumption of 0.1 litre of non-milk beverages per kilogram of body weight and the daily consumption of 418.4 kJ/kg bw from foods (equivalent to 0.05 kg/kg bw based on an estimated energy density of 8.4 kJ/g). These levels correspond to the daily consumption of 6 litres of non-milk beverages and 3 kg of food in a person with a body weight of 60 kg (typical adult) and a daily consumption of 1.5 litres of non-milk beverages and 750 g of food in a person with a body weight of 15 kg (typical 3-year-old child) (FAO/WHO, 2009).

- The level present in foods is assumed to be the highest maximum use level of the enzyme preparation, as reported by the manufacturer, of 1.54 g of TOS per kilogram of protein hydrolysate.
- The proportion of, respectively, solid foods and beverages that may contain the substance is set generally at 12.5% and 25% (100% for children).

Assuming the conservative scenario above and also that all processed foods and beverages contain, respectively, 10% and 5% of protein for hydrolysis, the Committee estimated that the theoretical maximum daily dietary exposures to serine protease (chymotrypsin) from *N. prasina* expressed in *B. licheniformis* would be approximately 2.9 mg of TOS per kilogram of body weight in adults ([1540 × 0.1 × 25% × 5%] + [1540 × 0.05 × 12.5% × 10%]) and 8.7 mg of TOS per kilogram of body weight in children ([1540 × 0.1 × 100% × 5%] + [1540 × 0.05 × 12.5% × 10%]).

3.2 General conclusions

The serine protease (chymotrypsin) enzyme preparation is expected to be inactivated in processed food. An estimate of the theoretical maximum dietary exposure to serine protease (chymotrypsin) was made by the Committee using the conservative budget method approach. Based on the level of TOS of 7.7% in the enzyme preparation and its uses in a variety of applications as ingredients in food and/or beverages at the maximum proposed use level per kilogram of processed protein, the Committee estimated theoretical maximum dietary exposures of 2.9 mg of TOS per kilogram of body weight per day for adults and 8.7 mg of TOS per kilogram of body weight per day for children.

The Committee noted that the above exposure estimates were too conservative, because they were made assuming that 100% of food products would be manufactured using the enzyme preparation and that 100% of the enzyme preparation would remain in the final food products. Therefore, the Committee concluded that a more refined estimate was necessary. Assuming that the serine protease (chymotrypsin) enzyme preparation was used to hydrolyse the entire daily human protein requirement of 1 g/kg bw (WHO/FAO/UNU, 2007) and using a maximum proposed use level of 1540 mg of TOS per kilogram of protein hydrolysate gives a dietary exposure estimate of 1.5 mg of TOS per kilogram of body weight per day. The Committee considered this estimate to be more relevant for the purpose of the safety assessment of the enzyme preparation, as it is based on human physiological protein requirements.

4. COMMENTS

4.1 Assessment of potential allergenicity

Serine protease (chymotrypsin) was evaluated for potential allergenicity according to the bioinformatics criteria recommended by FAO and WHO (FAO/WHO, 2001a). The amino acid sequence of serine protease (chymotrypsin) was compared with the amino acid sequences of known allergens. No matches of six contiguous amino acids were found between the *N. prasina* serine protease (chymotrypsin) and any allergenic proteins in the Structural Database of Allergenic Proteins. Also, no matches of 35% amino acid identity were found between the *N. prasina* serine protease (chymotrypsin) and any allergenic proteins using a sliding window of 80 amino acids. However, a 35% amino acid identity was found with Pla a 2 (*Platanus acerifolia* [London plane tree], which is not listed as a food allergen in the WHO-IUIS list) if the window length was extended beyond 80 amino acids and gaps were introduced. Nevertheless, multiple gap openings and gap extensions needed to obtain this identity suggest that the identity is most likely not biologically meaningful. Therefore, the Committee considered that oral intake of serine protease (chymotrypsin) is not anticipated to pose any risk of allergenicity.

4.2 Toxicological data

Toxicological studies were performed with the serine protease (chymotrypsin) enzyme preparation, which was produced according to the procedure used for commercial production and had an activity of 574.7 PROT per milligram of TOS. In a 13-week study of general toxicity in rats, no toxicologically relevant treatment-related effects were seen when the serine protease (chymotrypsin) enzyme preparation was administered daily by gavage at doses up to 500 mg of TOS per kilogram of body weight. The NOAEL was identified as the highest dose tested (i.e. 500 mg of TOS per kilogram of body weight per day). The serine protease (chymotrypsin) enzyme preparation was not mutagenic in a bacterial reverse mutation assay in vitro and was not clastogenic in an assay for chromosomal aberrations in human lymphocytes in vitro.

4.3 Assessment of dietary exposure

The serine protease (chymotrypsin) enzyme preparation is expected to be inactivated in processed food. An estimate of the theoretical maximum dietary exposure to serine protease (chymotrypsin) was made by the Committee using the conservative budget method approach. Based on the level of TOS of 7.7% in the enzyme preparation and its uses in a variety of applications as ingredients in food and/or beverages at the maximum proposed use levels per kilogram of processed protein, the Committee estimated theoretical maximum dietary exposures of 2.9 mg of TOS per kilogram of body weight per day for adults and 8.7 mg of TOS per kilogram of body weight per day for children.

The Committee noted that the above exposure estimates were too conservative, because they were made assuming that 100% of food products would be manufactured using the enzyme preparation and that 100% of the enzyme preparation would remain in the final food products. Therefore, the Committee concluded that a more refined estimate was necessary. Assuming that the serine protease (chymotrypsin) enzyme preparation was used to hydrolyse the entire daily human protein requirement of 1 g/kg bw (WHO/FAO/UNU, 2007) and using a maximum proposed use level of 1540 mg of TOS per kilogram of protein hydrolysate gives a dietary exposure estimate of 1.5 mg of TOS per kilogram of body weight per day. The Committee considered this estimate to be more relevant for the purpose of the safety assessment of the enzyme preparation, as it is based on human physiological protein requirements.

5. EVALUATION

Comparing the exposure estimate with the NOAEL from the 13-week study of oral toxicity in rats, the margin of exposure is approximately 350. The Committee allocated an acceptable daily intake (ADI) "not specified" for the serine protease (chymotrypsin) enzyme preparation from *N. prasina* expressed in the production strain *B. licheniformis*, used in the applications specified and in accordance with good manufacturing practice.

6. REFERENCES

EC (1998). *Report on methodologies for the monitoring of food additive intake across the European Union*. Final report submitted by the Task Coordinator, 16 January 1998. Brussels, Belgium, European Commission, Directorate General III Industry, Reports of a Working Group on Scientific Cooperation on Questions Relating to Food, Task 4.2 (SCOOP/INT/REPORT/2).

FAO/WHO (2001a). *Evaluation of allergenicity of genetically modified foods. Report of a Joint FAO/WHO Expert Consultation on Allergenicity of Foods Derived from Biotechnology, 22–25 January 2001*. Rome, Italy, Food and Agriculture Organization of the United Nations (http://www.who.int/foodsafety/publications/biotech/en/ec_jan2001.pdf).

FAO/WHO (2001b). *Guidelines for the preparation of working papers on intake of food additives for the Joint FAO/WHO Expert Committee on Food Additives*. Rome, Italy, Food and Agriculture Organization of the United Nations; Geneva, Switzerland, World Health Organization (http://www.who.int/foodsafety/chem/jecfa/en/intake_guidelines.pdf).

FAO/WHO (2009). Dietary exposure assessment of chemicals in food. In: *Principles and methods for the risk assessment of chemicals in food*. Rome, Italy, Food and Agriculture Organization of the United Nations; Geneva, Switzerland, World Health Organization (Environmental Health Criteria 240; http://whqlibdoc.who.int/ehc/WHO_EHC_240_9_eng_Chapter6.pdf).

FAO/WHO (2011). *Report of the Forty-third Session of the Codex Committee on Food Additives, Xiamen, China, 14–18 March 2011*. Rome, Italy, Food and Agriculture Organization of the United Nations, and Geneva, Switzerland, World Health Organization, Joint FAO/WHO Food Standards Programme, Codex Alimentarius Commission, Codex Committee on Food Additives (REP11/FA).

Friis E (2009). Sequence homology of *Nocardiopsis* sp. serine protease (iZyme B) to known toxins and WHO/FAO JECFA recommended allergen analysis of *Nocardiopsis* sp. serine protease (iZyme B). Unpublished study report no. LUNA 2009-14291-02 from Novozymes A/S, Bagsvaerd, Denmark. Submitted to WHO by Novozymes A/S, Bagsvaerd, Denmark.

Glerup P (2009). Serine endopeptidase PPA 26797: a 13-week oral (gavage) toxicity study in rats. Unpublished report of study no. 66063. Amended report no. 2007-42583-02 from LAB Research (Scantox), Ejby, Denmark. Submitted to WHO by Novozymes A/S, Bagsvaerd, Denmark.

Novozymes (2011). Chemical and technical assessment for an approval of the use of a protease enzyme preparation (iZyme B) produced by *Bacillus licheniformis* expressing the genes coding for chymotrypsin from serine protease gene from *Nocardiopsis prasina*. Submitted to WHO by Novozymes A/S, Bagsvaerd, Denmark, 8 December 2011.

Pedersen PB (2009). Serine endopeptidase PPA 26797: test for mutagenic activity with strains of *Salmonella typhimurium* and *Escherichia coli*. Unpublished report of study no. 20078045. Amended report no. 2007-38794-02 from Novozymes A/S, Bagsvaerd, Denmark. Submitted to WHO by Novozymes A/S, Bagsvaerd, Denmark.

Whitwell J (2009). Serine endopeptidase PPA 26797: induction of chromosome aberrations in cultured human peripheral blood lymphocytes. Unpublished report no. 1974/62. Amended report no. 2007-38802-02 from Covance Laboratories Ltd, Harrogate, England. Submitted to WHO by Novozymes A/S, Bagsvaerd, Denmark.

WHO/FAO/UNU (2007). *Protein and amino acid requirements in human nutrition. Report of a Joint WHO/FAO/UNU Expert Consultation, Geneva, 9–16 April 2002.* Geneva, Switzerland, World Health Organization (WHO Technical Report Series, No. 935; http://whqlibdoc.who.int/trs/WHO_TRS_935_eng.pdf).

SERINE PROTEASE (TRYPSIN) FROM FUSARIUM OXYSPORUM EXPRESSED IN FUSARIUM VENENATUM

First draft prepared by

S. Choudhuri[1], M. DiNovi[1], J.-C. Leblanc[2], I. Meyland[3], U. Mueller[4] and J. Srinivasan[1]

[1] Center for Food Safety and Applied Nutrition, Food and Drug Administration, College Park, Maryland, United States of America (USA)
[2] Agence nationale de sécurité sanitaire de l'alimentation, de l'environnement et du travail (ANSES), Maisons-Alfort, France
[3] Birkerød, Denmark
[4] Food Standards Australia New Zealand, Canberra, ACT, Australia

1. Explanation	51
1.1 Genetic modification	52
1.2 Chemical and technical considerations	52
2. Biological data	53
2.1 Biochemical aspects	53
2.2 Toxicological studies	54
2.2.1 Acute toxicity	54
2.2.2 Short-term studies of toxicity	54
2.2.3 Long-term studies of toxicity and carcinogenicity	56
2.2.4 Genotoxicity	56
2.2.5 Reproductive toxicity	56
2.3 Observations in humans	56
3. Dietary exposure	56
3.1 Assessment of dietary exposure	58
3.1.1 Screening budget method	58
3.2 General conclusions	58
4. Comments	59
4.1 Assessment of potential allergenicity	59
4.2 Toxicological data	60
4.3 Assessment of dietary exposure	60
5. Evaluation	60
6. References	61

1. EXPLANATION

At the request of the Codex Committee on Food Additives at its Forty-third Session (FAO/WHO, 2011), the Committee evaluated an enzyme preparation containing a serine protease with trypsin specificity (trypsin: Enzyme Commission number 3.4.21.4), which it had not evaluated previously. Serine protease (trypsin) catalyses the hydrolysis of peptide bonds in a protein, primarily at the carboxyl side of lysine (Lys-X) or arginine (Arg-X), where X is not proline. In this monograph, the expression "serine protease (trypsin)" refers to the serine protease (trypsin) enzyme and its amino acid sequence, and the expression "serine protease (trypsin) enzyme preparation" refers to the serine protease (trypsin) enzyme preparation as formulated for commercial use. The serine protease (trypsin) enzyme preparation is used as a food additive in the manufacture of partially or extensively hydrolysed proteins for applications in food and beverages, for protein fortification and for emulsification or flavour enhancement.

1.1 Genetic modification

Serine protease (trypsin) is produced from a genetically modified strain of *Fusarium venenatum* containing the serine protease (trypsin) gene from *F. oxysporum*. *Fusarium venenatum* is a fungus belonging to the class of hyphomycetales. It is a saprophyte found in the soil and is not considered to be a human pathogen (Joffe, 1986). Although the *Fusarium* species are known for their ability to produce mycotoxins, this *F. venenatum* strain is genetically modified to be non-toxigenic.

Prior to the introduction of the serine protease (trypsin) gene, the *F. venenatum* host strain was rendered incapable of producing trichothecenes and other related toxins by the deletion of the *tri5* gene encoding trichodiene synthase and replacing it with the acetamidase (*amdS*) gene from *A. nidulans* (Royer et al., 1999). The modified host strain was then transformed with an amplifiable plasmid deoxyribonucleic acid (DNA) fragment harbouring the serine protease (trypsin) gene from *F. oxysporum*. The individual transformed colonies were spore-purified, and a high-yielding transformant was selected for enzyme production. The final production strain is genetically stable and does not contain antibiotic resistance genes or other heterologous DNA. Batch analysis demonstrated that the serine protease (trypsin) enzyme preparations from *F. venenatum* were free of aflatoxin B1, T2 toxin, ochratoxin A, zearalenone, sterigmatocystin and diacetoxyscirpenol.

1.2 Chemical and technical considerations

Serine protease (trypsin) is manufactured by submerged, fed-batch, pure culture fermentation of a genetically modified *F. venenatum* production strain. The enzyme is secreted into the fermentation broth and is subsequently purified and concentrated. The enzyme concentrate is formulated with sodium benzoate, potassium sorbate, glycerol and water to achieve the desired activity and stability. The serine protease (trypsin) enzyme preparation contains commonly used food-grade materials and conforms to the General Specifications and Considerations for Enzyme Preparations Used in Food Processing (Annex 1, reference *184*).

Serine protease (trypsin) activity is measured in Kilo Microbial Trypsin Units (KMTU). One KMTU is defined as the amount of enzyme that releases 1 μmol of *p*-nitroaniline per minute from substrate (Ac-Arg-pNA) at a concentration of 1 mmol/l at pH 8.0 and at 37 °C. The mean protease activity of two unstandardized batches of enzyme concentrate was 78 KMTU per gram of enzyme concentrate. Total organic solids (TOS) consist of the enzyme of interest and residues of organic materials, such as proteins, peptides and carbohydrates, derived from the production organism and the manufacturing process. The serine protease (trypsin) enzyme preparation is typically used up to a level of 12 g/kg of protein in the product. The serine protease (trypsin) enzyme preparation is expected to be inactivated during processing.

2. BIOLOGICAL DATA

In this monograph, the expression "serine protease (trypsin)" is used when referring to the pure serine protease (trypsin) enzyme and its amino acid sequence, whereas the expression "serine protease (trypsin) enzyme preparation" is used when referring to the enzyme preparation used for various applications.

2.1 Biochemical aspects

The serine protease (trypsin) from *F. oxysporum* has been evaluated for potential allergenicity (Friis, 2012) using bioinformatics criteria recommended in the report of the Joint FAO/WHO Expert Consultation on Allergenicity of Foods Derived from Biotechnology (FAO/WHO, 2001a). According to the FAO/WHO allergenicity recommendation, the possibility of cross-reactivity between a query protein and a known allergen has to be considered if there is 1) more than 35% identity[1] in the amino acid sequence of the expressed protein using a sliding window of 80 amino acids and a suitable gap penalty or 2) an identity of six contiguous amino acids. Using these two criteria, the serine protease (trypsin) sequence from *F. oxysporum* was compared with the allergenic protein sequences in two databases: the Allermatch database and the Structural Database of Allergenic Proteins. These two databases allow for the bioinformatics analysis of protein allergenicity using the FAO/WHO recommendation, specifically the six contiguous amino acids identity match.

A similarity search using the Allermatch database (http://www.allermatch.org/) produced multiple matches showing a 35% or greater identity in a sliding window of 80 amino acids between the *F. oxysporum* serine protease (trypsin) and several allergenic proteins (Blo t 3, Der f 3, Der f 6, Der p 3, Der p 9, Eur m 3) that are not identified as food allergens in the WHO-International Union of Immunological Societies (WHO-IUIS) list. Similarly, a comparison using the Structural Database of Allergenic Proteins (http://fermi.utmb.edu/SDAP/) produced multiple matches showing 35% or greater identity in different sliding windows of 80 amino acids with the mite allergen Blo t 3. As all these allergenic proteins are serine proteases (trypsin, collagenolytic serine protease, chymotrypsin), some degree of amino acid sequence identity is expected. Further sequence analysis of these 80 amino acid–long windows revealed that the identities are driven by blocks of 4–8 amino acid–long sequences that are identical between the *F. oxysporum* serine protease (trypsin) and the allergen sequence being compared. A separate search for these 4–8 amino acid–long sequences using the National Center for Biotechnology Information protein database revealed that these sequences from various allergenic proteins are also widely distributed in various trypsin, non-trypsin and non-allergenic proteins in prokaryotes as well as lower and higher eukaryotes; some are also present in human trypsin and other human proteins.

---

[1] *Sequence identity* is quantified as per cent identity and is based on the number of identical amino acid residues in positions in the two sequences being aligned for comparison. Historically, the expression *sequence homology* has been frequently used to mean *sequence identity*, but *sequence homology* is truly an evolutionary term. Sequences are considered homologous if they are derived from a common ancestral sequence. Therefore, sequences are either homologous or not homologous, and there is no quantification of homology.

A second similarity search was done using the Allermatch database and the Structural Database of Allergenic Proteins to identify the six contiguous amino acid stretches of the F. oxysporum serine protease (trypsin) that are shared by allergenic proteins. The search produced multiple matches of six contiguous amino acids between the F. oxysporum serine protease (trypsin) and many allergenic proteins from the WHO-IUIS list that are not identified as food allergens. A separate search for these six contiguous amino acid sequences using the National Center for Biotechnology Information protein database revealed that they are widely distributed in various trypsin, trypsin-like, chymotrypsin, chymotrypsin-like and other serine proteases and in non-trypsin and non-allergenic proteins in prokaryotes as well as lower and higher eukaryotes; some are also present in human trypsin and other human proteins. Many of the six contiguous amino acid sequences of the F. oxysporum serine protease (trypsin) that are shared by the allergenic proteins Blo t 3, Bom p 4, Der f 3, Der p 3, Der p 9, Eur m 3 and Tyr p 3 are present in human trypsin as well. This suggests that the amino acid sequences that are shared by the F. oxysporum serine protease (trypsin) and these allergenic proteins are most likely not part of any allergenic epitopes.

Therefore, the totality of evidence suggests that despite multiple 35% or greater identity in a sliding window of 80 amino acids and multiple matches of six contiguous amino acids between the F. oxysporum serine protease (trypsin) and various allergenic proteins, the F. oxysporum serine protease (trypsin) does not appear to have the characteristics of a potential food allergen. Therefore, the Committee considered that oral intake of serine protease (trypsin) is not anticipated to pose a risk of allergenicity.

2.2 Toxicological studies

Several toxicological studies were conducted using the F. oxysporum serine protease (trypsin) enzyme preparation, such as a 13-week oral toxicity study and genotoxicity studies. The representative batch (PPF 26813) used for the toxicological studies had an activity of 117 KMTU per gram and a TOS content of 11%. The enzyme activity/TOS ratio of this batch was therefore 1.06 KMTU per milligram of TOS. Batch analyses demonstrated that the enzyme preparation was free from aflatoxin B1, ochratoxin A, sterigmatocystin, T2 toxin, zearalenone and diacetoxyscirpenol.

2.2.1 Acute toxicity

No information was available.

2.2.2 Short-term studies of toxicity

A 13-week oral toxicity study was performed under good laboratory practice (GLP) and conducted in accordance with Organisation for Economic Co-operation and Development (OECD) Test Guideline 408 (Repeated Dose 90-Day Oral Toxicity Study in Rodents) (Glerup, 2008). There were four groups of animals: one control group and three treatment groups. The dose levels of the serine protease (trypsin) enzyme preparation used were 58 mg of TOS per kilogram of body weight (bw) per day (low dose), 192 mg of TOS per kilogram of body weight per day (middle dose) and 581 mg of TOS per kilogram of body weight per day (high dose). An aqueous suspension of the serine protease (trypsin) enzyme preparation (batch PPF 26813) was administered daily to groups of 20 Sprague-Dawley rats (10 males and 10 females; approximately 6 weeks old at the start of the treatment) by gavage for 92 consecutive days. The control group received an equivalent amount of water by gavage.

Animals were evaluated with respect to general clinical observations, open field and stimuli-induced tests, body weight, feed consumption, water consumption, ophthalmoscopic examination, haematology, clinical chemistry, organ weights, macroscopic examination and microscopic examination. Blood samples were collected for haematology and clinical chemistry before the animals were euthanized at the end of treatment. A macroscopic examination of organs and tissues in situ was performed for all animals at termination, and any macroscopic change was recorded. Major organs and tissues were collected for histopathological examination from all control and high-dose animals, as well as from any animals that died before the scheduled sacrifice. Clinical signs of ill-health and any behavioural changes were recorded daily. All animals were evaluated for reactivity to different types of stimuli, grip strength and motor activity (open field test). Water consumption was recorded twice a week. Body weight, feed consumption and detailed clinical observations, such as skin/fur, eyes, mucous membranes, occurrence of secretions, excretions, autonomic activities (e.g. lacrimation, piloerection, pupil size, unusual respiratory pattern), changes in gait, posture, response to handling, clonic or tonic movements, stereotypies (e.g. excessive grooming, repetitive circling) and bizarre behaviour (e.g. self-mutilation, walking backwards), were recorded weekly. Open field and stimuli-induced tests were administered to all animals once during the last 2 weeks of the study, in order to determine grip strength, motor activity and response to auditory, visual and tactile stimuli. Ophthalmoscopic observations were performed for all animals before the initiation of treatment and then again for the control and the high-dose treatment groups after the termination of treatment.

One male rat from the mid-dose group died on day 35, and one female rat from the mid-dose group died on day 69. Subsequent necropsy of these animals could not determine the cause of death. Other than this, no toxicologically relevant, treatment-related adverse effects were observed with respect to stimuli-induced sensory reactivity and open field behaviour, body weight, organ weight, ophthalmoscopic examinations, feed consumption, water consumption, haematology, clinical chemistry, macroscopic examination and histopathology. Some statistically significant differences were observed between the control group and one or more of the treatment groups. For example, the body weights of high-dose males were lower (7%; $P < 0.05$) on day 14. Lower feed consumption (ranging from 9% to 14% lower than control) was noted in low-dose males in week 4, week 6 and weeks 8–13. As a result, the cumulative (week 1 through week 13) total feed consumption was lower in low-dose males by 10% ($P < 0.01$). On week 9, the feed consumption was also lower (8%; $P < 0.05$) in high-dose males. A lower feed consumption in low-dose males could be due to the fact that low-dose males tended to have the lowest mean body weight. Likewise, lower water consumption was noted in low-dose males (16% lower than controls) on day 84 ($P < 0.05$) and day 87 ($P < 0.01$), in mid-dose males on day 42 (14% lower than controls; $P < 0.05$) and day 84 (21% lower than controls; $P < 0.05$) and in high-dose males on day

87 (12% lower than controls; $P < 0.05$) of the study. Among the clinical chemistry parameters, the alanine aminotransferase activity was lower (16%; $P < 0.05$) in high-dose males. A lower absolute brain weight was observed in low-dose males (3% lower than control; $P < 0.05$) and in high-dose males (4.7% lower than control; $P < 0.01$). All these observations represent isolated data points showing statistical significance. The effects were not treatment related and did not occur in both sexes. In the case of decreased absolute brain weight, there was no corresponding decrease in relative brain weight; in fact, the relative brain weights were slightly increased. Therefore, it was concluded that the effects were not treatment related and hence were not toxicologically significant.

Overall, it can be concluded that no toxicologically relevant effects were seen in this study of general toxicity in rats when the serine protease (trypsin) enzyme preparation was administered daily by gavage for 92 consecutive days at doses up to 581 mg of TOS per kilogram of body weight per day. The no-observed-adverse-effect level (NOAEL) was identified as the highest dose tested, 581 mg of TOS per kilogram of body weight per day (Glerup, 2008).

2.2.3 Long-term studies of toxicity and carcinogenicity

No information was available.

2.2.4 Genotoxicity

The results of the bacterial reverse mutation assay and the in vitro chromosomal aberration assay with the serine protease (trypsin) enzyme preparation (batch PPF 26813) are summarized in Table 1. The bacterial reverse mutation assay was conducted in accordance with OECD Test Guideline 471 (Bacterial Reverse Mutation Test). The in vitro chromosomal aberration assay was conducted in accordance with OECD Test Guideline 473 (In Vitro Mammalian Chromosome Aberration Test; 1997). Both studies were certified for compliance with GLP and quality assurance. The results of these studies showed that the serine protease (trypsin) enzyme preparation was not genotoxic.

2.2.5 Reproductive toxicity

No information was available.

2.3 Observations in humans

No information was available.

3. DIETARY EXPOSURE

The enzyme is a serine protease (trypsin) from *Fusarium oxysporum* expressed in *Fusarium venenatum* that is used to hydrolyse proteins such as casein, whey, soya isolate, soya concentrate, wheat gluten and corn gluten. This enzyme is proposed to be used as a processing aid for production of partly or extensively hydrolysed proteins of vegetable and animal origin. Such protein hydrolysates may be used for various applications as ingredients in food and/or beverages

Table 1. Genotoxicity of the serine protease (trypsin) enzyme preparation

End-point	Test system	Concentration	Result	Reference
Reverse mutation[a]	*Salmonella typhimurium* TA98, TA100, TA1535 and TA1537 and *Escherichia coli* WP2*uvrA*	156–5000 μg test substance (in solution) per millilitre or plate[b] ± S9[c]	Negative	Pedersen (2007)
Chromosomal aberrations in vitro[d]	Human lymphocytes	1st experiment: 2813, 3750 and 5000 μg/ml ± S9 2nd experiment: 3200, 4000 and 5000 μg/ml ± S9	Negative	Whitwell (2007)

[a] Dose was based on dry matter of the test material. Positive control substances, 2-nitrofluorene, 9-aminoacridine, 2-aminoanthracene and *N*-methyl-*N'*-nitrosoguanidine, produced the expected increase in the number of revertants. In this assay, the test substance is considered positive (mutagenic) if it produces at least a doubling of the mean number of revertants per plate in one or more strains compared with the solvent control.

[b] All *Salmonella* strains were dosed per millilitre of test substance—i.e. exposed in liquid culture ("treat and plate assay")—whereas the *E. coli* was dosed per plate—i.e. exposed using the direct plate incorporation assay.

[c] S9 is the 9000 × *g* supernatant fraction of liver homogenate. It contains microsomes and cytosol and is devoid of mitochondria. The livers used to obtain the S9 fraction were obtained from rats treated with Aroclor 1254, which is a broad-spectrum inducer of cytochrome P450 enzymes. Hence, the S9 fraction acts as an exogenous metabolic activation system.

[d] Dose of test material was weighed out as received. Positive control chemicals, 4-nitroquinoline 1-oxide and cyclophosphamide, both produced a statistically significant increase in the incidence of cells with chromosomal aberrations. In this test, the test substance is considered positive (clastogenic) if a proportion of cells with chromosomal structural aberrations at one or more concentrations exceeds the historical negative control range in both replicate cultures; a statistically significant increase in the proportion of cells with chromosomal structural aberrations (excluding gaps) is observed at such concentrations; and there is a concentration-related increase in cells with chromosomal structural aberrations (excluding gaps).

The Committee evaluated one chemical and technical assessment submission received from a manufacturer of the enzyme preparation (Novozymes, 2011).

The serine protease (trypsin) enzyme preparation is used at the minimum levels necessary to achieve the required hydrolysis and according to requirements for normal production following good manufacturing practice. The optimum dosage depends on the desired effect and the specific circumstances in the processing plant. For protein hydrolysis, the recommended dosage by the manufacturer would be up to a maximum of 12 g of enzyme preparation per kilogram of processed protein. The enzyme preparation will be standardized with an approximate content of 4% TOS (total organic solids from the fermentation, mainly protein and carbohydrate components), giving an estimate of 480 mg of TOS per kilogram of protein hydrolysate.

3.1 Assessment of dietary exposure

Although it is expected that the enzyme would be inactivated and residues of the enzyme preparation would be removed in the production of final food products, but also because the resulting protein hydrolysates may be used for a variety of applications as ingredients in food and/or beverages, an upper-bound exposure assessment approach was used to estimate dietary exposure. This approach employed the maximizing assumptions that 100% of the food products would be manufactured using the enzyme preparation and that 100% of the enzyme preparation would remain in the final food products.

3.1.1 Screening by the budget method

The "budget method" is used to assess theoretical maximum daily dietary exposure. The budget method has been used as a screening method in assessing food additives by JECFA (FAO/WHO, 2001b) and for assessments within the European Union Scientific Cooperation Task 4.2 (EC, 1998).

The method relies on assumptions regarding 1) the level of consumption of foods and of non-milk beverages, 2) the level of presence of the substance in foods and in non-milk beverages and 3) the proportion of foods and of non-milk beverages that may contain the substance. More specifically:

- The levels of consumption of foods and beverages considered are maximum physiological levels of consumption—i.e. the daily consumption of 0.1 litre of non-milk beverages per kilogram of body weight and the daily consumption of 418.4 kJ/kg bw from foods (equivalent to 0.05 kg/kg bw based on an estimated energy density of 8.4 kJ/g). These levels correspond to the daily consumption of 6 litres of non-milk beverages and 3 kg of food in a person with a body weight of 60 kg (typical adult) and a daily consumption of 1.5 litres of non-milk beverages and 750 g of food in a person with a body weight of 15 kg (typical 3-year-old child) (FAO/WHO, 2009).

- The level present in foods is assumed to be the highest maximum use level of the enzyme preparation, as reported by the manufacturer, of 480 mg of TOS per kilogram of protein hydrolysate.

- The proportion of, respectively, solid foods and beverages that may contain the substance is set generally at 12.5% and 25% (100% for children).

Assuming the conservative scenario above and also that all processed foods and beverages contain, respectively, 10% and 5% of protein for hydrolysis, the Committee estimated that the theoretical maximum daily dietary exposures to serine protease (trypsin) from *F. oxysporum* expressed in *F. venenatum* would be approximately 0.9 mg of TOS per kilogram of body weight in adults ([480 × 0.1 × 25% × 5%] + [480 × 0.05 × 12.5% × 10%]) and 2.7 mg of TOS per kilogram of body weight in children ([480 × 0.1 × 100% × 5%] + [480 × 0.05 × 12.5% × 10%]).

3.2 General conclusions

The serine protease (trypsin) enzyme preparation is expected to be inactivated in processed food. An estimate of the theoretical maximum dietary exposure to the serine protease (trypsin) enzyme preparation was made by the Committee using the conservative budget method approach. Based on the level of TOS of 4% in the enzyme preparation and its uses in a variety of applications as ingredients in food and/or beverages at the maximum proposed use level per kilogram of processed protein, the Committee estimated theoretical maximum daily exposures of 0.9 mg of TOS per kilogram of body weight in adults and 2.7 mg of TOS per kilogram of body weight in children.

The Committee noted that the above exposure estimates were too conservative because they were made assuming that 100% of food products would be manufactured using the enzyme preparation and that 100% of the enzyme preparation would remain in the final food products. Therefore, the Committee concluded that a more refined estimate was necessary. Assuming that the serine protease (trypsin) enzyme preparation was used to hydrolyse the entire daily human protein requirement of 1 g/kg bw per day (WHO/FAO/UNU, 2007) and using a maximum proposed use level of 480 mg of TOS per kilogram of protein hydrolysate gives a dietary exposure estimate of 0.5 mg of TOS per kilogram of body weight per day. The Committee considered this estimate to be more relevant for the purpose of the safety assessment of the enzyme preparation, as it is based on human physiological protein requirements.

4. COMMENTS

4.1 Assessment of potential allergenicity

Serine protease (trypsin) was evaluated for potential allergenicity according to the bioinformatics criteria recommended by FAO and WHO (FAO/WHO, 2001a). A similarity search using the Allermatch database produced multiple matches showing a 35% or greater identity in a sliding window of 80 amino acids between the *F. oxysporum* serine protease (trypsin) and several allergenic proteins (Blo t 3, Der f 3, Der f 6, Der p 3, Der p 9, Eur m 3) that are not identified as food allergens in the WHO-IUIS list. Similarly, a comparison using the Structural Database of Allergenic Proteins produced multiple matches showing 35% or greater identity in different sliding windows of 80 amino acids with the mite allergen Blo t 3. A second similarity search was performed using the Allermatch database and the Structural Database of Allergenic Proteins to identify the six contiguous amino acid stretches of the *F. oxysporum* serine protease (trypsin) that are shared by allergenic proteins. The search produced multiple matches of six contiguous amino acids between the *F. oxysporum* serine protease (trypsin) and many allergenic proteins from the WHO-IUIS list that are not identified as food allergens. Further search and bioinformatics analysis using the National Center for Biotechnology Information protein database revealed that these sequences are widely distributed in various trypsin, trypsin-like, chymotrypsin, chymotrypsin-like and other serine proteases and in non-trypsin and non-allergenic proteins in prokaryotes as well as lower and higher eukaryotes. Many of the sequences of six contiguous amino acids of the *F. oxysporum* serine protease (trypsin) that are shared by the allergenic proteins Blo t 3, Bom p 4, Der f 3, Der p 3, Der p 9, Eur m 3 and Tyr p 3 are present in human trypsin as well. Therefore, the Committee considered that oral intake of serine protease (trypsin) is not anticipated to pose a risk of allergenicity.

4.2 Toxicological data

Toxicological studies were performed with the serine protease (trypsin) enzyme preparation using a batch that was representative of commercial material and had an activity of 1.06 KMTU per milligram of TOS. In a 13-week study of general toxicity in rats, no toxicologically relevant treatment-related effects were seen when the serine protease (trypsin) enzyme preparation was administered daily by gavage at doses up to 581 mg of TOS per kilogram of body weight. The NOAEL was identified as the highest dose tested (i.e. 581 mg of TOS per kilogram of body weight per day). The serine protease (trypsin) enzyme preparation was not mutagenic in a bacterial reverse mutation assay in vitro and was not clastogenic in an assay for chromosomal aberrations in human lymphocytes in vitro.

4.3 Assessment of dietary exposure

The serine protease (trypsin) enzyme preparation is expected to be inactivated in processed food. An estimate of the theoretical maximum dietary exposure to the serine protease (trypsin) enzyme preparation was made by the Committee using the conservative budget method approach. Based on the level of TOS of 4% in the enzyme preparation and its uses in a variety of applications as ingredients in food and/or beverages at the maximum proposed use level per kilogram of processed protein, the Committee estimated theoretical maximum daily exposures of 0.9 mg of TOS per kilogram of body weight in adults and 2.7 mg of TOS per kilogram of body weight in children.

The Committee noted that the above exposure estimates were too conservative because they were made assuming that 100% of food products would be manufactured using the enzyme preparation and that 100% of the enzyme preparation would remain in the final food products. Therefore, the Committee concluded that a more refined estimate was necessary. Assuming that the serine protease (trypsin) enzyme preparation was used to hydrolyse the entire daily human protein requirement of 1 g/kg bw per day (WHO/FAO/UNU, 2007) and using a maximum proposed use level of 480 mg of TOS per kilogram of protein hydrolysate gives a dietary exposure estimate of 0.5 mg of TOS per kilogram of body weight per day. The Committee considered this estimate to be more relevant for the purpose of the safety assessment of the enzyme preparation, as it is based on human physiological protein requirements.

5. EVALUATION

Comparing the dietary exposure estimate with the NOAEL from the 13-week study of oral toxicity in rats, the margin of exposure is approximately 1200. The Committee allocated an acceptable daily intake (ADI) "not specified" for the serine protease (trypsin) enzyme preparation from *F. oxysporum* expressed in the production strain *F. venenatum*, used in the applications specified and in accordance with good manufacturing practice.

6. REFERENCES

EC (1998). *Report on methodologies for the monitoring of food additive intake across the European Union*. Final report submitted by the Task Coordinator, 16 January 1998. Brussels, Belgium, European Commission, Directorate General III Industry, Reports of a Working Group on Scientific Cooperation on Questions Relating to Food, Task 4.2 (SCOOP/INT/REPORT/2).

FAO/WHO (2001a). *Evaluation of allergenicity of genetically modified foods. Report of a Joint FAO/WHO Expert Consultation on Allergenicity of Foods Derived from Biotechnology, 22–25 January 2001*. Rome, Italy, Food and Agriculture Organization of the United Nations (http://www.who.int/foodsafety/publications/biotech/en/ec_jan2001.pdf).

FAO/WHO (2001b). *Guidelines for the preparation of working papers on intake of food additives for the Joint FAO/WHO Expert Committee on Food Additives*. Rome, Italy, Food and Agriculture Organization of the United Nations; Geneva, Switzerland, World Health Organization (http://www.who.int/foodsafety/chem/jecfa/en/intake_guidelines.pdf).

FAO/WHO (2009). Dietary exposure assessment of chemicals in food. In: *Principles and methods for the risk assessment of chemicals in food*. Rome, Italy, Food and Agriculture Organization of the United Nations; Geneva, Switzerland, World Health Organization (Environmental Health Criteria 240; http://whqlibdoc.who.int/ehc/WHO_EHC_240_9_eng_Chapter6.pdf).

FAO/WHO (2011). *Report of the Forty-third Session of the Codex Committee on Food Additives, Xiamen, China, 14–18 March 2011*. Rome, Italy, Food and Agriculture Organization of the United Nations, and Geneva, Switzerland, World Health Organization, Joint FAO/WHO Food Standards Programme, Codex Alimentarius Commission, Codex Committee on Food Additives (REP11/FA).

Friis E (2012). WHO/FAO JECFA recommended allergen analysis of *Fusarium oxysporum* protease. Unpublished study report no. LUNA 2012-09879-01 from Novozymes A/S, Bagsvaerd, Denmark. Submitted to WHO by Novozymes A/S, Bagsvaerd, Denmark.

Glerup P (2008). 3-month toxicity study in rats. Unpublished study test report no. SP 387/TL1 of study no. 65860 from LAB Research (Scantox), Ejby, Denmark. Submitted to WHO by Novozymes A/S, Bagsvaerd, Denmark.

Joffe AZ (1986). Human infections associated with *Fusarium* species. In: *Fusarium species: their biology and toxicology*. John Wiley & Sons, pp. 293–298.

Novozymes (2011). Chemical and technical assessment for a protease enzyme preparation produced by a strain of *Fusarium venenatum* expressing the serine protease gene from *Fusarium oxysporum*. Submitted to WHO by Novozymes A/S, Bagsvaerd, Denmark.

Pedersen PB (2007). SP387/TL1, PPF26813: test for mutagenic activity with strains of *Salmonella typhimurium* and *Escherichia coli*. Unpublished study test report no. SP387/TL1, PPF26813 of study no. 20078062 from Novozymes A/S, Bagsvaerd, Denmark. Submitted to WHO by Novozymes A/S, Bagsvaerd, Denmark.

Royer JC et al. (1999). Deletion of the trichodiene synthase gene of *Fusarium venenatum*: two systems for repeated gene deletions. *Fungal Genetics and Biology*, 28:68–78.

Whitwell J (2007). SP387/TL1: induction of chromosome aberrations in cultured human peripheral blood lymphocytes. Unpublished report no. 1974/63-D6172 from Covance Laboratories Ltd, Harrogate, England. Submitted to WHO by Novozymes A/S, Bagsvaerd, Denmark.

WHO/FAO/UNU (2007). *Protein and amino acid requirements in human nutrition. Report of a Joint WHO/FAO/UNU Expert Consultation, Geneva, 9–16 April 2002*. Geneva, Switzerland, World Health Organization (WHO Technical Report Series, No. 935; http://whqlibdoc.who.int/trs/WHO_TRS_935_eng.pdf).

SAFETY EVALUATIONS OF GROUPS OF RELATED FLAVOURING AGENTS

INTRODUCTION

Assignment to structural class

Twelve groups of flavouring agents were evaluated using the Procedure for the Safety Evaluation of Flavouring Agents as outlined in Figure 1 (Annex 1, references *116, 122, 131, 137, 143, 149, 154, 160, 166, 173* and *178*). In applying the Procedure, the chemical is first assigned to a structural class as identified by the Committee at its forty-sixth meeting (Annex 1, reference *122*). The structural classes are as follows:

- *Class I.* Flavouring agents that have simple chemical structures and efficient modes of metabolism that would suggest a low order of toxicity by the oral route.
- *Class II.* Flavouring agents that have structural features that are less innocuous than those of substances in class I but are not suggestive of toxicity. Substances in this class may contain reactive functional groups.
- *Class III.* Flavouring agents that have structural features that permit no strong initial presumption of safety or may even suggest significant toxicity.

A key element of the Procedure involves determining whether a flavouring agent and the product(s) of its metabolism are innocuous and/or endogenous substances. For the purpose of the evaluations, the Committee used the following definitions, adapted from the report of its forty-sixth meeting (Annex 1, reference *122*):

- *Innocuous metabolic products* are defined as products that are known or readily predicted to be harmless to humans at the estimated dietary exposure to the flavouring agent.
- *Endogenous substances* are intermediary metabolites normally present in human tissues and fluids, whether free or conjugated; hormones and other substances with biochemical or physiological regulatory functions are not included. The estimated dietary exposure to a flavouring agent that is, or is metabolized to, an endogenous substance should be judged not to give rise to perturbations outside the physiological range.

Assessment of dietary exposure

Maximized survey-derived intake (MSDI)

Estimates of the dietary exposure to flavouring agents by populations are based on annual volumes of production. These data were derived from surveys in Europe, Japan and the USA. Manufacturers were requested to exclude use of flavouring agents in pharmaceutical, tobacco or cosmetic products when compiling these data. When using these production volumes to estimate dietary exposures, a correction factor of 0.8 is applied to account for under-reporting.

Figure 1. Procedure for the Safety Evaluation of Flavouring Agents

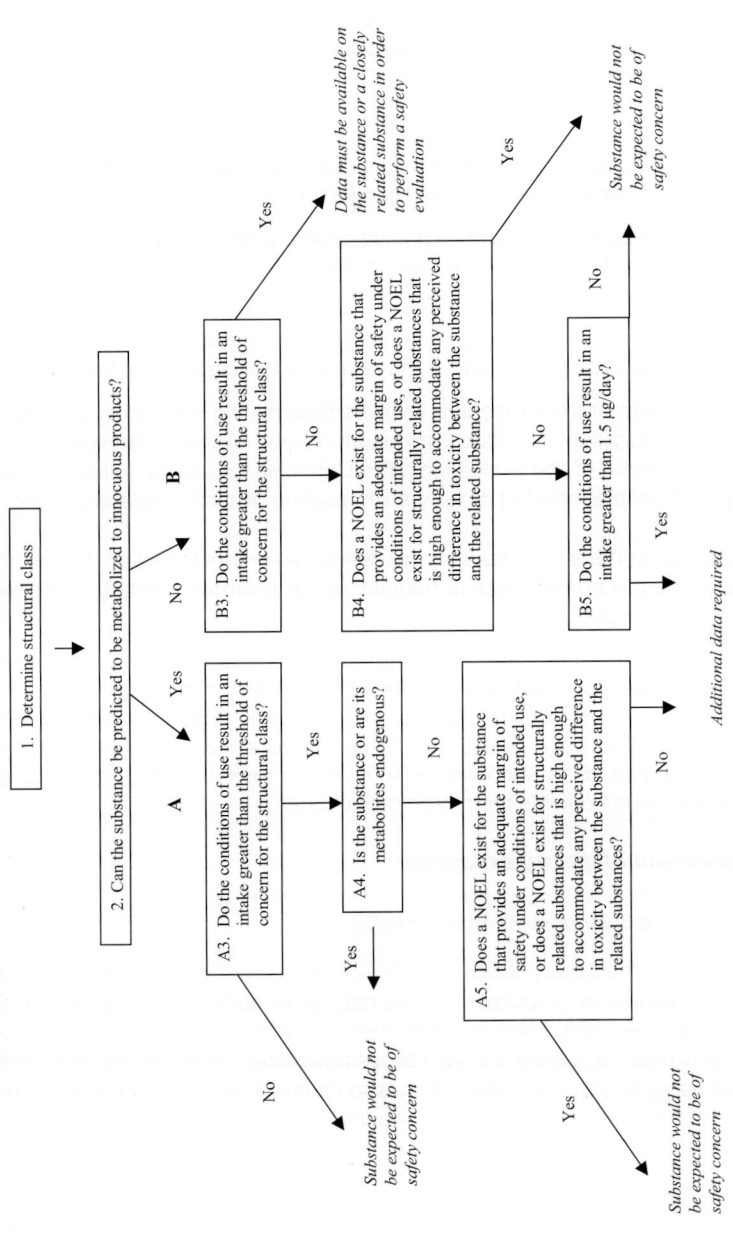

INTRODUCTION

$$\text{MSDI (µg/day)} = \frac{\text{annual volume of production (kg)} \times 10^9 \text{ (µg/kg)}}{\text{population of consumers} \times 0.8 \times 365 \text{ days}}$$

The population of consumers was assumed to be 32×10^6 in Europe, 13×10^6 in Japan and 31×10^6 in the USA.

Single portion exposure technique (SPET)

The SPET was developed by the Committee at its sixty-seventh meeting (Annex 1, reference *184*) to account for presumed patterns of consumer behaviour with respect to food consumption and the possible uneven distribution of dietary exposures among consumers of foods containing flavouring agents. It is based on reported use levels supplied by the industry. This single portion–derived estimate was designed to account for individuals' brand loyalty to food products and for niche products that would be expected to be consumed by only a small proportion of the population. Its use in the Procedure was endorsed at the sixty-ninth meeting of the Committee (Annex 1, reference *190*) to render the safety assessment more robust, replacing the sole use of MSDI estimates with the higher of the highest MSDI or the SPET estimate as the exposure estimate in the decision-tree. The Committee also agreed that it would not be necessary to re-evaluate flavouring agents that had already been assessed previously using the Procedure.

The SPET provides an estimate of dietary exposure for an individual who consumes a specific food product containing the flavouring agent every day. The SPET combines an average (or usual) added use level provided by the flavour industry with a standard portion size from 75 predefined food categories as described by the Committee at its sixty-seventh meeting. The standard portion is taken to represent the mean food consumption for consumers of these food categories. Among all the food categories with a reported use level, the calculated dietary exposure from the single food category leading to the highest dietary exposure from one portion is taken as the SPET estimate:

$$\text{SPET (µg/day)} = \text{standard portion size of food category } i \text{ (g/day)} \times \text{use level for food category } i \text{ (µg/g)}$$

The highest result is used in the evaluation.

The use level data provided by industry for each flavouring agent evaluated at this meeting and used in the SPET calculations are available on the WHO JECFA web site at http://www.who.int/foodsafety/chem/jecfa/publications/en/index.html.

Consideration of combined intakes from use as flavouring agents

The safety assessment of possible combined intakes of flavouring agents was based on the presence of common metabolites or a homologous series (as proposed at the sixty-eighth meeting; Annex 1, reference *187*) and using the MSDI exposure assessment (as proposed at the sixty-ninth meeting; Annex 1, reference *190*).

ALIPHATIC AND AROMATIC AMINES AND AMIDES (addendum)

First draft prepared by

B. Fields[1], M. DiNovi[2], J.-C. Leblanc[3], U. Mueller[1] and A. Renwick[4]

[1] Food Standards Australia New Zealand, Canberra, ACT, Australia
[2] Center for Food Safety and Applied Nutrition, Food and Drug Administration, College Park, Maryland, United States of America (USA)
[3] Agence nationale de sécurité sanitaire de l'alimentation, de l'environnement et du travail (ANSES), Maisons-Alfort, France
[4] School of Medicine, University of Southampton, Southampton, England

1. Evaluation	69
1.1 Introduction	69
1.2 Assessment of dietary exposure	70
1.3 Absorption, distribution, metabolism and elimination	71
1.4 Application of the Procedure for the Safety Evaluation of Flavouring Agents	71
1.5 Consideration of combined intakes from use as flavouring agents	80
1.6 Conclusion	80
2. Relevant background information	81
2.1 Explanation	81
2.2 Additional considerations on dietary exposure	81
2.3 Biological data	81
2.3.1 Biochemical data: absorption, distribution, metabolism and elimination	81
(a) 2-Aminoacetophenone (No. 2043)	81
(b) N-(2-Methylcyclohexyl)-2,3,4,5,6-pentafluorobenzamide (No. 2081)	81
(c) 3[(4-Amino-2,2-dioxido-1H-2,1,3-benzothiadiazin-5-yl)oxy]-2,2-dimethyl-N-propylpropanamide (No. 2082)	82
2.3.2 Toxicological studies	82
(a) Acute toxicity	82
(b) Short-term studies of toxicity	83
(c) Genotoxicity	89
3. References	89

1. EVALUATION

1.1 Introduction

The Committee evaluated an additional seven flavouring agents belonging to the group of aliphatic and aromatic amines and amides. The additional flavouring agents comprised one aniline (No. 2043), three menthyl amides (Nos 2078–2080) and three amides with alicyclic, aromatic or alkyl side-chains, including one with a pentafluorophenyl group (Nos 2077, 2081 and 2082). The evaluations were conducted using the Procedure for the Safety Evaluation of Flavouring Agents (see Figure 1, Introduction) (Annex 1, reference *131*). None of these flavouring agents has previously been evaluated by the Committee. Four of the seven flavouring agents in this group (Nos 2077 and 2080–2082) are reported to be flavour modifiers.

The Committee evaluated 37 other members of this group of flavouring agents at its sixty-fifth meeting (Annex 1, reference *178*). For 36 of these flavouring agents, the Committee concluded that they would not give rise to safety concerns based on estimated dietary exposures. For acetamide (No. 1592), the Committee considered it inappropriate for use as a flavouring agent or for food additive purposes based on the available data indicating carcinogenicity in mice and rats. For 27 flavouring agents, the dietary exposure estimates were based on anticipated annual volumes of production, and these evaluations were conditional pending submission of use levels or poundage data, which were provided at the sixty-ninth meeting (Annex 1, reference *190*). For the evaluation of 2-isopropyl-*N*-2,3-trimethylbutyramide (No. 1595), additional data available at the sixty-ninth meeting raised safety concerns, and the Committee concluded that the Procedure could not be applied to this flavouring agent until additional safety data became available.

At its sixty-eighth meeting, the Committee evaluated 12 additional members of this group of flavouring agents and concluded that all 12 were of no safety concern at estimated dietary exposures (Annex 1, reference *187*).

The Committee evaluated nine additional members of this group of flavouring agents at its seventy-third meeting (Annex 1, reference *202*). The Committee concluded that five of the nine flavouring agents did not raise any safety concerns at estimated dietary exposures. For one of the remaining four flavouring agents (No. 2007), the available data did not provide an adequate margin of exposure, and for the other three flavouring agents (Nos 2005, 2010 and 2011), no additional data were available. The Committee concluded that for these four flavouring agents, further data would be required to complete the safety evaluation.

One of the seven flavouring agents considered at the current meeting— namely, 2-aminoacetophenone (No. 2043)—is a natural component of food and has been detected in maize, corn chips, tuna, egg white, milk, soya milk, green tea, honey and beer (Kumazawa & Masuda, 1999; Kaneko, Kumazawa & Nishimura, 2011; Nijssen, van Ingen-Visscher & Donders, 2011; Scott-Thomas, Pearson & Chambers, 2011).

1.2   Assessment of dietary exposure

The total annual volumes of production of the seven aliphatic and aromatic amines and amides are approximately 362 kg in the USA, 0.1 kg in Europe and 0.1 kg in Japan (European Flavour and Fragrance Association, 2004; Japan Flavor and Fragrance Materials Association, 2005; Gavin, Williams & Hallagan, 2008; International Organization of the Flavor Industry, 2011). Approximately 83% of the total annual volume of production in the USA is accounted for by one flavouring agent in this group—namely, (2E,6E/Z,8E)-N-(2-methylpropyl)-2,6,8-decatrienamide (No. 2077).

Dietary exposures were estimated using the maximized survey-derived intake (MSDI) method and the single portion exposure technique (SPET). The highest estimates were all derived using the SPET and are reported in Table 1. The estimated daily dietary exposure is highest for (2E,6E/Z,8E)-N-(2-methylpropyl)-2,6,8-decatrienamide (No. 2077) (4500 µg, the SPET value obtained from non-alcoholic beverages). For the other flavouring agents, daily dietary exposures as SPET or MSDI estimates range from 0.01 to 3000 µg, with the SPET yielding the highest estimate in each case.

Annual volumes of production of this group of flavouring agents as well as the daily dietary exposures calculated using both the MSDI method and the SPET are summarized in Table 2.

1.3 Absorption, distribution, metabolism and elimination

The metabolism of aliphatic and aromatic amines and amides was described in the reports of the sixty-fifth, sixty-eighth and seventy-third meetings of the Committee (Annex 1, references *178, 187* and *202*).

In general, aliphatic and aromatic amines and amides are rapidly absorbed from the gastrointestinal tract and metabolized by deamination, hydrolysis or oxidation to polar metabolites that are readily eliminated in the urine. Aliphatic amides have been reported to undergo hydrolysis in mammals; however, the rate of hydrolysis is dependent on the chain length and may involve a number of different enzymes.

In relation to the additional flavouring agents considered at the current meeting of the Committee, only limited information regarding metabolic pathways is available for specific substances.

Published studies indicate that 2-aminoacetophenone (No. 2043) is a minor intermediate of tryptophan metabolism that is further metabolized to anthranilic acid, which is excreted in urine or involved in subsequent metabolic pathways.

Distribution and metabolism studies were provided for N-(2-methylcyclohexyl)-2,3,4,5,6-pentafluorobenzamide (No. 2081). Widespread tissue distribution of parent compound and/or metabolites was observed following oral administration to rats. Minimal metabolite formation was observed following incubation with rat liver microsomes. Those metabolites that were identified were hydroxylation products.

A pharmacokinetic study on 3[(4-amino-2,2-dioxido-1H-2,1,3-benzothiadiazin-5-yl)oxy]-2,2-dimethyl-N-propylpropanamide (No. 2082) indicated rapid absorption following oral administration to rats; however, oral bioavailability was less than 10%. Minimal metabolite formation was observed following incubation with rat and human liver microsomes. Identified metabolites were hydroxylation products.

1.4 Application of the Procedure for the Safety Evaluation of Flavouring Agents

Step 1. In applying the Procedure for the Safety Evaluation of Flavouring Agents to the additional flavouring agents in this group, the Committee assigned one

Table 1. Summary of the results of the safety evaluations of aliphatic and aromatic amines and amides used as flavouring agents[a,b,c]

Flavouring agent	No.	CAS No. and structure	Step A3/B3[d] Does estimated dietary exposure exceed the threshold of concern?	Follow-on from step B3[e] Are additional data available for flavouring agent with an estimated dietary exposure exceeding the threshold of concern? Step A4/B4[e] Adequate margin of exposure for the flavouring agent or a related substance?	Comments on predicted metabolism	Related structure name (No.) and structure (if applicable)	Conclusion based on current estimated dietary exposure
Structural class I							
2-Aminoacetophenone	2043	551-93-9	A3: No, SPET: 10	NR	Note 1		No safety concern
Structural class III							
(2E,6E/Z,8E)-N-(2-Methylpropyl)-2,6,8-decatrienamide	2077	25394-57-4	B3: Yes, SPET: 4500	The NOEL of 572 mg/kg bw per day in a 28-day study in rats (Moore, 2002) is 7600 times the estimated daily dietary exposure to No. 2077 when used as a flavouring agent.	Note 2		No safety concern

ALIPHATIC AND AROMATIC AMINES AND AMIDES (addendum)

Name	No.	CAS No.	B3/SPET	Comments	Structure	Conclusion
(2S,5R)-N-[4-(2-Amino-2-oxoethyl)phenyl]-5-methyl-2-(propan-2-yl)-cyclohexanecarboxamide	2078	1119711-29-3	B3: Yes, SPET: 3000	The NOAEL of 300 mg/kg bw per day for the structurally related N-p-benzeneacetonitrile menthanecarboxamide (No. 2009) in a 90-day study in rats (Eapen, 2006) is 6000 times the estimated daily dietary exposure to No. 2078 when used as a flavouring agent.	Note 2	N-p-Benzeneacetonitrile menthanecarboxamide (No. 2009)	No safety concern
(1R,2S,5R)-N-(4-Methoxyphenyl)-5-methyl-2-(1-methylethyl)-cyclohexanecarboxamide	2079	68489-09-8	B3: Yes, SPET: 625	The NOAEL of 300 mg/kg bw per day for the structurally related N-p-benzeneacetonitrile menthanecarboxamide (No. 2009) in a 90-day study in rats (Eapen, 2006) is 29 000 times the estimated daily dietary exposure to No. 2079 when used as a flavouring agent.	Note 2	N-p-Benzeneacetonitrile menthanecarboxamide (No. 2009)	No safety concern

continued

Table 1 (continued)

Flavouring agent	No.	CAS No. and structure	Step A3/B3[a] Does estimated dietary exposure exceed the threshold of concern?	Follow-on from step B3[a] Are additional data available for flavouring agent with an estimated dietary exposure exceeding the threshold of concern? Step A4/B4[e] Adequate margin of exposure for the flavouring agent or a related substance?	Comments on predicted metabolism	Related structure name (No.) and structure (if applicable)	Conclusion based on current estimated dietary exposure
N-Cyclopropyl-5-methyl-2-isopropylcyclohexane-carboxamide	2080	73435-61-7	B3: Yes, SPET: 3000	The NOEL of 8 mg/kg bw per day for the structurally related N-ethyl 2-isopropyl-5-methylcyclohexanecarboxamide (No. 1601) in a 28-day study in rats (Miyata, 1995) is 160 times the SPET estimate (3000 µg/day) and 480 000 times the MSDI estimate (1 µg/day) when No. 2080 is used as a flavouring agent.	Note 2	N-Ethyl 2-isopropyl-5-methylcyclohexanecarbox-amide (No. 1601)	No safety concern
N-(2-Methylcyclohexyl)-2,3,4,5,6-pentafluorobenzamide	2081	1003050-32-5	B3: No, SPET: 50	B4: Yes. The NOAEL of 130 mg/kg bw per day in a 28-day study in rats (Dunster, Watson & Brooks, 2009) is 160 000 times the estimated daily dietary exposure to No. 2081 when used as a flavouring agent.	Note 3		No safety concern

ALIPHATIC AND AROMATIC AMINES AND AMIDES (addendum)

3-[(4-Amino-2,2-dioxido-1H-2,1,3-benzothiadiazin-5-yl)oxy]-2,2-dimethyl-N-propylpropanamide	2082	1093200-92-0	B3: Yes, SPET: 1250	The NOEL of 20 mg/kg bw per day in a 90-day study in rats (Dong, 2009b) is 960 times the estimated daily dietary exposure to No. 2082 when used as a flavouring agent.	Note 3	No safety concern

bw, body weight; CAS, Chemical Abstracts Service; NOAEL, no-observed-adverse-effect level; NOEL, no-observed-effect level; NR, not required for evaluation because consumption of the flavouring agent was determined to be of no safety concern at step A3 of the Procedure

[a] Fifty-eight flavouring agents in this group were previously evaluated by the Committee (Annex 1, references *178*, *187* and *202*).
[b] *Step 1:* One flavouring agent is in structural class I (No. 2043), and six flavouring agents (Nos 2077–2082) are in structural class III.
[c] *Step 2:* Flavouring agent No. 2043 is expected to be metabolized to innocuous products. The remaining six flavouring agents (Nos 2077–2082) cannot be predicted to be metabolized to innocuous products.
[d] The thresholds for human dietary exposure for structural classes I and III are 1800 and 90 μg/person per day, respectively. All dietary exposure values are expressed in μg/day. The dietary exposure value listed represents the highest estimated dietary exposure calculated using either the SPET or the MSDI method. The SPET gave the highest estimated dietary exposure in each case.
[e] The margins of exposure were calculated based on the estimated dietary exposure calculated using the SPET. In cases where the resulting margin of exposure was relatively low, a comparison with the MSDI was also made.

Notes:
1. 2-Aminoacetophenone is a product of tryptophan metabolism and is expected to be transformed to anthranilic acid, which is excreted in the urine or involved in subsequent metabolic pathways.
2. Amides are expected to undergo hydrolysis and/or oxidation and enter into known pathways of metabolism.
3. Minimal metabolism of Nos 2081 and 2082 was observed in vitro; small amounts of hydroxylation products were formed for each flavouring agent.

Table 2. Annual volumes of production and daily dietary exposures for aliphatic and aromatic amines and amides used as flavouring agents in Europe, the USA and Japan

Flavouring agent (No.)	Most recent annual volume of production (kg)[a]	Dietary exposure				Annual consumption via natural occurrence in foods (kg)[d]	Consumption ratio[e]
		MSDI[b]		SPET[c]			
		µg/day	µg/kg bw per day	µg/day	µg/kg bw per day		
2-Aminoacetophenone (2043)							
Europe	0.1	0.01	0.0002				
USA	0.05	0.01	0.0001				
Japan	0.1	0.04	0.0006	10	0.2	396	7920
(2E,6E/Z,8E)-N-(2-Methylpropyl)-2,6,8-decatrienamide (2077)							
Europe	ND	ND	ND				
USA	300	33	1	4500	75	+	NA
Japan	ND	ND	ND				
(2S,5R)-N-[4-(2-Amino-2-oxoethyl)phenyl]-5-methyl-2-(propan-2-yl)cyclohexanecarboxamide (2078)							
Europe	ND	ND	ND				
USA	50	6	0.1	3000	50	–	NA
Japan	ND	ND	ND				
(1R,2S,5R)-N-(4-Methoxyphenyl)-5-methyl-2-(1-methylethyl)-cyclohexanecarboxamide (2079)							
Europe	ND	ND	ND				
USA	0.1	0.01	0.0001	625	11	–	NA
Japan	ND	ND	ND				

N-Cyclopropyl-5-methyl-2-isopropylcyclohexanecarboxamide (2080)			
Europe	ND	ND	
USA	10	1	0.02
Japan	ND	ND	
	3000	50	–	NA
N-(2-Methylcyclohexyl)-2,3,4,5,6-pentafluorobenzamide (2081)			
Europe	ND	ND	
USA	1	0.1	0.002
Japan	ND	ND	
	50	1	–	NA
3[(4-Amino-2,2-dioxido-1H-2,1,3-benzothiadiazin-5-yl)oxy]-2,2-dimethyl-N-propylpropanamide (2082)			
Europe	ND	ND	
USA	1	0.1	0.002
Japan	ND	ND	
	1250	21	–	NA
Total			
Europe	0.1		
USA	362		
Japan	0.1		

NA, not applicable; ND, no data reported; +, reported to occur naturally in foods, but no quantitative data; –, not reported to occur naturally in foods

[a] From European Flavour and Fragrance Association (2004), Japan Flavor and Fragrance Materials Association (2005), Gavin, Williams & Hallagan (2008) and International Organization of the Flavor Industry (2011). Values greater than 0 kg but less than 0.1 kg were reported as 0.1 kg.
[b] MSDI (μg/person per day) calculated as follows:
(annual volume, kg) × (1 × 10^9 μg/kg)/(population × survey correction factor × 365 days), where population (10%, "eaters only") = 32 × 10^6 for Europe, 31 × 10^6 for the USA and 13 × 10^6 for Japan; and where survey correction factor = 0.8 for the surveys in the USA, Europe and Japan, representing the assumption that only 80% of the annual flavour volume was reported in the poundage surveys (European Flavour and Fragrance Association, 2004; Japan Flavor and Fragrance Materials Association, 2005; Gavin, Williams & Hallagan, 2008; International Organization of the Flavor Industry, 2011).

continued

Table 2 (continued)

MSDI (µg/kg bw per day) calculated as follows:
(µg/person per day)/body weight, where body weight = 60 kg. Slight variations may occur from rounding.
c SPET (µg/person per day) calculated as follows:
(standard food portion, g/day) × (average use level) (International Organization of the Flavor Industry, 2011). The dietary exposure from the single food category leading to the highest dietary exposure from one portion is taken as the SPET estimate.
SPET (µg/kg bw per day) calculated as follows:
(µg/person per day)/body weight, where body weight = 60 kg. Slight variations may occur from rounding.
d Quantitative data for the USA reported by Stofberg & Grundschober (1987).
e The consumption ratio is calculated as follows:
(annual consumption via natural occurrence in foods in the USA, kg)/(most recent reported annual volume of production as a flavouring agent in the USA, kg).

flavouring agent (No. 2043) to structural class I. The remaining six flavouring agents (Nos 2077–2082) were assigned to structural class III (Cramer, Ford & Hall, 1978).

Step 2. One of the flavouring agents (No. 2043) in this group is predicted to be metabolized to innocuous products. The evaluation of this flavouring agent therefore proceeded via the A-side of the Procedure. The remaining flavouring agents (Nos 2077–2082) in this group could not be predicted to be metabolized to innocuous products. Therefore, the evaluation of these flavouring agents proceeded via the B-side of the Procedure.

Step A3. The highest dietary exposure of 2-aminoacetophenone (No. 2043) is below the threshold of concern (i.e. 1800 μg/person per day for class I). This flavouring agent would not be expected to be of safety concern at current estimated dietary exposures.

Step B3. The highest dietary exposures for five flavouring agents in structural class III are above the threshold of concern (i.e. 90 μg/person per day for class III). Accordingly, for all of these flavouring agents, data are required on the flavouring agent or a closely related substance in order to perform a safety evaluation. For one of the flavouring agents in structural class III (No. 2081), the highest estimated dietary exposure is below the threshold of concern. Accordingly, evaluation of this flavouring agent proceeded to step B4.

Step B4. For N-(2-methylcyclohexyl)-2,3,4,5,6-pentafluorobenzamide (No. 2081), the no-observed-adverse-effect level (NOAEL) of 130 mg/kg of body weight (bw) per day in a 28-day study in rats (Dunster, Watson & Brooks, 2009) provides a margin of exposure of 160 000 in relation to the highest estimated dietary exposure to No. 2081 (SPET = 50 μg/day) when used as a flavouring agent.

Consideration of flavouring agents with high exposure evaluated via the B-side of the decision-tree:

Toxicity data on flavouring agent No. 2009, previously evaluated by the Committee (Annex 1, reference *202*), were used in support of the safety evaluation of two flavouring agents (Nos 2078 and 2079). Toxicity data on flavouring agent No. 1601, previously evaluated by the Committee (Annex 1, reference *178*), were used in support of the safety evaluation of flavouring agent No. 2080. Toxicity data on flavouring agent No. 2077 were previously evaluated by the Committee (Annex 1, reference *178*).

For (2*E*,6*E*/*Z*,8*E*)-*N*-(2-methylpropyl)-2,6,8-decatrienamide (No. 2077), the no-observed-effect level (NOEL) of 572 mg/kg bw per day in a 28-day study in rats (Moore, 2002) provides a margin of exposure of 7600 in relation to the highest estimated dietary exposure to No. 2077 (SPET = 4500 μg/day) when used as a flavouring agent.

For (2*S*,5*R*)-*N*-[4-(2-amino-2-oxoethyl)phenyl]-5-methyl-2-(propan-2-yl)-cyclohexanecarboxamide (No. 2078), the NOAEL of 300 mg/kg bw per day for the structurally related substance *N*-*p*-benzeneacetonitrile menthanecarboxamide (No. 2009) in a 90-day study in rats (Eapen, 2006) is 6000 times the highest estimated dietary exposure to No. 2078 (SPET = 3000 μg/day) when used as a flavouring agent.

For (1*R*,2*S*,5*R*)-*N*-(4-methoxyphenyl)-5-methyl-2-(1-methylethyl)cyclohexanecarboxamide (No. 2079), the NOAEL of 300 mg/kg bw per day for the structurally related substance *N-p*-benzeneacetonitrile menthanecarboxamide (No. 2009) in a 90-day study in rats (Eapen, 2006) is 29 000 times the highest estimated dietary exposure to No. 2079 (SPET = 625 µg/day) when used as a flavouring agent.

For *N*-cyclopropyl-5-methyl-2-isopropylcyclohexanecarboxamide (No. 2080), the NOEL of 8 mg/kg bw per day for the structurally related substance *N*-ethyl-2-isopropyl-5-methylcyclohexanecarboxamide (No. 1601) in a 28-day study in rats (Miyata, 1995) is 160 times the SPET estimate (3000 µg/day) and 480 000 times the MSDI estimate (1 µg/day) when No. 2080 is used as a flavouring agent.

For 3[(4-amino-2,2-dioxido-1H-2,1,3-benzothiadiazin-5-yl)oxy]-2,2-dimethyl-*N*-propylpropanamide (No. 2082), the NOEL of 20 mg/kg bw per day in a 90-day study in rats (Dong, 2009b) provides a margin of exposure of 960 in relation to the highest estimated dietary exposure to No. 2082 (SPET = 1250 µg/day) when used as a flavouring agent.

The Committee concluded that the five flavouring agents with high exposure evaluated via the B-side of the decision-tree (Nos 2077–2080 and 2082) would not pose a safety concern at current estimated dietary exposures.

Table 1 summarizes the evaluations of the seven additional flavouring agents belonging to the group of aliphatic and aromatic amines and amides (Nos 2043 and 2077–2082).

1.5 Consideration of combined intakes from use as flavouring agents

The highest MSDI values for members of the current group are 33 µg/day (No. 2077), 6 µg/day (No. 2078) and 1 µg/day (No. 2080). The seven additional flavouring agents in this group have diverse structures, with various potential sites of metabolism, and are not likely to be metabolized to common products. The Committee concluded that under the conditions of use as flavouring agents, the combined intake of the flavouring agents in this group would not saturate metabolic pathways, and the combined intakes, including those of previously evaluated members of the group, would not raise safety concerns.

1.6 Conclusion

In the previous evaluations of members of this group of flavouring agents, studies of acute toxicity, short-term studies of toxicity, long-term studies of toxicity and carcinogenicity, and studies of genotoxicity and reproductive toxicity were available. Previously evaluated data from short-term studies of toxicity were used to support the safety evaluations of four members of the current group. Data from short-term studies of toxicity on two members of the current group (Nos 2081 and 2082) were evaluated at the present meeting. The Committee concluded that none of the seven flavouring agents evaluated at the present meeting, which are additions to the group of aliphatic and aromatic amines and amides evaluated previously, raise any safety concerns at current estimated dietary exposures.

2. RELEVANT BACKGROUND INFORMATION

2.1 Explanation

This monograph addendum summarizes key aspects relevant to the safety evaluation of seven aliphatic and aromatic amines and amides, which are additions to a group of 58 flavouring agents evaluated previously by the Committee at its sixty-fifth, sixty-eighth and seventy-third meetings (Annex 1, references *178*, *187* and *202*).

2.2 Additional considerations on dietary exposure

Annual volumes of production and estimated dietary exposures calculated using both the MSDI method and the SPET for each flavouring agent are reported in Table 2. There is no additional information on estimated dietary exposures.

2.3 Biological data

2.3.1 Biochemical data: absorption, distribution, metabolism and elimination

General information on the metabolism of aliphatic and aromatic amines and amides was previously provided in the reports of the sixty-fifth, sixty-eighth and seventy-third meetings (Annex 1, references *178*, *187* and *202*). Specific information relevant to the flavouring agents considered in this report is provided below.

(a) 2-Aminoacetophenone (No. 2043)

2-Aminoacetophenone is a minor intermediate metabolite of tryptophan formed via the kynurenine pathway. The proposed biosynthetic route involves formation of 2-aminoacetophenone from kynurenine (3-(*o*-aminobenzoyl)alanine) (Kaseda, Noguchi & Kido, 1973). In humans, 2-aminoacetophenone has been identified as a tryptophan metabolite. It is ultimately metabolized to anthranilic acid, which is readily excreted in the urine (Kochen et al., 1975).

(b) N-(2-Methylcyclohexyl)-2,3,4,5,6-pentafluorobenzamide (No. 2081)

In a qualitative tissue distribution study, single oral doses of No. 2081 labelled at the amide carbon with ^{14}C were administered via corn oil gavage to Sprague-Dawley rats (five of each sex) at 50 mg/kg bw. The distribution of radioactivity was investigated using whole-body autoradiography. Radioactivity representing No. 2081 and/or its metabolites was absorbed and widely distributed throughout the body. Tissue radioactivity appeared greatest during the first 3 hours after dosing, with the majority of tissues containing levels qualitatively similar to that observed in blood. Highest concentrations were associated with the excretory organs (i.e. liver, kidney, gastrointestinal tract and urinary bladder). By 72 hours after dosing, radioactivity was still present (at low levels) in the gastrointestinal tract, liver, kidney, lung, blood, skin, tongue, Harderian gland and lacrimal gland. No sex-related differences in the distribution of the parent compound and/or its metabolites were observed (Penketh et al., 2009).

In an in vitro metabolism study, No. 2081 (1, 3, 30 or 100 μmol/l) labelled at the amide carbon with ^{14}C was incubated with rat liver microsomes for 10, 30, 60 and 120 minutes. A total of 16 fractions corresponding to potential metabolites were observed via liquid chromatography–tandem mass spectrometry. The major metabolites were identified as hydroxylation products. The mass spectral data indicated that hydroxylation takes place on the methylcyclohexyl moiety, but the low levels of metabolites produced prevented further analysis to specifically identify the positions of hydroxylation (Foster, 2009).

(c) *3[(4-Amino-2,2-dioxido-1H-2,1,3-benzothiadiazin-5-yl)oxy]-2,2-dimethyl-N-propylpropanamide (No. 2082)*

In an in vitro metabolism study, small amounts of two metabolites (designated M1 and M2) were observed when No. 2082 (10 μmol/l) was separately incubated with rat or human liver microsomes for 2 hours. M1 and M2 were identified as single hydroxylation metabolites of the parent compound; however, the positions of hydroxylation could not be determined. No other metabolites were identified in the study (Kong, Yuan & Wang, 2009).

In a metabolic stability study, No. 2082 (2 μmol/l) was separately incubated with rat or human liver microsomes for 60 minutes. The percentage of the test material remaining unchanged after 60 minutes was 99% and 100% for rat and human liver microsomes, respectively (Yuan, 2008).

In a pharmacokinetic study, No. 2082 was administered to fasted male Sprague-Dawley rats (three per group) via single intravenous injection (2.5 mg/kg bw) or single oral gavage (5, 10 or 20 mg/kg bw). Mean pharmacokinetic parameters are presented in Table 3. The area under the plasma concentration–time curve ($AUC_{0-\infty}$) and peak plasma concentration (C_{max}) increased less than proportionally to oral dose. The time to C_{max} (T_{max}) and the half-life ($t_{½}$) did not vary markedly with oral dose. Oral bioavailability was low and decreased with dose, ranging from 9.6% to 3.9%. Pharmacokinetic parameters were also determined for a metabolite of unstated identity. Relative oral exposure to this metabolite, as determined by AUC ratios, increased with dose, but was only 0.02–0.09% of that of the parent compound (Walburn, 2009).

2.3.2 *Toxicological studies*

(a) Acute toxicity

Oral acute toxicity studies have been conducted on two of the seven flavouring agents in this group. For (1R,2S,5R)-N-(4-methoxyphenyl)-5-methyl-2-(1-methylethyl)cyclohexanecarboxamide (No. 2079), no deaths occurred following gavage administration to rats at doses up to 2000 mg/kg bw, the highest dose tested (Bradshaw, 2008). For 3[(4-amino-2,2-dioxido-1H-2,1,3-benzothiadiazin-5-yl)oxy]-2,2-dimethyl-N-propylpropanamide (No. 2082), no deaths occurred following gavage administration to rats at doses up to 50 mg/kg bw, the highest dose tested (Arulnesan, 2008).

Table 3. Mean pharmacokinetic parameters for 3[(4-amino-2,2-dioxido-1H-2,1,3-benzothiadiazin-5-yl)oxy]-2,2-dimethyl-N-propylpropanamide in male rats

Parameter	Intravenous 2.5 mg/kg bw	Oral 5 mg/kg bw	Oral 10 mg/kg bw	Oral 20 mg/kg bw
$AUC_{0-\infty}$ (ng·h/ml)	1760 ± 289	337 ± 31	453 ± 45	553 ± 29
$AUC_{0-\infty}$/dose ([ng·h/ml]/mg/kg bw)	704 ± 116	67.3 ± 6.1	45.3 ± 4.5	27.7 ± 1.5
C_{max} (ng/ml)	NA	75 ± 23	103 ± 34	111 ± 22
C_{max}/dose ([ng/ml]/mg/kg bw)	NA	15.0 ± 4.6	10.3 ± 3.4	5.5 ± 1.1
CL (ml/min per kilogram body weight)	24.1 ± 3.7	ND	ND	ND
$t_{\frac{1}{2}}$ (h)	0.78 ± 0.04	3.7 ± 1.6	3.0 ± 1.1	4.5 ± 0.2
V_{ss} (ml/kg bw)	860 ± 120	ND	ND	ND
T_{max} (h)	NA	1.3 ± 0.6	1.0 ± 0.0	1.3 ± 0.6
F (%)	100	9.6	6.4	3.9

CL, clearance; F, bioavailability; NA, not applicable; ND, not determined; V_{ss}, volume of distribution at steady state
Source: Walburn (2009)

(b) Short-term studies of toxicity

Short-term studies of toxicity of (2E,6E/Z,8E)-N-(2-methylpropyl)-2,6,8-decatrienamide (No. 2077) have been previously evaluated by the Committee as a structurally related substance in support of a previously evaluated group of amides (Annex 1, reference 178). Short-term studies of toxicity of N-(2-methylcyclohexyl)-2,3,4,5,6-pentafluorobenzamide (No. 2081) and 3-[4-amino-2,2-dioxido-1H-2,1,3-benzothiadiazin-5-yl)oxy]-2,2-dimethyl-N-propylpropanamide (No. 2082) are described below and summarized in Table 4.

(i) N-(2-Methylcyclohexyl)-2,3,4,5,6-pentafluorobenzamide (No. 2081)

In a 14-day dose range–finding study conducted according to good laboratory practice (GLP), Han Wistar rats (three of each sex per group) were fed diets containing 0 (control), 3000, 5000, 7500 or 15 000 mg of No. 2081 per kilogram. Target dose levels corresponding to these dietary concentrations were 0, 200, 333, 500 and 1000 mg/kg bw per day, respectively. Clinical observations were performed once daily. Body weights and feed consumption were recorded on days 1, 4, 8, 11 and 15. On completion of the dosing period, all surviving animals were killed and subjected to internal and external macroscopic examination. Animals that died during the dosing period were also necropsied.

Dose-related reductions in feed consumption were observed in all treatment groups for both sexes. For all rats consuming the 15 000 mg/kg diet, this feed

Table 4. Results of short-term studies of toxicity of aliphatic and aromatic amines and amides used as flavouring agents

No.	Flavouring agent	Species; sex	No. of test groups[a] / no. per group[b]	Route	Duration (days)	NOEL/NOAEL (mg/kg bw per day)	Reference
2077	(2E,6E/Z,8E)-N-(2-Methylpropyl)-2,6,8-decatrienamide	Rat; M, F	3/10	Diet	28	572[c]	Moore (2002)[d]
2081	N-(2-Methylcyclohexyl)-2,3,4,5,6-pentafluorobenzamide	Rat; M, F	3/10	Diet	28	130[e]	Dunster, Watson & Brooks (2009)
2082	3[(4-Amino-2,2-dioxido-1H-2,1,3-benzothiadiazin-5-yl)-oxy]-2,2-dimethyl-N-propylpropanamide	Rat; M, F	3/40	Diet	90	20[c]	Dong (2009b)

F, female; M, male

[a] Total number of test groups does not include control animals.
[b] Total number per test group includes both male and female animals.
[c] NOEL.
[d] Moore (2002) was evaluated by the Committee at its sixty-fifth meeting (Annex 1, reference *178*) to support the safety evaluation of three structurally related flavouring agents (Nos 1596, 1597 and 1598). The test article was an extract of the flowering plant *Heliopsis longipes* ("gold root extract"), which was reported to contain approximately 50% *N*-isobutyl-2,6,8-decatrienamide. The three carbon–carbon double bonds were drawn in the *trans* configuration, which would correspond to (2*E*,6*E*,8*E*)-*N*-(2-methylpropyl)-2,6,8-decatrienamide. Analytical data provided for the current meeting on a preparation of gold root extract indicated a composition of 56% (2*E*,6*Z*,8*E*)-*N*-(2-methylpropyl)-2,6,8-decatrienamide (i.e. the central double bond is in the *cis* configuration). The NOEL of 572 mg/kg bw per day was established previously by the Committee using a content of 50%. This NOEL has been retained for the current evaluation.
[e] NOAEL.

Table 5. Feed consumption and changes in body weight for rats fed diets containing No. 2081

Parameter	Day(s)	Dietary concentration (mg/kg)				
		0	3000	5000	7500	15 000
Males						
Mean body weight (g)	1	170	251	239	168	178
	15ª	246	278	233	147	132
Body weight change (%)		45	11	−3	−13	−26
Mean feed consumption (g/day)	1–4	18	12 (66%)ᶜ	4 (22%)	3 (17%)	1 (6%)
	1–15ᵇ	21	18 (86%)	18 (86%)	11 (52%)	—ᵇ
Females						
Mean body weight (g)	1	131	149	168	137	141
	15ª	165	166	172	125	108
Body weight change (%)		26	11	2	−9	−23
Mean feed consumption (g/day)	1–4	14	9 (64%)	4 (29%)	3 (21%)	4 (29%)
	1–15ᵇ	15	12 (80%)	11 (73%)	11 (73%)	—ᵇ

ª Day 8 for animals receiving 15 000 mg/kg.
ᵇ Rats receiving 15 000 mg/kg were sacrificed on day 8.
ᶜ Feed consumption relative to control (%).

aversion was almost complete, so that all six (three of each sex) starved animals needed to be sacrificed in extremis on day 8. One female in the 15 000 mg/kg feed group had hunched posture on day 6. There were no clinical signs for the remaining animals. The reduced feed consumption resulted in body weight losses or reduced body weight gains, as shown in Table 5.

No macroscopic abnormalities were detected in animals fed the 15 000 mg/kg diet following early termination. All males in the 5000 and 7500 mg/kg diet groups had small seminal vesicles at necropsy. One female in the 7500 mg/kg diet group had hydronephrosis in the right kidney. All animals on the 5000 mg/kg diet had sloughing on the glandular region of the stomach. No macroscopic abnormalities were observed in animals receiving the 3000 mg/kg diet. The study report concluded that appropriate dietary concentrations for a 28-day study would be 750, 1500 and 3000 mg/kg (Dunster, Watson & Brooks, 2009).

In a subsequent 28-day study conducted according to GLP and Organisation for Economic Co-operation and Development (OECD) Test Guideline 407, Han Wistar rats (five of each sex per group) were fed diets containing 0 (control), 750, 1500 or 3000 mg of No. 2081 per kilogram. These dietary concentrations resulted in mean achieved doses of 0, 67, 130 and 262 mg/kg bw per day, respectively (averaged over both sexes for the duration of the study). Clinical signs, functional observations (behavioural assessments, motor activity, forelimb/hindlimb grip strength and sensory reactivity), body weight, feed consumption and water consumption were monitored throughout the study. However, water consumption was monitored for overt changes only by visual inspection of the bottles during the first 2 weeks; water was then weighed daily for the remainder of the study. Haematology and blood chemistry were evaluated for all animals at the end of the study. All animals were subjected to gross necropsy examination, including measurement of weights of selected organs. Histopathological evaluation of selected tissues from control animals and the high-dose group was performed.

There were no unscheduled deaths during the study period. There were no treatment-related clinical signs of toxicity, behavioural changes or effects on sensory reactivity. There were no toxicologically relevant effects on functional performance. Males from all treatment groups showed statistically significant reductions in overall mobile activity; however, there were no effects on overall general activity. In the absence of clinical signs of toxicity and a lack of effects on all other functional parameters investigated, the reductions in overall mobile activity are not considered to be toxicologically relevant. The male 3000 mg/kg diet group showed statistically significant ($P < 0.01$) decreased body weight gain during week 1 (59% of mean body weight gain of controls) and week 4 (53% of controls); however, reduced body weight gain over the 28 days of dosing (78% of controls) was not statistically significant. Females from this treatment group also showed reduced body weight gain during these weeks; however, statistical significance was not achieved. Feed consumption for high-dose males was 80% and 93% of that of controls for weeks 1 and 1–4, respectively; however, these reductions were not statistically significant. Feed consumption for high-dose females was 81% and 93% of that of controls for weeks 1 and 1–4, respectively; these reductions were also not statistically significant.

The 750 and 1500 mg/kg diet groups for both sexes showed no statistically significant differences from controls in body weight gain or feed consumption. There were no statistically significant intergroup differences in water consumption over days 15–28. However, males but not females treated with 3000 mg/kg diet consumed an average 2.6 ml per rat per day (or approximately 13% per day) more than the control group. Males treated with 3000 mg/kg diet showed a statistically significant reduction in erythrocyte count ($P < 0.05$) and statistically significant increases in neutrophil count ($P < 0.05$), clotting time ($P < 0.01$), activated partial thromboplastin time (APTT; $P < 0.01$) and alkaline phosphatase activity ($P < 0.01$). APTT was also increased in males at 1500 mg/kg diet ($P < 0.05$). Females treated with 3000 mg/kg diet showed a statistically significant increase in total protein ($P < 0.05$), albumin ($P < 0.05$) and total cholesterol levels ($P < 0.01$).

At necropsy, no treatment-related macroscopic anomalies were observed. There were also no treatment-related histopathological changes. Females treated with 3000 mg/kg diet showed a statistically significant reduction in absolute spleen weight and a statistically significant increase in relative spleen weight; however, all of the individual values were within the normal range for this rat strain. In the absence of any histopathological changes, the intergroup difference was considered to be of no toxicological relevance. Livers from females from all treatment groups showed statistically significant increases in absolute weight (by 7–15%) and relative (to body weight) weight (by 13–25%). In the absence of any hepatic histopathological correlates, this finding is not considered to be toxicologically relevant.

Considering the findings of the 14-day dose range–finding study and the 28-day study together suggests that the most important attribute of the test compound when present in the diet is to cause pronounced feed aversion, resulting in a dose-related reduction in body weight gain. This feed aversion is almost complete at 15 000 mg/kg diet, so that the rats were starved and needed to be sacrificed. The concentration at which this feed aversion effect was no longer observed was 1500 mg/kg diet. A few statistically significant changes in some clinical chemistry and haematology parameters at 3000 mg/kg diet were not associated with any gross pathology or histopathology findings and may be secondary to reduced body weight gain resulting from feed aversion. The NOAEL for this study is therefore considered to be 1500 mg/kg diet, which gave an achieved dose of 130 mg/kg bw per day (Dunster, Watson & Brooks, 2009).

>	(ii)	3-[4-Amino-2,2-dioxido-1H-2,1,3-benzothiadiazin-5-yl)oxy]-2,2-dimethyl-N-propylpropanamide (No. 2082)

In a 28-day dose range–finding study conducted according to GLP and based on United States Food and Drug Administration (USFDA) Redbook guidelines, Sprague-Dawley rats (five of each sex per group) were administered feed calculated to provide average doses of 0 (control), 10, 30 and 100 mg of No. 2082 per kilogram of body weight per day. The animals were observed twice daily for mortality, morbidity, injury and availability of feed and water. Clinical observations and body weights were recorded weekly. Feed consumption was measured twice weekly. Blood samples for clinical chemistry evaluations were collected prior to necropsy. At study termination, necropsy examinations were performed, organ weights were recorded and liver and pituitary were examined microscopically.

No mortality occurred prior to the scheduled sacrifice. There were no test article–related clinical signs or effects on body weight, feed consumption or clinical chemistry. There were no test article–related macroscopic or microscopic findings. There were dose-dependent increases in the absolute and relative (to brain) weights of the pituitary gland for males and females, as shown in Table 6. In females, relative (to body weight) pituitary gland weights were also increased in a dose-dependent manner.

For absolute pituitary weight, the increases were statistically significant in males receiving 100 mg/kg bw per day and in all female treatment groups. For brain weight–relative pituitary gland weights, the increases were statistically significant in males receiving 100 mg/kg bw per day and in females receiving 30 and 100 mg/kg bw per day. Increases in body weight–relative pituitary weights were statistically significant in males at 100 mg/kg bw per day and in females at 30 mg/kg bw per day. There was no microscopic correlate, such as hyperplasia and/or hypertrophy, for these pituitary gland findings, which casts doubt on their toxicological relevance. By comparing the control pituitary weights in this study with those in the subsequent 13-week study (Dong, 2009b) that was completed only a few months later, it becomes clear that the apparent increases in pituitary weights relative to controls for females are due to an anomalously low mean value observed for the control group in the

Table 6. Pituitary gland weights in rats fed No. 2082 in the diet

Parameter	Dose (mg/kg bw per day)			
	0	10	30	100
Males				
Pituitary gland weight (mg)	14.3	15.0	16.8	21.4*
Increase relative to control group (%)	—	5	17	50
Pituitary gland weight/body weight (%)	0.0044	0.0046	0.0051	0.0064*
Pituitary gland weight/brain weight (%)	0.0079	0.0080	0.0092	0.0113*
Females				
Pituitary gland weight (mg)	17.2	19.4*	20.4**	20.6**
Increase relative to control group (%)	—	13	19	20
Pituitary gland weight/body weight (%)	0.0078	0.0081	0.0091*	0.0088
Pituitary gland weight/brain weight (%)	0.0097	0.0105	0.0113**	0.0114**

* $P < 0.05$; ** $P < 0.01$

dose range–finding study. In males receiving 100 mg/kg bw per day, the mean pituitary weight was 21.4 mg; however, this value is skewed by a high pituitary weight of 31.7 mg observed for one male, whereas the range for the remaining four males was 17.4–21.7 mg.

It is concluded that there are no toxicologically relevant effects in this study, and the NOEL is therefore considered to be 100 mg/kg bw per day. However, the study director chose a maximum dose of 20 mg/kg bw per day for a subsequent 13-week study on the basis of this apparent dose-related increase in pituitary weight (Dong, 2009a).

In a 13-week dietary study conducted according to GLP and based on USFDA Redbook guidelines, Sprague-Dawley rats (20 of each sex per group) were provided feed delivering 0 (control), 5, 10 or 20 mg of No. 2082 per kilogram of body weight per day. Observations for morbidity, mortality, injury and the availability of feed and water were conducted twice daily. Detailed clinical examinations were conducted on all study animals once weekly. Functional observational battery evaluations were conducted on all study animals at the end of week 12. Body weights and feed consumption were measured weekly. Ophthalmoscopic examinations were conducted before commencement of the study and prior to terminal necropsy. Blood samples for haematology and clinical chemistry evaluations were collected prior to commencement, at week 6 and prior to necropsy. Blood samples for hormone assays (adrenocorticotropic hormone, thyroid stimulating hormone, corticosterone, follicle stimulating hormone and luteinizing hormone) were collected prior to necropsy. Urine samples were collected prior to necropsy. At study termination, necropsy examinations were performed, organ weights were recorded and selected tissues were examined microscopically.

No mortality and no clinical signs of toxicity were observed during the 13-week study period. There were no test article–related adverse effects in the detailed clinical observations, the functional observational battery evaluations, ophthalmoscopic examinations, body weight, feed consumption, haematology, clinical chemistry, hormone evaluations or urine analysis. There were no test article–related macroscopic findings, organ weight findings or microscopic findings in any of the treatment groups in this study. In particular, there were no effects on pituitary weights (absolute, body weight relative or brain weight relative).

Based on a lack of treatment-related findings in this study, the NOEL was considered to be the highest dose tested, 20 mg/kg bw per day (Dong, 2009b).

(c) Genotoxicity

In vitro and in vivo genotoxicity studies have been conducted on four flavouring agents in this group. These studies are summarized in Table 7. Negative results in in vitro bacterial mutation assays, both with and without metabolic activation, were obtained for the four flavouring agents (Nos 2043, 2079, 2081 and 2082). One flavouring agent was tested for the induction of unscheduled deoxyribonucleic acid (DNA) synthesis in vitro, with negative results (No. 2043). Chromosomal aberration assays conducted in mammalian cells with one of the flavouring agents in this group were negative (No. 2082). Micronucleus assays conducted in mice with two of the flavouring agents were also negative (Nos 2081 and 2082).

3. REFERENCES

Arulnesan N (2008). Acute oral toxicity study with S6973 in rats. Unpublished report submitted to WHO by the International Organization of the Flavor Industry, Brussels, Belgium.

Bowden JP, Chung KT, Andrews AW (1976). Mutagenic activity of tryptophan metabolites produced by rat intestinal microflora. *Journal of the National Cancer Institute*, 57:921–924.

Bowles AJ (2008). Reverse mutation assay "Ames test" using *Salmonella typhimurium*. Unpublished report submitted to WHO by the International Organization of the Flavor Industry, Brussels, Belgium.

Bradshaw J (2008). Acute oral toxicity in the rat—fixed dose method. Unpublished report submitted to WHO by the International Organization of the Flavor Industry, Brussels, Belgium.

Cardoso R (2009a). Chromosome aberration test of S6973 in Chinese hamster ovary cells. Unpublished report submitted to WHO by the International Organization of the Flavor Industry, Brussels, Belgium.

Cardoso R (2009b). In vivo mouse micronucleus test of S6973. Unpublished report submitted to WHO by the International Organization of the Flavor Industry, Brussels, Belgium.

Cramer GM, Ford RA, Hall RL (1978). Estimation of toxic hazard—a decision tree approach. *Food and Cosmetics Toxicology*, 16:255–276.

Dong G (2009a). S6973: a 28-day dietary administration toxicity study in rats. Unpublished report no. 1646-001 from MPI Research, Inc. Submitted to WHO by the International Organization of the Flavor Industry, Brussels, Belgium.

Dong G (2009b). S6973: a 13-week dietary administration toxicity study in rats. Unpublished report submitted to WHO by the International Organization of the Flavor Industry, Brussels, Belgium.

Dunster J, Watson P, Brooks P (2009). PFMC benzamide: twenty-eight day repeated dose oral (dietary) toxicity study in the rat. Unpublished report submitted to WHO by the International Organization of the Flavor Industry, Brussels, Belgium.

Eapen AK (2006). A 90-day oral (dietary) toxicity study of GR-72-0180 in rats with a 28-day recovery period. Unpublished report submitted to WHO by the International Organization of the Flavor Industry, Brussels, Belgium.

Table 7. Results of genotoxicity studies with aliphatic and aromatic amines and amides used as flavouring agents

No.	Flavouring agent	End-point	Test object	Concentration/dose	Results	Reference
In vitro						
2043	2-Aminoacetophenone	Reverse mutation	*Salmonella typhimurium* TA98, TA100, TA1535, TA1537 and TA1538	50–500 µg/plate	Negative[a,b]	Bowden, Chung & Andrews (1976)
2043	2-Aminoacetophenone	Reverse mutation	*S. typhimurium* C3076, D3052, G46, TA98, TA100, TA1535, TA1537 and TA1538 and *Escherichia coli* WP2 and WP2uvrA	0.1–1000 µg/ml agar	Negative[a,b]	Thompson et al. (1983)
2043	2-Aminoacetophenone	Unscheduled DNA synthesis	Rat hepatocytes	0.5–1000 nmol/ml	Negative	Thompson et al. (1983)
2079	(1*R*,2*S*,5*R*)-*N*-(4-Methoxyphenyl)-5-methyl-2-(1-methylethyl)-cyclohexanecarboxamide	Reverse mutation	*S. typhimurium* TA98, TA100, TA102, TA1535 and TA1537	50, 150, 500, 1500 and 5000 µg/plate	Negative[a,b]	Bowles (2008)
2081	*N*-(2-Methylcyclohexyl)-2,3,4,5,6-pentafluorobenzamide	Reverse mutation	*S. typhimurium* TA98, TA100, TA1535 and TA1537 and *E. coli* WP2uvrA	500, 750, 1000, 2000 and 4000 µg/plate	Negative[a,b,c]	Kirby (2008)
2082	3[(4-Amino-2,2-dioxido-1H-2,1,3-benzothiadiazin-5-yl)oxy]-2,2-dimethyl-*N*-propylpropanamide	Reverse mutation	*S. typhimurium* TA98 and TA100	62, 190, 560, 1670 and 5000 µg/plate	Negative[a,b]	Zhang (2008)
2082	3[(4-Amino-2,2-dioxido-1H-2,1,3-benzothiadiazin-5-yl)oxy]-2,2-dimethyl-*N*-propylpropanamide	Reverse mutation	*S. typhimurium* TA98, TA100, TA1535 and TA1537 and *E. coli* WP2uvrA	200, 440, 990, 2200 and 5000 µg/plate	Negative[a,d]	Maniatis (2009)

ALIPHATIC AND AROMATIC AMINES AND AMIDES (addendum)

2082	3[(4-Amino-2,2-dioxido-1H-2,1,3-benzothiadiazin-5-yl)oxy]-2,2-dimethyl-N-propylpropanamide	Chromosomal aberration	Chinese hamster ovary WB$_L$ cells	625, 1250, 2500 and 5000 µg/ml	Negative[e,f]	Cardoso (2009a)
In vivo						
2081	N-(2-Methylcyclohexyl)-2,3,4,5,6-pentafluorobenzamide	Micronucleus induction	CD-1 mouse bone marrow cells	500, 1000 and 2000 mg/kg bw[g]	Negative[h]	Shore (2008)
2082	3[(4-Amino-2,2-dioxido-1H-2,1,3-benzothiadiazin-5-yl)oxy]-2,2-dimethyl-N-propylpropanamide	Micronucleus induction	CD-1 mouse bone marrow cells	2000 mg/kg bw[g]	Negative[h]	Cardoso (2009b)

[a] With and without metabolic activation.
[b] Plate incorporation method.
[c] Precipitate observed at 5000 µg/plate in a preliminary assay.
[d] Plate incorporation and preincubation methods.
[e] Three separate exposure conditions were tested: 1) 3-hour exposure in the absence of metabolic activation; 2) 3-hour exposure in the presence of metabolic activation; and 3) 18-hour exposure in the absence of metabolic activation.
[f] Moderate cytotoxicity was observed at 5000 µg/ml in the presence of metabolic activation.
[g] Administered by gavage.
[h] Bone marrow was harvested at 24 and 48 hours.

European Flavour and Fragrance Association (2004). European inquiry on volume use. Private communication to the Flavor and Extract Manufacturers Association, Washington, DC, USA. Submitted to WHO by the International Organization of the Flavor Industry, Brussels, Belgium.

Foster J (2009). PFMC benzamide. In vitro metabolism using rat liver microsomes. Unpublished report submitted to WHO by the International Organization of the Flavor Industry, Brussels, Belgium.

Gavin CL, Williams MC, Hallagan JB (2008). *Flavor and Extract Manufacturers Association of the United States 2005 poundage and technical effects update survey.* Washington, DC, USA, Flavor and Extract Manufacturers Association of the United States.

International Organization of the Flavor Industry (2011). Interim inquiry on volume use and added use levels for flavoring agents to be presented at the 76th meeting of JECFA. Private communication to the Flavor and Extract Manufacturers Association, Washington, DC, USA. Submitted to WHO by the International Organization of the Flavor Industry, Brussels, Belgium.

Japan Flavor and Fragrance Materials Association (2005). Japanese inquiry on volume use. Private communication to the Flavor and Extract Manufacturers Association, Washington, DC, USA. Submitted to WHO by the International Organization of the Flavor Industry, Brussels, Belgium.

Kaneko S, Kumazawa K, Nishimura O (2011). Studies on the key aroma compounds in soy milk made from three different soybean cultivars. *Journal of Agricultural and Food Chemistry*, 59(22):12 204–12 209.

Kaseda H, Noguchi T, Kido R (1973). Biosynthetic routes to 2-aminoacetophenone and 2-amino-3-hydroxyacetophenone. *Journal of Biochemistry*, 74(1):127–133.

Kirby P (2008). Evaluation of a test article in the *Salmonella typhimurium/Escherichia coli* plate incorporation mutation assay in the presence and absence of induced rat liver S-9. Unpublished report submitted to WHO by the International Organization of the Flavor Industry, Brussels, Belgium.

Kochen W et al. (1975). Tryptophan-stoffwechseluntersuchungen bei unbehandelten phenylketonurikern. *Zeitschrift für klinische Chemie und klinische Biochemie*, 13(1):1–12.

Kong L, Yuan R, Wang R (2009). In vitro biotransformation of S6973 in Sprague-Dawley rat and human liver microsomes. Unpublished report submitted to WHO by the International Organization of the Flavor Industry, Brussels, Belgium.

Kumazawa K, Masuda H (1999). Identification of potent odorants in Japanese green tea (Sencha). *Journal of Agricultural and Food Chemistry*, 47(12):5169–5172.

Maniatis T (2009). Bacterial reverse mutation test of S6973. Unpublished report submitted to WHO by the International Organization of the Flavor Industry, Brussels, Belgium.

Miyata N (1995). Summary of 28-day repeated-dose oral toxicity study of WS-3. Unpublished report submitted to WHO by the International Organization of the Flavor Industry, Brussels, Belgium.

Moore GE (2002). 28-day dietary toxicity study in rodents. Product identification: gold root extract. Unpublished report submitted to WHO by the International Organization of the Flavor Industry, Brussels, Belgium.

Nijssen LM, van Ingen-Visscher CA, Donders JJH (2011). Volatile Compounds in Food (VCF) database, version 13.1. Zeist, the Netherlands, TNO Triskelion (http://www.vcf-online.nl/VcfHome.cfm).

Penketh S et al. (2009). PFMC benzamide. Whole-body autoradiography in rats. Unpublished report submitted to WHO by the International Organization of the Flavor Industry, Brussels, Belgium.

Scott-Thomas A, Pearson J, Chambers S (2011). Potential sources of 2-aminoacetophenone to confound the *Pseudomonas aeruginosa* breath test, including analysis of a food challenge study. *Journal of Breath Research*, 5(4):046002.

Shore K (2008). In vivo test for chemical induction of micronucleated polychromatic erythrocytes in mouse bone marrow cells. Unpublished report submitted to WHO by the International Organization of the Flavor Industry, Brussels, Belgium.

Stofberg J, Grundschober F (1987). Consumption ratio and food predominance of flavoring materials. *Perfumer & Flavorist*, 12:27–68.

Thompson CZ et al. (1983). The induction of bacterial mutation and hepatocyte unscheduled DNA synthesis by monosubstituted anilines. *Environmental Mutagenesis*, 5(6):803–811.

Walburn J (2009). Pharmacokinetic study of S6973 in male Sprague-Dawley rats following a single bolus intravenous or oral dose. Unpublished report submitted to WHO by the International Organization of the Flavor Industry, Brussels, Belgium.

Yuan R (2008). Metabolic stability test of S6973 in rat and human liver microsomes. Unpublished report submitted to WHO by the International Organization of the Flavor Industry, Brussels, Belgium.

Zhang B (2008). TA98 and TA100 reverse mutation test of S6973. Unpublished report submitted to WHO by the International Organization of the Flavor Industry, Brussels, Belgium.

ALIPHATIC AND AROMATIC ETHERS (addendum)

First draft prepared by

I.G. Sipes[1], M. DiNovi[2] and P. Sinhaseni[3]

[1] Department of Pharmacology, College of Medicine, University of Arizona, Tucson, Arizona, United States of America (USA)
[2] Center for Food Safety and Applied Nutrition, Food and Drug Administration, College Park, Maryland, USA
[3] Community Risk Analysis Research and Development Center, Bangkok, Thailand

1. Evaluation 95
   1.1 Introduction 95
   1.2 Assessment of dietary exposure 96
   1.3 Absorption, distribution, metabolism and elimination 96
   1.4 Application of the Procedure for the Safety Evaluation of Flavouring Agents 105
   1.5 Consideration of combined intakes from use as flavouring agents 106
   1.6 Consideration of secondary components 106
   1.7 Conclusions 107
2. Relevant background information 107
   2.1 Explanation 107
   2.2 Additional considerations on dietary exposure 107
   2.3 Biological data 107
      2.3.1 Biochemical data: hydrolysis, absorption, distribution, metabolism and elimination 107
      2.3.2 Toxicological studies 107
         (a) Acute toxicity 108
         (b) Short-term studies of toxicity 108
         (c) Genotoxicity 110
         (d) Reproductive toxicity 114
3. References 114

1. EVALUATION

1.1 Introduction

The Committee evaluated a group of 10 aliphatic and aromatic ethers used as flavouring agents. The evaluations were conducted according to the Procedure for the Safety Evaluation of Flavouring Agents (see Figure 1, Introduction) (Annex 1, reference *131*). None of these agents have previously been evaluated.

The Committee previously evaluated 29 other members of this group of flavouring agents at its sixty-first meeting (Annex 1, reference *166*), including benzyl butyl ether (No. 1253) and dibenzyl ether (No. 1256), which were evaluated only for specifications at the twenty-fourth meeting (Annex 1, reference *53*). The Committee concluded that all 29 flavouring agents in that group were of no safety concern at estimated dietary exposures.

Seven of the 10 flavouring agents (Nos 2133–2135, 2137–2139 and 2142) in this group have been reported to occur naturally and can be found in strawberry, lychee, *Salvia* species, dill blossom, dill herb, clary sage, grape brandy, arctic bramble, black currant (buds), *Cinnamomum*, elder flower, elderberry juice, grapefruit juice, green tea (roasted), fresh tomato and white wine (Nofal, Ho & Chang, 1982; Nijssen, van Ingen-Visscher & Donders, 2011).

1.2 Assessment of dietary exposure

The total annual volumes of production of the 10 aliphatic and aromatic ethers are approximately 20 kg in the USA, 0.1 kg in Europe and 371 kg in Japan (European Flavour and Fragrance Association, 2004; Japan Flavor and Fragrance Materials Association, 2005; Gavin, Williams & Hallagan, 2008; International Organization of the Flavor Industry, 2011). Two flavouring agents in this group, butyl β-naphthyl ether (No. 2141) and linalool oxide pyranoid (No. 2135), with annual volumes of production of 202 kg and 159 kg, respectively, account for approximately 97% of the total annual volume of production in Japan. Approximately 95% of the reported volume of production in the USA is accounted for by 3,6-dimethyl-2,3,3a,4,5,7a-hexahydrobenzofuran (No. 2133).

Dietary exposures were estimated using the maximized survey-derived intake (MSDI) method and the single portion exposure technique (SPET), with the highest values reported in Table 1. The estimated daily dietary exposure is highest for digeranyl ether (No. 2142) (10 000 µg, the SPET value obtained from gelatines and puddings). For the other flavouring agents, the estimated daily dietary exposures range from 0.01 to 5000 µg, with the SPET yielding the highest estimates.

Annual volumes of production of this group of flavouring agents as well as the daily dietary exposures calculated using both the MSDI method and the SPET are summarized in Table 2.

1.3 Absorption, distribution, metabolism and elimination

Information on the hydrolysis, absorption, distribution, metabolism and elimination of flavouring agents belonging to the group of aliphatic and aromatic ethers has previously been described in the report of the sixty-first meeting (Annex 1, reference *166*).

The straight-chain aliphatic ethers (Nos 2134, 2138, 2139 and 2142) are predicted to undergo *O*-dealkylation in vivo to yield the corresponding alcohol and aldehyde, which subsequently undergo complete oxidation via the fatty acid pathway and tricarboxylic acid cycle. Alternatively, the aliphatic ether may undergo ω-1 oxidation to yield polar hydroxylated metabolites. These are conjugated with glucuronic acid and excreted or are further oxidized and/or excreted. The alicyclic ethers (Nos 2133, 2135, 2137 and 2140) are predicted to largely undergo ring hydroxylation by cytochrome P450. The resulting metabolites undergo conjugation with glucuronic acid, followed by excretion in the urine. The aromatic ethers (Nos 2136 and 2141) are predicted to undergo ring hydroxylation, *O*-demethylation

ALIPHATIC AND AROMATIC ETHERS (addendum)

Table 1. Summary of the results of the safety evaluations of aliphatic and aromatic ethers used as flavouring agents[a,b,c]

Flavouring agent	No.	CAS No. and structure	Step A3[d] Does estimated dietary exposure exceed the threshold of concern?	Step A5[e] Adequate margin of exposure for the flavouring agent or a related substance?	Comments on predicted metabolism	Related structure name (No.) and structure (if applicable)	Conclusion based on current estimated dietary exposure
Structural class II							
3,6-Dimethyl-2,3,3a,4,5,7a-hexahydrobenzofuran	2133	70786-44-6	No, SPET: 180	NR	Note 1		No safety concern
Ethyl linalyl ether	2134	72845-33-1	No, SPET: 0.1	NR	Note 2		No safety concern

continued

Table 1 (continued)

Flavouring agent	No.	CAS No. and structure	Step A3[1] Does estimated dietary exposure exceed the threshold of concern?	Step A5[5] Adequate margin of exposure for the flavouring agent or a related substance?	Comments on predicted metabolism	Related structure name (No.) and structure (if applicable)	Conclusion based on current estimated dietary exposure
Linalool oxide pyranoid	2135	14049-11-7	Yes, SPET: 600	Yes. The NOEL of 2.5 mg/kg bw per day in a 90-day study in rats for the structurally related tetrahydro-4-methyl-2-(2-methylpropen-1-yl)-pyran (No. 1237) (Posternak, Linder & Vodoz, 1969) is 250 times the estimated dietary exposure to No. 2135 calculated using the SPET and 3300 times compared with the MSDI (45 μg/day) when No. 2135 is used as a flavouring agent.	Note 1	Tetrahydro-4-methyl-2-(2-methylpropen-1-yl)pyran (No. 1237)	No safety concern
Nerolidol oxide	2137	1424-83-5	Yes, SPET: 2500	No	Note 1		Additional data required to complete evaluation

ALIPHATIC AND AROMATIC ETHERS (addendum)

Flavouring agent	No.	CAS no.	Does intake exceed the threshold for human intake?	Comments on predicted metabolism and toxicity		Conclusion based on current estimated dietary exposure
Methyl hexyl ether	2138	4747-07-3	Yes, SPET: 5000	Yes. The NOEL of 900 mg/kg bw per day in a 90-day study in rats for the structurally related methyl tert-butyl ether (Robinson, Bruner & Olson, 1990) is 11 000 times the estimated dietary exposure to No. 2138 when used as a flavouring agent.	Note 2 Methyl tert-butyl ether	No safety concern
Myrcenyl methyl ether	2139	24202-00-4	No, SPET: 63	NR	Note 2	No safety concern
Digeranyl ether	2142	31147-36-1	Yes, SPET: 10 000	Yes. The NOEL of 50 mg/kg bw per day in a 196-day study in rats for the structurally related geraniol (No. 1223) (Hagan et al., 1967) is 300 times the estimated dietary exposure to No. 2142 calculated using the SPET and 300 million times compared with the MSDI (0.01 µg/day) when used as a flavouring agent.	Note 2 Geraniol (No. 1223)	No safety concern

continued

Table 1 (continued)

Flavouring agent	No.	CAS No. and structure	Step A3[1] Does estimated dietary exposure exceed the threshold of concern?	Step A5[a] Adequate margin of exposure for the flavouring agent or a related substance?	Comments on predicted metabolism	Related structure name (No.) and structure (if applicable)	Conclusion based on current estimated dietary exposure
Structural class III							
Isoamyl phenethyl ether	2136	56011-02-0	Yes, SPET: 150	Yes. The NOEL of 196 mg/kg bw per day in a 90-day study in rats for the structurally related dibenzyl ether (No. 1256) (Burdock & Ford, 1992) is 78 000 times the estimated dietary exposure to No. 2136 when used as a flavouring agent.	Note 3	Dibenzyl ether (No. 1256)	No safety concern
5-Isopropyl-2,6-diethyl-2-methyltetrahydro-2H-pyran	2140	1120363-98-5	Yes, SPET: 2500	Yes. The NOEL of 2.5 mg/kg bw per day in a 90-day study in rats for the structurally related tetrahydro-4-methyl-2-(2-methylpropen-1-yl)-pyran (No. 1237) (Posternak, Linder & Vodoz, 1969) is 60 times the estimated dietary exposure to No. 2140 calculated using the SPET and 15 million times compared with the MSDI (0.01 μg/day) when used as a flavouring agent.	Note 1	Tetrahydro-4-methyl-2-(2-methylpropen-1-yl)pyran (No. 1237)	No safety concern

ALIPHATIC AND AROMATIC ETHERS (addendum)

Butyl β-naphthyl ether	2141	10484-56-7	Yes, SPET: 400	Yes. The NOEL of 5.1 mg/kg bw per day in a 90-day study in rats for the structurally related β-naphthyl ethyl ether (No. 1258) (Oser, Carson & Oser, 1965) is 770 times the estimated dietary exposure to No. 2141 calculated using the SPET and 5300 times compared with the MSDI (58 μg/day) when used as a flavouring agent.	Note 3	β-Naphthyl ethyl ether (No. 1258)	No safety concern

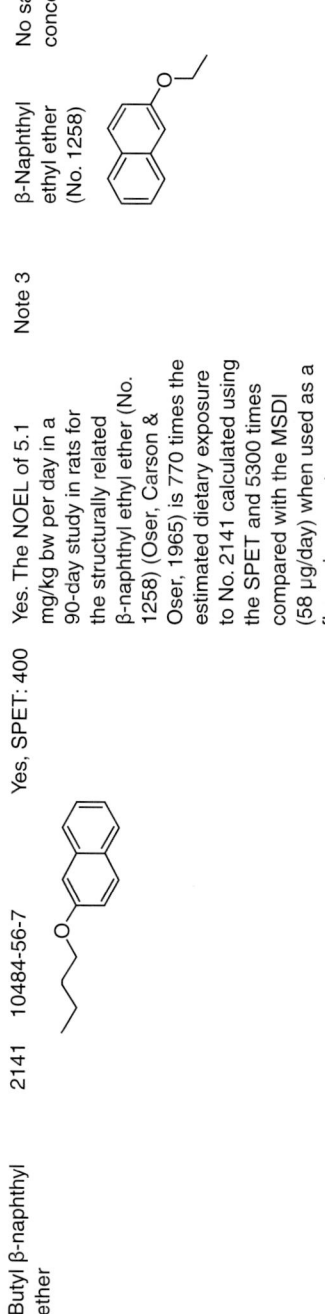

bw, body weight; CAS, Chemical Abstracts Service; NOEL, no-observed-effect level; NR, not required for evaluation because dietary exposure to the flavouring agent was determined to be of no safety concern at step A3 of the Procedure

[a] Twenty-nine flavouring agents in this group were previously evaluated by the Committee (Annex 1, reference 166).
[b] *Step 1*: Seven flavouring agents in this group (Nos 2133–2135, 2137–2139 and 2142) are in structural class II. Three flavouring agents in this group (Nos 2136, 2140 and 2141) are in structural class III (Cramer, Ford & Hall, 1987).
[c] *Step 2*: All of the flavouring agents in this group can be predicted to be metabolized to innocuous products.
[d] The thresholds for human dietary exposure for structural classes II and III are 540 and 90 μg/person per day, respectively. All dietary exposures are expressed in μg/day. The dietary exposure value listed represents the highest estimated dietary exposure calculated using either the SPET or the MSDI method. The SPET gave the highest estimated dietary exposure in each case.
[e] The margins of exposure were calculated based on the estimated dietary exposure calculated using the SPET. In cases where the resulting margin of exposure was relatively low, a comparison with the MSDI was also made.

Notes:
1. Alicyclic ethers are expected to undergo ring hydroxylation by cytochrome P450, conjugation with glucuronic acid and then excretion in the urine.
2. The straight-chain aliphatic ethers are expected to undergo O-dealkylation to yield the corresponding alcohol and aldehyde, which subsequently undergo complete oxidation in the fatty acid pathway and tricarboxylic acid cycle.
3. The aromatic ethers are expected to undergo ring hydroxylation, O-demethylation or side-chain oxidation, depending upon the position of the substituents, followed by conjugation with glucuronic acid, sulfate or glycine.

Table 2. Annual volumes of production and daily dietary exposures for aliphatic and aromatic ethers used as flavouring agents in Europe, the USA and Japan

Flavouring agent (No.)	Most recent annual volume of production (kg)[a]	Dietary exposure			SPET[c]		Natural occurrence in foods
		MSDI[b]					
		µg/day	µg/kg bw per day	µg/day	µg/kg bw per day		
3,6-Dimethyl-2,3,3a,4,5,7a-hexahydrobenzofuran (2133)							
Europe	ND	ND		180	3		+[d]
USA	19	2	0.03				
Japan	ND	ND					
Ethyl linalyl ether (2134)							
Europe	ND	ND		0.1	0.002		+[d]
USA	ND	ND					
Japan	0.1	0.03	0.0005				
Linalool oxide pyranoid (2135)							
Europe	ND	ND		600	10		+[d]
USA	ND	ND					
Japan	159	45	0.8				
Isoamyl phenethyl ether (2136)							
Europe	ND	ND		150	3		–
USA	ND	ND					
Japan	10	3	0.05				

ALIPHATIC AND AROMATIC ETHERS (addendum)

Nerolidol oxide (2137)					
Europe	ND	ND			
USA	ND	ND	2500	42	+[d]
Japan	0.2	0.05	0.001		
Methyl hexyl ether (2138)					
Europe	0.1	0.01	0.0002		
USA	0.1	0.01	0.0002	5000	83	+[d]
Japan	ND	ND	ND		
Myrcenyl methyl ether (2139)					
Europe	ND	ND	ND		
USA	ND	ND	ND	63	1	+[d]
Japan	0.1	0.03	0.0005		
5-Isopropyl-2,6-diethyl-2-methyltetrahydro-2H-pyran (2140)					
Europe	ND	ND	ND		
USA	0.1	0.01	0.0002	2500	42	–
Japan	ND	ND	ND		
Butyl β-naphthyl ether (2141)					
Europe	ND	ND	ND		
USA	ND	ND	ND	400	7	–
Japan	202	58	1		
Digeranyl ether (2142)					
Europe	ND	ND	ND	10 000	167	+[e]

continued

Table 2 *(continued)*

Flavouring agent (No.)	Most recent annual volume of production (kg)[a]	Dietary exposure				Natural occurrence in foods
		MSDI[b]		SPET[c]		
		μg/day	μg/kg bw per day	μg/day	μg/kg bw per day	
USA	0.1	0.01	0.0002			
Japan	ND	ND	ND			
Total						
Europe	0.1					
USA	20					
Japan	371					

ND, no data reported; +, reported to occur naturally in foods, but no quantitative data; –, not reported to occur naturally in foods

[a] From European Flavour and Fragrance Association (2004), Japan Flavor and Fragrance Materials Association (2005), Gavin, Williams & Hallagan (2008) and International Organization of the Flavor Industry (2011). Values greater than 0 kg but less than 0.1 kg were reported as 0.1 kg.

[b] MSDI (μg/person per day) calculated as follows:
(annual volume, kg) × (1 × 10^9 μg/kg)/(population × survey correction factor × 365 days), where population (10%, "eaters only") = 32 × 10^6 for Europe, 31 × 10^6 for the USA and 13 × 10^6 for Japan; and where survey correction factor = 0.8 for the surveys in Europe, the USA and Japan, representing the assumption that only 80% of the annual flavour volume was reported in the poundage surveys (European Flavour and Fragrance Association, 2004; Japan Flavor and Fragrance Materials Association, 2005; Gavin, Williams & Hallagan, 2008; International Organization of the Flavor Industry, 2011).
MSDI (μg/kg bw per day) calculated as follows:
(μg/person per day)/body weight, where body weight = 60 kg. Slight variations may occur from rounding.

[c] SPET (μg/person per day) calculated as follows:
(standard food portion, g/day) × (highest usual use level) (International Organization of the Flavor Industry, 2011). The dietary exposure from the single food category leading to the highest dietary exposure from one portion is taken as the SPET estimate.
SPET (μg/kg bw per day) calculated as follows:
(μg/person per day)/body weight, where body weight = 60 kg. Slight variations may occur from rounding.

[d] Nijssen, van Ingen-Fisscher & Donders (2011).

[e] Nofal, Ho & Chang (1982).

or side-chain oxidation (depending upon the position of the substituents). These metabolites can undergo conjugation with glucuronic acid, sulfate or glycine, the products of which are readily excreted.

1.4 Application of the Procedure for the Safety Evaluation of Flavouring Agents

Step 1. In applying the Procedure for the Safety Evaluation of Flavouring Agents to the 10 flavouring agents in this group of aliphatic and aromatic ethers, the Committee assigned 7 flavouring agents to structural class II (Nos 2133–2135, 2137–2139 and 2142) and 3 flavouring agents to structural class III (Nos 2136, 2140 and 2141) (Cramer, Ford & Hall, 1978).

Step 2. All the flavouring agents in this group are expected to be metabolized to innocuous products. The evaluation of all flavouring agents in this group therefore proceeded via the A-side of the Procedure.

Step A3. The highest dietary exposures to three of the flavouring agents (Nos 2133, 2134 and 2139) in structural class II are below the threshold of concern (i.e. 540 µg/person per day for class II). According to the Procedure, these flavouring agents raise no safety concern at their current estimated dietary exposures. The estimated dietary exposures to four of the flavouring agents (Nos 2135, 2137, 2138 and 2142) are above the threshold of concern for structural class II (i.e. 540 µg/person per day for class II). The three flavouring agents in structural class III (Nos 2136, 2140 and 2141) have estimated dietary exposures above the threshold of concern (i.e. 90 µg/person per day for class III). The evaluation of the seven flavouring agents with estimated dietary exposures above the thresholds of concern proceeded to step A4.

Step A4. Because all of the flavouring agents with estimated dietary exposures above the thresholds of concern, and their metabolites, are not endogenous, the evaluation proceeded to step A5.

Step A5. For linalool oxide pyranoid (No. 2135), the no-observed-effect level (NOEL) of 2.5 mg/kg of body weight (bw) per day for the structurally related substance tetrahydro-4-methyl-2-(2-methylpropen-1-yl)pyran (No. 1237) from a 90-day dietary study in rats (Posternak, Linder & Vodoz, 1969) provides a margin of exposure of 250 in relation to the dietary exposure to No. 2135 as calculated using the SPET (600 µg/person per day) or 3300 in relation to the MSDI (45 µg/day) when No. 2135 is used as a flavouring agent.

For isoamyl phenethyl ether (No. 2136), the NOEL of 196 mg/kg bw per day for the structurally related substance dibenzyl ether (No. 1256) from a 90-day dietary study in rats (Burdock & Ford, 1992) provides a margin of exposure of 78 000 in relation to the dietary exposure to No. 2136 calculated using the SPET (150 µg/day) when No. 2136 is used as a flavouring agent.

For nerolidol oxide (No. 2137), a no-observed-adverse-effect level (NOAEL) for this flavouring agent or a structurally related substance was not available. The Committee therefore concluded that additional data would be necessary to complete the safety evaluation.

For methyl hexyl ether (No. 2138), the NOEL of 900 mg/kg bw per day for the structurally related substance methyl *tert*-butyl ether from a 90-day study in rats (Robinson, Bruner & Olson, 1990) provides a margin of exposure of 11 000 in relation to the dietary exposure to No. 2138 (SPET = 5000 µg/day) when used as a flavouring agent.

For 5-isopropyl-2,6-diethyl-2-methyltetrahydro-2H-pyran (No. 2140), the NOEL of 2.5 mg/kg bw per day for the structurally related substance tetrahydro-4-methyl-2-(2-methylpropen-1-yl)pyran (No. 1237) from a 90-day dietary study in rats (Posternak, Linder & Vodoz, 1969) provides a margin of exposure of 60 in relation to the dietary exposure to No. 2140 as the SPET value (2500 µg/day) or 15 million in relation to the MSDI (0.01 µg/day) when No. 2140 is used as a flavouring agent.

For butyl β-naphthyl ether (No. 2141), the NOEL of 5.1 mg/kg bw per day for the structurally related β-naphthyl ethyl ether (No. 1258) from a 90-day study in rats (Oser, Carson & Oser, 1965) provides a margin of exposure of 770 relative to the SPET value (400 µg/day) or 5300 in relation to the MSDI (58 µg/day) when No. 2141 is used as a flavouring agent.

For digeranyl ether (No. 2142), the NOEL of 50 mg/kg bw per day from a 196-day dietary study in rats (Hagan et al., 1967) for the structurally related substance, and predicted metabolite, geraniol (No. 1223; a mixture of 3,7-dimethyl-2,6-octadienol and 3,7-dimethyl-1,6-octadienol) provides a margin of exposure of 300 relative to the SPET value (10 000 µg/day) or 300 million in relation to the MSDI (0.01 µg/day) when No. 2142 is used as a flavouring agent.

The Committee therefore concluded that 9 of the 10 additional flavouring agents (Nos 2133–2136 and 2138–2142) in this group of aliphatic and aromatic ethers would not pose a safety concern at current estimated dietary exposures. Table 1 summarizes these evaluations.

1.5 Consideration of combined intakes from use as flavouring agents

Flavouring agents Nos 2142, 1257, 1258 and 1259 are ethers of 2-naphthol, which could be a common metabolite, but the combined intake of these is less than the threshold of concern for structural class III (i.e. 90 µg/person per day). Flavouring agents Nos 2136, 1252, 1253 and 1254 are benzyl or phenethyl ethers, but the combined intake of these is also less than the threshold of concern for structural class III. The other flavouring agents in this group have diverse structures, with various potential sites of metabolism, and are not likely to be metabolized to common products. The Committee concluded that combined intakes of the flavouring agents in this group would not raise safety concerns.

1.6 Consideration of secondary components

One flavouring agent in this group (No. 2135) has a minimum assay value of less than 95% (see Annex 5). The secondary component of linalool oxide pyranoid (No. 2135) is linalool (No. 356), which was previously evaluated (Annex 1, reference *137*) and considered not to present a safety concern at estimated dietary exposures.

1.7 Conclusions

In the previous evaluation of flavouring agents in this group of aliphatic and aromatic ethers, studies of metabolism and acute toxicity, short-term and long-term studies of toxicity, and studies of genotoxicity were available. The results of those studies did not raise safety concerns. The additional data from acute toxicity studies (Nos 1237, 2136 and 2140), short-term studies of toxicity (No. 1239), studies of reproductive toxicity (No. 1243) and genotoxicity studies (Nos 2133 and 2140 and several flavouring agents previously evaluated) considered at this meeting support the previous evaluation (Annex 1, reference *166*) and raised no safety concerns.

The Committee concluded that 9 of these 10 flavouring agents, which are additions to the group of aliphatic and aromatic ethers evaluated previously, would not give rise to safety concerns at current estimated dietary exposures. For one flavouring agent (No. 2137), additional data are required to complete the evaluation.

2. RELEVANT BACKGROUND INFORMATION

2.1 Explanation

This monograph summarizes the key data relevant to the safety evaluation of a group of 10 aliphatic and aromatic ethers used as flavouring agents (Table 1), which are additions to the group of aliphatic and aromatic ethers evaluated previously.

2.2 Additional considerations on dietary exposure

Seven of the 10 flavouring agents (Nos 2133–2135, 2137–2139 and 2142) in this group have been reported to occur naturally and can be found in strawberry, lychee, *Salvia* species, dill blossom, dill herb, clary sage, grape brandy, arctic bramble, black currant (buds), *Cinnamomum*, elder flower, elderberry juice, grapefruit juice, green tea (roasted), fresh tomato and white wine (Nofal, Ho & Chang, 1982; Nijssen, van Ingen-Visscher & Donders, 2011) (see Table 2).

2.3 Biological data

2.3.1 Biochemical data: hydrolysis, absorption, distribution, metabolism and elimination

No relevant information additional to that available and described in the monograph of the sixty-first meeting (Annex 1, reference *166*) was available on the absorption, distribution, metabolism or elimination of flavouring agents belonging to the group of aliphatic and aromatic ethers.

2.3.2 Toxicological studies

Results of additional short-term studies on the toxicity of cycloionone (No. 1239), previously evaluated by the Committee (Annex 1, reference *166*), have become available since the sixty-first meeting. These are summarized below. Results of acute toxicity studies for 2 flavouring agents (Nos 2136 and 2140) in the present group of 10 flavouring agents and for 1 previously evaluated flavouring agent (No. 1237) are presented. Results of long-term studies of toxicity have not been reported for any of the 10 additional flavouring agents evaluated in this group, but results of a reproductive study on p-methylanisole (No. 1243) are presented. Results of additional genotoxicity studies on a number of these flavouring agents are also described below.

(a) Acute toxicity

The oral median lethal doses (LD_{50} values) for isoamyl phenethyl ether (No. 2136) have been reported to range from 300 to greater than 5000 mg/kg bw in rats (Moreno, 1979; Hope, 1982; Vaeth, 2005). In Sprague-Dawley rats, an oral LD_{50} value of greater than 2000 mg/kg bw has been reported for 5-isopropyl-2,6-diethyl-2-methyltetrahydro-2H-pyran (No. 2140) (Bradshaw, 2008). LD_{50} values of 4735 mg/kg bw for female rats and greater than 2000 mg/kg bw for male rats have also been reported for tetrahydro-4-methyl-2(2-methylpropen-1-yl)pyran (No. 1237) (Leuschner, 2001), confirming the low toxicity of aliphatic and aromatic ethers. Results of these acute oral toxicity studies are summarized in Table 3.

(b) Short-term studies of toxicity

Results of short-term studies of the toxicity of aliphatic and aromatic ethers used as flavouring agents are summarized in Table 4 and described below.

In a 14-day dietary feeding study, Sprague-Dawley rats (five of each sex per group) were maintained on a control diet or a diet containing cycloionone (No. 1239). The average daily consumption was estimated to be 36.6 and 33.7 mg/kg bw per day for the males and females, respectively. Body weights were recorded on days 1, 8 and 15. Rats were observed daily to assess viability, clinical observations and behaviour patterns. After euthanasia and necropsy on day 15, tissues and organs of the thoracic and abdominal cavities were subjected to gross evaluation. Histological examinations were performed on the liver and kidneys.

All test animals survived and appeared healthy, with no signs of gross toxicity. There were no significant differences in body weight gain or feed consumption between test and control groups. Necropsy findings at terminal sacrifice were unremarkable, and no significant differences in organ to body weight ratios were noted between test and control groups. Microscopic evaluation of the kidneys and livers did not reveal any treatment-related histomorphological alterations.

The NOEL in this study was 36.6 and 33.7 mg/kg bw per day, the only dose tested, in males and females, respectively (Wnorowski, 2006a).

In a 28-day repeated-dose oral gavage toxicity study, Sprague-Dawley rats (five of each sex per dose) were maintained on diets containing cycloionone (No. 1239) at doses of 0, 30, 120, 400 or 1000 mg/kg bw per day. Throughout the study, the animals were observed daily for signs of gross toxicity and mortality and weekly for feed consumption and body weight measurements. During the fourth

Table 3. Results of acute oral toxicity studies with aliphatic and aromatic ethers used as flavouring agents

No.	Flavouring agent	Species; sex	LD_{50} (mg/kg bw)	Reference
2136	Isoamyl phenethyl ether	Rat; NR	>5000	Moreno (1979)
2136	Isoamyl phenethyl ether	Rat; NR	>5000	Hope (1982)
2136	Isoamyl phenethyl ether	Rat; M, F	300–2000	Vaeth (2005)
2140	5-Isopropyl-2,6-diethyl-2-methyltetrahydro-2H-pyran	Rat; M, F	>2000	Bradshaw (2008)
1237	Tetrahydro-4-methyl-2-(2-methylpropen-1-yl)pyran	Rat; M, F	>2000 (M); 4735 (F)	Leuschner (2001)

F, female; M, male; NR, not reported

Table 4. Results of short-term studies of the toxicity of aliphatic and aromatic ethers used as flavouring agents

No.	Flavouring agent	Species; sex	No. of test groups[a] / no. per group[b]	Route	Duration (days)	NOEL/NOAEL (mg/kg bw per day)	Reference
1239	Cycloionone	Rat; M, F	1/10	Diet	14	36.6 (M); 33.7 (F)[c]	Wnorowski (2006a)
1239	Cycloionone	Rat; M, F	4/10	Gavage	28	120[d]	Wnorowski (2006b)

F, female; M, male

[a] Total number of test groups does not include control animals.
[b] Total number per test group includes both male and female animals.
[c] NOEL, the only dose tested.
[d] NOAEL.

week of exposure, a functional observational battery was performed, including assessment of sensory reactivity, grip strength and locomotor activity. Prior to termination of the study, blood was collected and evaluated with respect to serum chemistry and haematology parameters. Gross necropsies were performed on all animals, and selected tissues were collected, weighed and preserved. Histological examinations were performed on the tissues of the control and 1000 mg/kg bw per day groups. Pathological evaluations were extended to the livers, kidneys, thyroids and testicles of animals in the 120 and 400 mg/kg bw per day treatment groups.

No fatalities occurred during the study. Between days 3 and 6, one 1000 mg/kg bw per day female developed facial staining, ventral staining and reduced faecal volume. Otherwise, all rats appeared healthy, and no signs of gross toxicity, adverse pharmacological effects or abnormal behaviour were noted during the study. Although the females treated with 1000 mg/kg bw per day experienced an average body weight loss during the last week of the study, there were no statistically significant differences in body weight gain or feed consumption between any of the test and control groups. Statistical evaluation of organ to body weight ratios revealed significant increases ($P < 0.01$) in relative liver weights in both males and females from the 400 and 1000 mg/kg bw per day groups compared with the control group. Relative kidney weights in the males from the 1000 mg/kg bw per day group were increased compared with the control group. No other statistically significant differences in organ to body weight ratios or haematology parameters were noted between the test and control groups. Statistical evaluation of clinical chemistry values revealed significant differences compared with controls in total serum protein levels for males in the 400 and 1000 mg/kg bw per day treatment groups, globulin levels for females in the 120 mg/kg bw per day treatment group and males and females in the 1000 mg/kg bw per day treatment groups, alkaline phosphatase activity for males in the 400 mg/kg bw per day treatment group and males and females in the 1000 mg/kg bw per day treatment groups, gamma-glutamyl transpeptidase activity and blood urea nitrogen level for females in the 1000 mg/kg bw per day treatment group, and albumin and blood glucose levels for males in the 1000 mg/kg bw per day treatment group. When compared with controls, females treated with 120 mg/kg bw per day exhibited a statistically significant ($P < 0.05$) increase in globulin level, but this effect was not observed in the females dosed with 30 or 400 mg/kg bw per day or in the males treated with 30, 120 or 400 mg/kg bw per day. Histopathological evaluation revealed lesions in the liver, kidneys, thyroids and testicles of the animals in the 1000 mg/kg bw per day treatment group. Lesions were also observed in the kidneys, testicles and thyroids of rats from the 400 mg/kg bw per day group.

Based on the results of the histological evaluations and the supporting clinical chemistry for this 28-day study, the author concluded that the NOAEL for cycloionone in rats is 120 mg/kg bw per day (Wnorowski, 2006b).

(c)   Genotoxicity

Genotoxicity studies have been reported for 3,6-dimethyl-2,3,3a,4,5,7a-hexahydrobenzofuran (No. 2133) and for 5-isopropyl-2,6-diethyl-2-methyltetrahydro-2H-pyran (No. 2140), as well as for the previously evaluated eucalyptol (No. 1234), 2,2,6-trimethyl-6-vinyltetrahydropyran (No. 1236), tetrahydro-4-methyl-2-(2-methylpropen-1-yl)pyran (No. 1237) and dibenzyl ether (No. 1256). The results of these studies are presented in Table 5 and summarized below.

(i)   In vitro

3,6-Dimethyl-2,3,3a,4,5,7a-hexahydrobenzofuran (No. 2133) was tested for mutagenicity to bacteria by incubation with *Salmonella typhimurium* tester strains TA98, TA100, TA102, TA1535 and TA1537 either alone or with an exogenous induced rat liver bioactivation system using the plate incorporation or preincubation method. At five concentrations up to 316 μg/plate, the test substance produced no increase in revertant mutants in any strain, whereas appropriate control substances established the responsiveness of the tester strains (Stien, 2005).

5-Isopropyl-2,6-diethyl-2-methyltetrahydro-2H-pyran (No. 2140) was tested for mutagenicity to bacteria by incubation with *S. typhimurium* tester strains TA98,

Table 5. Studies of genotoxicity with aliphatic and aromatic ethers used as flavouring agents

No.	Flavouring agent	End-point	Test object	Concentration	Results	Reference
In vitro						
2133	3,6-Dimethyl-2,3,3a,4,5,7a-hexahydrobenzofuran	Reverse mutation	Salmonella typhimurium TA98, TA100, TA102, TA1535 and TA1537	3.16, 10, 31.6, 100 and 316 µg/plate	Negative[a,b]	Stien (2005)
2133	3,6-Dimethyl-2,3,3a,4,5,7a-hexahydrobenzofuran	Reverse mutation	S. typhimurium TA98, TA100, TA102, TA1535 and TA1537	3.16, 10, 31.6, 100 and 316 µg/plate	Negative[a,c]	Stien (2005)
2140	5-Isopropyl-2,6-diethyl-2-methyltetrahydro-2H-pyran	Reverse mutation	S. typhimurium TA98, TA100, TA102, TA1535 and TA1537	50, 150, 500, 1500 and 5000 µg/plate	Negative[a]	Thompson (2008)
1234	Eucalyptol	Comet assay	Mouse lymphoma cells	1151, 2302, 4605 and 9210 µg/ml[d]	Negative	Ribeiro, Marques & Salvadori (2006)
1236	2,2,6-Trimethyl-6-vinyltetrahydropyran	Reverse mutation	S. typhimurium TA98, TA100, TA102, TA1535 and TA1537	0.15, 0.5, 1.5, 5, 15, 50, 150, 500, 1500 and 5000 µg/plate	Negative[a,e]	Thompson (2004)
1236	2,2,6-Trimethyl-6-vinyltetrahydropyran	Reverse mutation	S. typhimurium TA98, TA100, TA102, TA1535 and TA1537	0.15, 0.5, 1.5, 5, 15, 50, 150, 500, 1500 and 5000 µg/plate	Negative[a,c]	Thompson (2004)
1237	Tetrahydro-4-methyl-2-(2-methylpropen-1-yl)pyran	Reverse mutation	S. typhimurium TA100	3, 10, 33, 100, 333, 1000, 3330 and 5000 µg/plate	Negative[a,c,e]	Verspeek-Rip (2002)
1237	Tetrahydro-4-methyl-2-(2-methylpropen-1-yl)pyran	Reverse mutation	S. typhimurium TA100	3, 10, 33, 100, 333, 1000, 3330 and 5000 µg/plate	Negative[a,b,e]	Verspeek-Rip (2002)
1237	Tetrahydro-4-methyl-2-(2-methylpropen-1-yl)pyran	Reverse mutation	S. typhimurium TA98, TA102, TA1535 and TA1537	3, 10, 33, 100, 333 and 1000 µg/plate	Negative[a,b,e]	Verspeek-Rip (2002)
1237	Tetrahydro-4-methyl-2-(2-methylpropen-1-yl)pyran	Reverse mutation	S. typhimurium TA98, TA102, TA1535 and TA1537	3, 10, 33, 100, 333 and 1000 µg/plate	Negative[a,c,e]	Verspeek-Rip (2002)

continued

Table 5 (continued)

No.	Flavouring agent	End-point	Test object	Concentration	Results	Reference
1237	Tetrahydro-4-methyl-2-(2-methylpropen-1-yl)pyran	Reverse mutation	S. typhimurium TA97, TA98, TA100, TA102 and TA1535	16, 50, 160, 500 and 1600 µg/plate	Negative[a]	Scheerbaum (2001)
1256	Dibenzyl ether	Reverse mutation	S. typhimurium TA98 and TA100	19.8 and 198 µg/plate (0.01 and 1 mmol/l)[f]	Negative[a]	Kubo, Urano & Utsumi (2002)

[a] With and without metabolic activation.
[b] Plate incorporation method.
[c] Preincubation method.
[d] Concentrations based on a density of 0.921 g/ml at 25 °C.
[e] Toxicity observed under some conditions of the assay.
[f] Calculated using relative molecular mass of 198.27.

TA100, TA102, TA1535 and TA1537 either alone or with an exogenous induced rat liver bioactivation system. At five concentrations up to 5000 µg/plate, the test substance produced no increase in revertant mutants in any strain, whereas appropriate control substances established the responsiveness of the tester strains (Thompson, 2008).

Negative results for deoxyribonucleic acid (DNA) damage were obtained in an in vitro mouse lymphoma single-cell gel (comet) assay for eucalyptol (No. 1234) at concentrations equivalent to 0, 1151, 2302, 4605 and 9210 µg/ml (based on a density of 0.921 g/ml at 25 °C). Test material concentrations of 2302 µg/ml and higher were toxic to the cells. However, DNA breaks were not induced. Eucalyptol was determined to be a strong cytotoxicant, but, under the conditions of the assay, it did not increase the level of DNA damage to mammalian cells (Ribeiro, Marques & Salvadori, 2006).

2,2,6-Trimethyl-6-vinyltetrahydropyran (No. 1236) was tested for mutagenicity to bacteria by incubation with *S. typhimurium* tester strains TA98, TA100, TA102, TA1535 and TA1537 either alone or with an exogenous induced rat liver bioactivation system using the plate incorporation and preincubation methods. At 10 concentrations up to 5000 µg/plate, the test substance produced no increase in revertant mutants in any strain, whereas appropriate control substances established the responsiveness of the tester strains (Thompson, 2004).

Tetrahydro-4-methyl-2-(2-methylpropen-1-yl)pyran (No. 1237) was tested for mutagenicity to bacteria by incubation with *S. typhimurium* tester strains TA98, TA100, TA102, TA1535 and TA1537 either alone or with an exogenous induced rat liver bioactivation system using the plate incorporation and preincubation methods. Tetrahydro-4-methyl-2-(2-methylpropen-1-yl)pyran was toxic to TA98, TA100, TA1537 and TA1535, but not to TA102, in the absence of metabolic activation in the plate incorporation test and was toxic to all strains at 1000 µg/plate and higher in the preincubation test. At eight concentrations up to 5000 µg/plate for tester strain TA100 and at six concentrations up to 1000 µg/plate for tester strains TA98, TA102, TA1535 and TA1537, the test substance produced no increase in revertant mutants in any strain, whereas appropriate control substances established the responsiveness of the tester strains (Verspeek-Rip, 2002). Additionally, tetrahydro-4-methyl-2-(2-methylpropen-1-yl)pyran was tested for mutagenicity to bacteria by incubation with *S. typhimurium* tester strains TA97, TA98, TA100, TA102 and TA1535 either alone or with an exogenous induced rat liver bioactivation system. At five concentrations up to 1600 µg/plate, the test substance produced no increase in revertant mutants in any strain, whereas appropriate control substances established the responsiveness of the tester strains (Scheerbaum, 2001).

Dibenzyl ether (No. 1256) was tested for mutagenicity to bacteria by incubation with *S. typhimurium* tester strains TA98 and TA100 either alone or with an exogenous induced rat liver bioactivation system. At concentrations of 19.8 and 198 µg/plate, the test substance produced no increase in revertant mutants in any strain, whereas appropriate control substances established the responsiveness of the tester strains (Kubo, Urano & Utsumi, 2002).

(d) Reproductive toxicity

A reproductive toxicity study has been reported in which Wistar rats (10 of each sex per dose) were administered p-methylanisole (p-cresyl methyl ether; No. 1243) at doses of 100, 300 or 1000 mg/kg bw per day by oral gavage. The study was conducted 2 weeks pre-mating and during the mating period (maximum of 2 weeks) in both sexes and during the entire gestation period, 4 days of lactation and approximately 1 week thereafter in females. The F_0 animals were examined for their reproductive performance, including determination of the number of implantation sites and the calculation of post-implantation loss. Feed consumption of the F_0 parents was determined regularly during the pre-mating period in both sexes and at gestation days 0–7, 7–14 and 14–20 and lactation days 1–4 in the dams. Body weights of F_0 animals were determined once per week, in males throughout the study and in females during pre-mating and mating. During gestation and lactation, F_0 females were weighed on gestation days 0, 7, 14 and 20, on the parturition day and on postnatal day 4. Pups were sexed and examined for macroscopically evident changes on the day of birth and weighed on postnatal days 1 and 4. Seven days after parturition, four of the female pups and all of the parental females were sacrificed and examined. The other pups were subjected to euthanasia 4 days after parturition for gross necropsy. The male animals were euthanized 29 days after the initial dose. Organ weights were recorded, and a histopathological examination was performed.

At 1000 mg/kg bw per day, the F_0 parental animals displayed reductions in feed consumption and body weight, apathy, ataxia and unsteady gait after treatment. The enlarged livers harvested at necropsy were characterized as being due to centrilobular hypertrophy. Reproductive end-points observed were increases in the incidence of post-implantation loss, stillborn pups, total litter loss and insufficient maternal care. At this high dose, none of the pups survived to postnatal day 4. At 300 mg/kg bw per day, the F_0 rats displayed total litter loss, insufficient maternal care, decreased maternal feed consumption during lactation, increased numbers of litters with stillborn pups as well as increased numbers of stillborn pups, and reduced live birth index. In the 300 mg/kg bw per day group, the following test material–related adverse effects or findings were reported in F_1 pups: reduced viability and decreased pup body weights. In the 100 mg/kg bw per day group, no test material–related adverse findings were observed in F_1 pups.

The NOAEL for developmental toxicity was determined to be 100 mg/kg bw per day, based on prenatal and postnatal offspring mortality at 300 and 1000 mg/kg bw per day as well as reduced pup body weights at 300 mg/kg bw per day. The NOAEL for general systemic toxicity of the test material was determined to be 100 mg/kg bw per day for the F_0 parental animals (Schneider et al., 2010).

3. REFERENCES

Bradshaw J (2008). Acute oral toxicity in the rat—fixed dose method. Unpublished report no. 2082/0174 from Harlan Laboratories Ltd, Derbyshire, England. Submitted to WHO by the International Organization of the Flavor Industry, Brussels, Belgium.

Burdock GA, Ford RA (1992). Safety evaluation of dibenzyl ether. *Food and Chemical Toxicology*, 30(7):559–566.

Cramer GM, Ford RA, Hall RL (1978). Estimation of toxic hazard—a decision tree approach. *Food and Cosmetics Toxicology*, 16(3):255–276.

European Flavour and Fragrance Association (2004). European inquiry on volume use. Private communication to the Flavor and Extract Manufacturers Association of the United States, Washington, DC, USA. Submitted to WHO by the International Organization of the Flavor Industry, Brussels, Belgium.

Gavin CL, Williams MC, Hallagan JB (2008). *Flavor and Extract Manufacturers Association of the United States 2005 poundage and technical effects update survey.* Washington, DC, USA, Flavor and Extract Manufacturers Association of the United States.

Hagan EC et al. (1967). Food flavourings and compounds of related structure. II. Subacute and chronic toxicity. *Food and Cosmetics Toxicology*, 5(2):141–157.

Hope J (1982). Acute oral toxicity test: anther. Unpublished report no. M.T. 2169 from Unilever Research Laboratory, Bedfordshire, England. Submitted to WHO by the International Organization of the Flavor Industry, Brussels, Belgium.

International Organization of the Flavor Industry (2011). Interim inquiry on volume use and added use levels for flavoring agents to be presented at the 76th JECFA meeting. Private communication to the Flavor and Extract Manufacturers Association of the United States, Washington, DC, USA. Submitted to WHO by the International Organization of the Flavor Industry, Brussels, Belgium.

Japan Flavor and Fragrance Materials Association (2005). Japanese inquiry on volume use. Private communication to the Flavor and Extract Manufacturers Association of the United States, Washington, DC, USA. Submitted to WHO by the International Organization of the Flavor Industry, Brussels, Belgium.

Kubo T, Urano K, Utsumi H (2002). Mutagenicity characteristics of 255 environmental chemicals. *Journal of Health Science*, 48(6):545–554.

Leuschner J (2001). Acute toxicity study of Rosenoxid-D by oral administration to Sprague Dawley rats. Unpublished report no. 13877/00 from LPT Laboratory of Pharmacology and Toxicology KG, Hamburg, Germany. Submitted to WHO by the International Organization of the Flavor Industry, Brussels, Belgium.

Moreno OM (1979). Acute toxicity in rats and acute dermal toxicity in rabbits. Unpublished report no. MB 78-3426 from MB Research Laboratories, Inc., Spinnerstown, PA, USA. Submitted to WHO by the International Organization of the Flavor Industry, Brussels, Belgium.

Nijssen LM, van Ingen-Visscher CA, Donders JJH (2011). Volatile Compounds in Food (VCF) database, version 13.1. Zeist, the Netherlands, TNO Triskelion (http://www.vcf-online.nl/VcfHome.cfm).

Nofal MA, Ho C, Chang SS (1982). Major volatile constituents of Egyptian rose absolute. *Perfumer & Flavorist*, 7(4):23–26.

Oser BL, Carson S, Oser M (1965). Toxicological tests on flavouring matters. *Food and Cosmetics Toxicology*, 3(4):563–569.

Posternak JM, Linder A, Vodoz CA (1969). Toxicological tests on flavoring matters. *Food and Cosmetics Toxicology*, 7(4):405–407.

Ribeiro DA, Marques MEA, Salvadori DMF (2006). In vitro cytotoxic and non-genotoxic effects of Gutta-Percha solvents on mouse lymphoma cells by single cell gel (comet) assay. *Brazilian Dental Journal*, 17(3):228–232.

Robinson M, Bruner RH, Olson GR (1990). Fourteen- and ninety-day oral toxicity studies of methyl tertiary-butyl ether in Sprague-Dawley rats. *International Journal of Toxicology*, 9(5):525–540.

Scheerbaum D (2001). Rose oxide rac: reverse mutation assay (Ames test) with *Salmonella typhimurium*. Unpublished report no. USO75593 from the Dr. U. Noack Laboratorium für Angewandte Biologie, Sarstedt, Germany. Submitted to WHO by the International Organization of the Flavor Industry, Brussels, Belgium.

Schneider S et al. (2010). *p*-Cresolmethylether reproduction/developmental toxicity screening test in Wistar rats: oral administration (gavage). Unpublished report no. 90R0506/09068 from BASF SE, Ludwigshafen, Germany. Submitted to WHO by the International Organization of the Flavor Industry, Brussels, Belgium.

Stien J (2005). Mutagenicity study of dill-ether in the *Salmonella typhimurium* reverse mutation assay (in vitro). Unpublished report no. 18432/10/04 from LPT Laboratory of Pharmacology and Toxicology KG, Hamburg, Germany. Submitted to WHO by the International Organization of the Flavor Industry, Brussels, Belgium.

Thompson PW (2004). Limetol (batch # 9000523633): reverse mutation assay "Ames test" using *Salmonella typhimurium*. Unpublished report no. 1895/005 from SafePharm Laboratories Ltd, Derbyshire, England. Submitted to WHO by the International Organization of the Flavor Industry, Brussels, Belgium.

Thompson PW (2008). PI 24640: reverse mutation assay "Ames test" using *Salmonella typhimurium*. Unpublished report no. 2082/0175 from Harlan Laboratories Ltd, Derbyshire, England. Submitted to WHO by the International Organization of the Flavor Industry, Brussels, Belgium.

Vaeth A (2005). Dillether acute oral toxicity study in the rat. Unpublished report no. 02718 from Frey-Tox GmbH, Germany. Submitted to WHO by the International Organization of the Flavor Industry, Brussels, Belgium.

Verspeek-Rip CM (2002). Evaluation of the mutagenic activity of rose oxide co in the *Salmonella typhimurium* reverse mutation assay. Unpublished report no. 359448 from NOTOX B.V., 's-Hertogenbosch, the Netherlands. Submitted to WHO by the International Organization of the Flavor Industry, Brussels, Belgium.

Wnorowski G (2006a). Repeat dose oral toxicity: 28-day rodent study—cycloionone. Unpublished report no. 5860 from Product Safety Labs, East Brunswick, NJ, USA. Submitted to WHO by the International Organization of the Flavor Industry, Brussels, Belgium.

Wnorowski G (2006b). 14-day dietary toxicity study: rats—cycloionone, lot #2/97. Unpublished report no. 5209 from Product Safety Labs, East Brunswick, NJ, USA. Submitted to WHO by the International Organization of the Flavor Industry, Brussels, Belgium.

ALIPHATIC LINEAR α,β-UNSATURATED ALDEHYDES, ACIDS AND RELATED ALCOHOLS, ACETALS AND ESTERS (addendum)

First draft prepared by

G.M. Williams[1], M.J. DiNovi[2] and A.G. Renwick[3]

[1] New York Medical College, Valhalla, New York, United States of America (USA)
[2] Center for Food Safety and Applied Nutrition, Food and Drug Administration, College Park, Maryland, USA
[3] School of Medicine, University of Southampton, Southampton, England

1. Evaluation 117
   1.1 Introduction 117
   1.2 Assessment of dietary exposure 118
   1.3 Absorption, distribution, metabolism and elimination 118
   1.4 Application of the Procedure for the Safety Evaluation of
       Flavouring Agents 118
   1.5 Consideration of combined intakes from use as flavouring
       agents 123
   1.6 Conclusions 123
2. Relevant background Information 123
   2.1 Explanation 123
   2.2 Additional considerations on dietary exposure 123
   2.3 Biological data 124
       2.3.1 Biochemical data: hydrolysis, absorption, distribution,
             metabolism and elimination 124
       2.3.2 Toxicological studies 125
           (a) Acute toxicity 125
           (b) Genotoxicity 126
           (c) Cardiotoxicity 130
3. References 131

1. EVALUATION

1.1 Introduction

The Committee evaluated a group of flavouring agents consisting of five aliphatic linear α,β-unsaturated aldehydes, acids and related alcohols, acetals and esters. This group included trans-2-nonenyl acetate (No. 2163), propyl sorbate (No. 2164), cis-2-octenol (No. 2165), trans-2-tridecenol (No. 2166) and ethyl 2-hexenoate (mixture of isomers) (No. 2167). 2-Phenyl-4-methyl-2-hexenal (No. 2069) was submitted for evaluation, but the Committee considered that it did not belong to this group of flavouring agents, and the compound was therefore not further considered. The evaluations were conducted according to the Procedure for the Safety Evaluation of Flavouring Agents (see Figure 1, Introduction) (Annex 1, reference *131*). None of these flavouring agents have previously been evaluated by the Committee.

The Committee previously evaluated 37 other members of this group of flavouring agents at its sixty-third meeting (Annex 1, reference *173*). The Committee concluded that all 37 flavouring agents in that group were of no safety concern at estimated dietary exposures.

The Committee also evaluated 22 additional members of this group of flavouring agents at its sixty-ninth meeting (Annex 1, reference *190*). The Committee concluded that the 22 additional flavouring agents in that group were of no safety concern at estimated dietary exposures.

Two of the five flavouring agents (Nos 2165 and 2167) in this group have been reported to occur naturally and can be found in chicken, ginger, mushroom, raspberry, blackberry, boysenberry, apple, grapes, guava, feyoa, mangifera, passiflora and wine (Nijssen, van Ingen-Visscher & Donders, 2011).

1.2 Assessment of dietary exposure

The total annual volumes of production of the five aliphatic linear α,β-unsaturated aldehydes, acids and related alcohols, acetals and esters are approximately 0.1 kg in the USA and 113 kg in Japan (European Flavour and Fragrance Association, 2004; Japan Flavor and Fragrance Materials Association, 2005; Gavin, Williams & Hallagan, 2008; International Organization of the Flavor Industry, 2011). Two flavouring agents in this group, propyl sorbate (No. 2164) with 80 kg and ethyl 2-hexenoate (mixture of isomers) (No. 2167) with 33 kg, account for approximately 99% of the total annual volume of production in Japan.

Dietary exposures were estimated using both the maximized survey-derived intake (MSDI) method and the single portion exposure technique (SPET), with the highest values reported in Table 1. The estimated daily dietary exposure is highest for propyl sorbate (No. 2164) (300 μg, the SPET value obtained from non-alcoholic beverages). For the other flavouring agents, the estimated daily dietary exposures range up to 25 μg.

Annual volumes of production of this group of flavouring agents as well as the daily dietary exposures calculated using both the MSDI method and the SPET are summarized in Table 2.

1.3 Absorption, distribution, metabolism and elimination

Information on the hydrolysis, absorption, distribution, metabolism and elimination of flavouring agents belonging to the group of aliphatic linear α,β-unsaturated aldehydes, acids and related alcohols, acetals and esters has previously been described in the report of the sixty-third meeting (Annex 1, reference *173*). The alkenols and alkenoic acid esters in this group are expected to be hydrolysed and completely metabolized by the fatty acid β-oxidation pathway or the tricarboxylic acid cycle.

1.4 Application of the Procedure for the Safety Evaluation of Flavouring Agents

Step 1. In applying the Procedure for the Safety Evaluation of Flavouring Agents to the five additional flavouring agents in this group of aliphatic linear

Table 1. Summary of the results of the safety evaluations of aliphatic linear α,β-unsaturated aldehydes, acids and related alcohols, acetals and esters used as flavouring agents[a,b,c]

Flavouring agent	No.	CAS No. and structure	Step A3[1] Does estimated dietary exposure exceed the threshold of concern?	Comments on predicted metabolism	Conclusion based on current estimated dietary exposure
Structural class I					
trans-2-Nonenyl acetate	2163	30418-89-4	No, SPET: 2	Note 1	No safety concern
Propyl sorbate	2164	10297-72-0	No, SPET: 300	Note 1	No safety concern
cis-2-Octenol	2165	26001-58-1	No, SPET: 20	Note 2	No safety concern
trans-2-Tridecenol	2166	74962-98-4	No, SPET: 3	Note 2	No safety concern

continued

Table 1 (continued)

Flavouring agent	No.	CAS No. and structure	Step A3[c] Does estimated dietary exposure exceed the threshold of concern?	Comments on predicted metabolism	Conclusion based on current estimated dietary exposure
Ethyl 2-hexenoate (mixture of isomers)	2167	1552-67-6	No, SPET: 25	Note 1	No safety concern

CAS, Chemical Abstracts Service

[a] Fifty-nine flavouring agents in this group were previously evaluated by the Committee (Annex 1, references *173* and *190*).
[b] *Step 1*: All of the flavouring agents in this group (Nos 2163–2167) are in structural class I.
[c] *Step 2*: All of the flavouring agents in this group can be predicted to be metabolized to innocuous products.
[d] The threshold for human dietary exposure for structural class I is 1800 μg/person per day. All dietary exposure values are expressed in μg/day. The dietary exposure value listed represents the highest estimated dietary exposure calculated using either the SPET or the MSDI method. The SPET gave the highest estimated dietary exposure in each case.

Notes:
1. Hydrolysed to corresponding alcohols and acids, followed by complete metabolism in the fatty acid pathway or the tricarboxylic acid cycle.
2. The alcohol group would be oxidized to the corresponding carboxylic acid and completely metabolized by the fatty acid pathway.

Table 2. Annual volumes of production of aliphatic linear α,β-unsaturated aldehydes, acids and related alcohols, acetals and esters used as flavouring agents in Europe, the USA and Japan

Flavouring agent (No.)	Most recent annual volume of production (kg)[a]	Dietary exposure				Natural occurrence in foods
		MSDI[b]		SPET[c]		
		μg/day	μg/kg bw per day	μg/day	μg/kg bw per day	
trans-2-Nonenyl acetate (2163)						
Europe	ND	ND	ND	2	0.03	−
USA	0.1	0.01	0.0002			
Japan	ND	ND	ND			
Propyl sorbate (2164)						
Europe	ND	ND	ND	300	5	−
USA	ND	ND	ND			
Japan	80	23	0.4			
cis-2-Octenol (2165)						
Europe	ND	ND	ND	20	0.3	+
USA	ND	ND	ND			
Japan	0.1	0.03	0.0005			
trans-2-Tridecenol (2166)						
Europe	ND	ND	ND	3	0.04	−
USA	ND	ND	ND			
Japan	0.3	0.1	0.002			
Ethyl 2-hexenoate (mixture of isomers) (2167)						
Europe	ND	ND	ND	25	0.4	+
USA	ND	ND	ND			
Japan	33	9	0.2			

continued

Table 2 (continued)

Flavouring agent (No.)	Most recent annual volume of production (kg)[a]	Dietary exposure				Natural occurrence in foods
		MSDI[b]		SPET[c]		
		µg/day	µg/kg bw per day	µg/day	µg/kg bw per day	
Total						
Europe	ND					
USA	0.1					
Japan	113					

bw, body weight; ND, no data reported; +, reported to occur naturally in foods (Nijssen, van Ingen-Visscher & Donders, 2011), but no quantitative data; –, not reported to occur naturally in foods

[a] From European Flavour and Fragrance Materials Association (2004), Japan Flavor and Fragrance Materials Association (2005), Gavin, Williams & Hallagan (2008) and International Organization of the Flavor Industry (2011). Values greater than 0 kg but less than 0.1 kg were reported as 0.1 kg.
[b] MSDI (µg/person per day) calculated as follows:
(annual volume, kg) × (1 × 10⁹ µg/kg)/(population × survey correction factor × 365 days), where population (10%, "eaters only") = 32 × 10⁶ for Europe, 31 × 10⁶ for the USA and 13 × 10⁶ for Japan; and where survey correction factor = 0.8 for the surveys in Europe, the USA and Japan, representing the assumption that only 80% of the annual flavour volume was reported in the poundage surveys (European Flavour and Fragrance Association, 2004; Japan Flavor and Fragrance Materials Association, 2005; Gavin, Williams & Hallagan, 2008; International Organization of the Flavor Industry, 2011).
MSDI (µg/kg bw per day) calculated as follows:
(µg/person per day)/body weight, where body weight = 60 kg. Slight variations may occur from rounding.
[c] SPET (µg/person per day) calculated as follows:
(standard food portion, g/day) × (highest usual use level) (International Organization of the Flavor Industry, 2011). The dietary exposure from the single food category leading to the highest dietary exposure from one portion is taken as the SPET estimate.
SPET (µg/kg bw per day) calculated as follows:
(µg/person per day)/body weight, where body weight = 60 kg. Slight variations may occur from rounding.

α,β-unsaturated aldehydes, acids and related alcohols, acetals and esters, the Committee assigned all five to structural class I (Nos 2163–2167) (Cramer, Ford & Hall, 1978).

Step 2. All flavouring agents in this group are expected to be metabolized to innocuous products. The evaluation of all flavouring agents in this group therefore proceeded via the A-side of the Procedure.

Step A3. The estimated dietary exposures to the five flavouring agents in structural class I are below the threshold of concern (i.e. 1800 μg/person per day for class I). According to the Procedure, none of the five flavouring agents raise safety concern at their current estimated dietary exposures.

Table 1 summarizes the evaluations of the five aliphatic linear α,β-unsaturated aldehydes, acids and related alcohols, acetals and esters (Nos 2163–2167) in this group.

1.5   Consideration of combined intakes from use as flavouring agents

The highest MSDI for any member of this group is 23 μg/person per day (No. 2164), which is less than 2% of the threshold of concern, 1800 μg/person per day. Consideration of combined intakes is not deemed necessary, because the additional flavouring agents would not contribute significantly to the combined intake of this flavouring group.

1.6   Conclusions

In the previous evaluations of the aliphatic linear α,β-unsaturated aldehydes, acids and related alcohols, acetals and esters, studies of acute toxicity, short-term studies of toxicity, long-term studies of toxicity and carcinogenicity, and studies of genotoxicity and reproductive toxicity were available. None raised safety concerns. New data on acute toxicity and genotoxicity were available at the present meeting, and these supported the previous safety evaluations (Annex 1, references *173* and *190*).

The Committee concluded that these five flavouring agents, which are additions to the group of aliphatic linear α,β-unsaturated aldehydes, acids and related alcohols, acetals and esters evaluated previously, would not give rise to safety concerns at current estimated dietary exposures.

2.   RELEVANT BACKGROUND INFORMATION

2.1   Explanation

This monograph summarizes the key data relevant to the safety evaluation of five flavouring agents that are additions to the group of aliphatic linear α,β-unsaturated aldehydes, acids and related alcohols, acetals and esters evaluated previously (see Table 1).

2.2   Additional considerations on dietary exposure

Two of the five flavouring agents (Nos 2165 and 2167) in this group have been reported to occur naturally and can be found in chicken, ginger, mushroom, raspberry, blackberry, boysenberry, apple, grapes, guava, feyoa, mangifera, passiflora and wine (Nijssen, van Ingen-Visscher & Donders, 2011).

2.3 Biological data

2.3.1 Biochemical data: hydrolysis, absorption, distribution, metabolism and elimination

Information on the hydrolysis, absorption, distribution, metabolism and elimination of flavouring agents belonging to the group of aliphatic linear α,β-unsaturated aldehydes, acids and related alcohols, acetals and esters has previously been described in the report of the sixty-third meeting (Annex 1, reference *173*). The alkenols and alkenoic acid esters (Nos 2163–2167) in this group are expected to be rapidly hydrolysed. The unsaturated alcohols are expected to be successively oxidized to the corresponding aldehydes and carboxylic acids, which are metabolized by the fatty acid β-oxidation pathway and tricarboxylic acid cycle, followed by excretion (Nelson & Cox, 2008). If substances are hydrolysed before absorption, the resulting aliphatic alcohols and carboxylic acids are rapidly absorbed in the gastrointestinal tract.

In addition to undergoing aldehyde group oxidation, α,β-unsaturated 2-phenylaldehyde derivatives are predicted to be subject to alkenyl group glutathione conjugation. The structurally related substance 2-phenylpropenal (atropaldehyde) readily forms glutathione conjugates at a half-life of formation of approximately 1 minute when incubated in vitro with free glutathione at pH 8.0 (Thompson, Kinter & Macdonald, 1996). The in vitro half-lives of formation were 37.8, 11.3, 17.7 and 26.4 seconds for control and the glutathione *S*-transferases GSTM1-1, GSTP1-1 and GSTA1-1, respectively (Dieckhaus et al., 2001). The formation of the glutathione conjugate was catalysed most efficiently by GSTM1-1, then GSTP1-1, followed by GSTA1-1, with second-order rate constants of 0.275 ± 0.035, 0.164 ± 0.005 and 0.042 ± 0.005 per micromole per litre per second, respectively (Dieckhaus et al., 2001). In addition, 2-phenylpropenal is a metabolite of the antiepileptic drug felbamate, which has been studied in six adult male Sprague-Dawley rats administered a single 800 mg/kg body weight (bw) dose of felbamate by gavage. In this study, 2-phenylpropenal was identified in the 18-hour urine. One healthy male human volunteer was also given a tablet containing 600 mg (9 mg/kg bw, as determined by the study authors) of felbamate, and the 8-hour urine analysis revealed reduced and oxidized mercapturic acid conjugates of 2-phenylpropenal (Thompson, Gulden & Macdonald, 1997). Therefore, aliphatic linear unsaturated derivatives are expected to form glutathione conjugates and be rapidly eliminated by the body as mercapturic acid derivatives.

4-Hydroxy α,β-unsaturated acids are products of lipid peroxidation, which is a viable pathway for this group of aliphatic linear α,β-unsaturated aldehydes, acids and related alcohols, acetals and esters (Annex 1, reference *174*). Until recently, it was presumed that aliphatic linear α,β-unsaturated aldehydes, acids and related alcohols, acetals and esters are primarily metabolized through β-oxidation. It has been proposed, based on liver perfusion experiments in rats followed by metabolite identification, that 4-hydroxy acids are metabolized through two pathways. The major pathway involves the formation of 4-hydroxy-4-phosphoacyl-coenzyme A derivatives, followed by isomerization, leading to 3-hydroxyacyl-coenzyme A derivatives and subsequently four rounds of β-oxidation; the minor pathway involves direct β-oxidation followed by α-oxidation, then two rounds of β-oxidation (Zhang et al., 2009).

2.3.2 Toxicological studies

Short-term and long-term studies of toxicity have not been reported for any of the five flavouring agents evaluated in this group. Additional studies of toxicity for previously evaluated flavouring agents have become available since the sixty-ninth meeting (Annex 1, reference *190*). New studies for 2-dodecenal (No. 1350), 2-hexenal (No. 1353) and 2-nonenal (No. 1362) are summarized below.

(a) Acute toxicity

The acute toxicity of 2-dodecenal (No. 1350) was assessed in three experiments using single oral doses of 0, 100, 1000 or 2000 mg/kg bw administered to two mice of each sex per dose. Mice were monitored at intervals of approximately 1, 2–4, 6, 24, 30 and 48 hours after dosing, and no toxic reactions were observed. At the 2–4 hour observation of the 100 mg/kg bw dosed mice, ruffled fur was noted (Honarvar et al., 2007b). A median lethal dose (LD_{50} value) for this study was determined to be greater than 2000 mg/kg bw.

Five studies of the acute toxicity of 2-hexenal (No. 1353) were performed, in each of which two male and two female NMRI mice were administered single oral doses of 2-hexenal at 0, 100, 1000, 1250, 1500 or 2000 mg/kg bw in corn oil. Mice were evaluated at 1, 2–4, 6, 24, 30 and 48 hours after dosing. No effects were observed with the 100 and 1000 mg/kg bw doses. Administration of 1250 mg/kg bw resulted in a reduction of spontaneous activity, changes in abdominal position, eyelid closure, ruffled fur and apathy. The same observations were seen at 1500 and 2000 mg/kg bw, but tremors were also observed. At 1250 mg/kg bw, one male died between 2 and 4 hours and one male and one female died at 24 hours; the last female died at 48 hours after administration. Administration of 2-hexenal at 1500 mg/kg bw resulted in the death of both males and both females 6 hours after administration. At the 2000 mg/kg bw dose, both males were dead at 24 hours; the two females were normal at 24 hours and remained normal upon observation at 30 and 48 hours (Honarvar et al., 2007a). An LD_{50} value for this study was determined to be less than 1250 mg/kg bw.

The acute toxicity of 2-nonenal (No. 1362) was assessed in four experiments using single oral doses of 2-nonenal at 0, 100, 500, 1000 or 2000 mg/kg bw administered to two mice of each sex by oral gavage in corn oil. Mice were monitored at intervals of approximately 1, 2–4, 6, 24, 30 and 48 hours after administration. No toxic reactions were observed below the 2000 mg/kg bw dose. At 2000 mg/kg bw, one male and one female displayed reduced spontaneous activity, and all mice administered this dose had ruffled fur (Honarvar et al., 2008). An LD_{50} for this study was determined to be greater than 2000 mg/kg bw.

The results of the acute oral toxicity studies with these three flavouring agents are summarized in Table 3.

Table 3. Results of acute oral toxicity studies with aliphatic linear α,β-unsaturated aldehydes, acids and related alcohols, acetals and esters used as flavouring agents

No.	Flavouring agent	Species; sex	LD_{50} (mg/kg bw)	Reference
1350	2-Dodecenal	Mouse; M, F	>2000	Honarvar et al. (2007b)
1353	2-Hexenal	Mouse; M, F	<1250	Honarvar et al. (2007a)
1362	2-Nonenal	Mouse; M, F	>2000	Honarvar et al. (2008)

F, female; M, male

(b) Genotoxicity

Studies of genotoxicity in vitro and in vivo with previously evaluated members of this group of aliphatic linear α,β-unsaturated aldehydes, acids and related alcohols, acetals and esters are summarized in Table 4 and described below.

(i) In vitro

Studies were performed to investigate the mutagenic potential of three aliphatic linear α,β-unsaturated aldehydes, 2-dodecenal (No. 1350), 2-hexenal (No. 1353) and 2-nonenal (No. 1362), previously evaluated by the Committee.

An Ames mutagenicity assay was conducted in *S. typhimurium* strains TA98, TA100, TA102, TA1535 and TA1537 by the plate incorporation and preincubation methods in the presence and absence of metabolic activation by rat liver S9 for *trans*-2-dodecenal (No. 1350) (Sokolowski et al., 2007b; Bhatia, Politano & Api, 2010) and *trans*-2-hexenal (No. 1353) (Sokolowski et al., 2007a; Bhatia, Politano & Api, 2010).

trans-2-Dodecenal (No. 1350) was tested for induction of gene mutations in *Salmonella typhimurium* strains TA98, TA100, TA102, TA1535 and TA1537 by the plate incorporation and preincubation methods, with and without S9 liver microsomal activation. In a pre-experiment using the plate incorporation test, concentrations of 0.1, 0.3, 1, 3, 10, 33 and 100 µg/plate without S9 mix and 1, 3, 10, 33, 100, 333 and 1000 µg/plate with S9 mix were tested. In the main plate incorporation assay, concentrations of 0, 3, 10, 33, 100, 333, 1000, 2500 and 5000 µg/plate were tested, and concentrations of 0, 0.3, 1, 3, 10, 33, 100, 333 and 1000 µg/plate were used in the main preincubation experiments. Cytotoxic effects were observed in all strains with and without S9 metabolic activation. No significant increases in revertant colony numbers were observed in any strain at any concentration in the presence or absence of metabolic activation (Sokolowski et al., 2007b; Bhatia, Politano & Api, 2010).

An Ames assay of *trans*-2-hexenal (No. 1353) was conducted using the plate incorporation test method in *S. typhimurium* strains TA98, TA100, TA102, TA1535 and TA1537, with and without S9 rat liver microsomal activation, at 3, 10, 33, 100, 333, 1000, 2500 and 5000 µg/plate. Cytotoxicity was observed in the TA98, TA100, TA102 and TA1535 strains at 2500 µg/plate and above with and without S9 mix and in the TA1537 strain at 1000 µg/plate and above without S9 mix and at 2500 µg/plate and above with S9 mix. No substantial increases in revertant colony numbers

Table 4. Studies of genotoxicity with aliphatic linear α,β-unsaturated aldehydes, acids and related alcohols, acetals and esters used as flavouring agents

No.	Flavouring agent	End-point	Test object	Concentration/dose	Results	Reference
In vitro						
1350	2-Dodecenal	Reverse mutation	*Salmonella typhimurium* TA98, TA100, TA102, TA1535 and TA1537	0.1, 0.3, 1, 3, 10, 33, 100, 333, 1000, 2500 and 5000 µg/plate	Negative[a,b,c]	Sokolowski et al. (2007b)
1350	2-Dodecenal	Reverse mutation	*S. typhimurium* TA98, TA100, TA102, TA1535 and TA1537	0.1, 0.3, 1, 3, 10, 33, 100, 333, 1000, 2500 and 5000 µg/plate	Negative[a,b,c]	Bhatia, Politano & Api (2010)
1350	2-Dodecenal	Reverse mutation	*S. typhimurium* TA98, TA100, TA102, TA1535 and TA1537	0.3, 1, 3, 10, 33, 100, 333 and 1000 µg/plate	Negative[a,c,d]	Sokolowski et al. (2007b)
1350	2-Dodecenal	Reverse mutation	*S. typhimurium* TA98, TA100, TA102, TA1535 and TA1537	0.3, 1, 3, 10, 33, 100, 333 and 1000 µg/plate	Negative[a,c,d]	Bhatia, Politano & Api (2010)
1353	2-Hexenal	Reverse mutation	*S. typhimurium* TA98, TA100, TA102, TA1535 and TA1537	3, 10, 33, 100, 333, 1000, 2500 and 5000 µg/plate	Negative[a,b,c]	Sokolowski et al. (2007a)
1353	2-Hexenal	Reverse mutation	*S. typhimurium* TA98, TA100, TA102, TA1535 and TA1537	3, 10, 33, 100, 333, 1000, 2500 and 5000 µg/plate	Negative[a,b,c]	Bhatia, Politano & Api (2010)
1353	2-Hexenal	Reverse mutation	*S. typhimurium* TA98, TA100, TA102, TA1535 and TA1537	1, 3, 10, 33, 100, 333, 1000 and 2500 µg/plate	Negative[a,c,d]	Sokolowski et al. (2007a)
1353	2-Hexenal	Reverse mutation	*S. typhimurium* TA98, TA100, TA102, TA1535 and TA1537	1, 3, 10, 33, 100, 333, 1000 and 2500 µg/plate	Negative[a,c,d]	Bhatia, Politano & Api (2010)
In vivo						
1350	2-Dodecenal	Micronucleus induction	NMRI mice; M, F	500, 1000 and 2000 mg/kg bw	Negative[e,f]	Honarvar et al. (2007b)
1350	2-Dodecenal	Micronucleus induction	NMRI mice; M, F	500, 1000 and 2000 mg/kg bw	Negative[e,f]	Bhatia, Politano & Api (2010)

continued

Table 4 (continued)

No.	Flavouring agent	End-point	Test object	Concentration/dose	Results	Reference
1353	2-Hexenal	Micronucleus induction	NMRI mice; M, F	250, 500 and 1000 mg/kg bw	Negative[e,f]	Honarvar et al. (2007a)
1353	2-Hexenal	Micronucleus induction	NMRI mice; M, F	250, 500 and 1000 mg/kg bw	Negative[e,f]	Bhatia, Politano & Api (2010)
1362	2-Nonenal	Micronucleus induction	NMRI mice; M, F	500, 1000 and 2000 mg/kg bw	Negative[e,f]	Honarvar et al. (2008)
1362	2-Nonenal	Micronucleus induction	NMRI mice; M, F	500, 1000 and 2000 mg/kg bw	Negative[e,f]	Bhatia, Politano & Api (2010)

F, female; M, male

[a] With and without metabolic activation.
[b] Plate incorporation.
[c] Cytotoxic effects observed.
[d] Preincubation.
[e] Bone marrow was sampled 24 hours after treatment at all doses and additionally at 48 hours at the highest dose.
[f] Administered via corn oil gavage.

were observed in any test strain, at any concentration of *trans*-2-hexenal, or in the presence or absence of metabolic activation. Under the test conditions, *trans*-2-hexenal was considered to be non-mutagenic (Sokolowski et al., 2007a; Bhatia, Politano & Api, 2010).

The preincubation test method was also used in the same five strains of *S. typhimurium* with and without S9 at 2-hexenal concentrations of 0, 1, 3, 10, 33, 100, 333, 1000 and 2500 µg/plate. Cytotoxicity was observed in strains TA98, TA100, TA102, TA1535 and TA1537 at 333 µg/plate and above without S9 mix and at 1000 µg/plate and above with S9 mix. A minor but dose-dependent increase in revertant colony numbers was observed in the TA100 strain without S9 mix, which was 2.1 times the control background at 100 µg/plate. At higher concentrations, the numbers of revertants were reduced due to the toxic effects. To verify these results, an additional preincubation study was performed with 25, 50, 75, 100, 150 and 200 µg/plate with and without S9 in the TA100 strain only. An increase in the number of revertant colonies was observed, which was more than 2 times the control level in tester strain TA100 at 100 µg/plate; however, this was not statistically significant and therefore not considered mutagenic (Sokolowski et al., 2007a; Bhatia, Politano & Api, 2010).

(ii)   In vivo

A study was conducted to determine the potential of *trans*-2-hexenal (No. 1353) administered by oral gavage to male and female NMRI mice (five of each sex per dose) at doses of 0, 250, 500 or 1000 mg/kg bw to induce micronuclei in bone marrow erythrocytes. The mice were examined for acute toxicity at 1, 2–4, 6, 24 and 48 hours after administration of *trans*-2-hexenal. The bone marrow was sampled 24 hours after dosing in the negative and positive control groups and the 250, 500 and 1000 mg/kg bw dose groups. An additional bone marrow sample was collected after 48 hours in the 1000 mg/kg bw dose group. The bone marrow smears were stained with Giemsa stain, and at least 2000 polychromatic erythrocytes (PCEs) were analysed per animal for micronuclei. Cytotoxic effects were determined by the ratio of PCEs to total erythrocytes for each animal. The 1000 mg/kg bw dose of *trans*-2-hexenal resulted in clinical signs of toxicity, including reduction of spontaneous activity, abdominal position changes, eyelid closure and ruffled fur. The number of PCEs was also decreased compared with the vehicle control, indicating a cytotoxic effect in the bone marrow. No toxic reactions were observed at 250 or 500 mg/kg bw. No increase in the frequency of micronuclei was observed in comparison with the corresponding vehicle controls after the administration of 2-hexenal at any dose level. Under the conditions of this study, 2-hexenal was classified as non-mutagenic in the mouse micronucleus assay (Honarvar et al., 2007a; Bhatia, Politano & Api, 2010).

Similarly, studies were conducted to determine the potential of *trans*-2-dodecenal (No. 1350) and *trans*-2-nonenal (No. 1362) at doses of 0, 500, 1000 and 2000 mg/kg bw to induce micronuclei in PCEs of male and female NMRI mice (five of each sex per dose). Mice were examined for acute toxicity at intervals of approximately 1, 2–4, 6, 24 and 48 hours after dosing. Sampling of the bone marrow for Giemsa staining was performed 24 hours after dosing, and an additional 48-hour sample was taken for the high-dose group. PCEs and total erythrocytes were counted to determine cytotoxicity, and PCEs were analysed for micronuclei. Toxic reactions were not observed, and there were no statistically significant increases in the frequency of the detected micronuclei at any dose or time period with *trans*-2-dodecenal (Honarvar et al., 2007b) or *trans*-2-nonenal (Honarvar et al., 2008; Bhatia, Politano & Api, 2010).

(c)   Cardiotoxicity

A study was performed to assess the effect of *trans*-2-hexenal (No. 1353) on the myocardium of mice. Male ICR mice at 8 weeks of age were administered *trans*-2-hexenal at 0, 0.1, 1, 10 or 50 mg/kg bw per week by intragastric instillation in corn oil for 2 or 4 weeks. Cardiac function, myocardial morphology, cardiomyocyte apoptosis and the cytochrome *c*–mediated caspase-activated apoptotic pathway were measured. The formation of 2-hexenal–protein adducts in the heart was measured by an enzyme-linked immunosorbent assay using an antibody to a hexenal–lysine conjugate.

2-Hexenal–protein adducts were significantly increased at 1, 10 and 50 mg/kg bw per week. The 0.1 mg/kg bw per week dosed mice showed a significant increase in adducts only at the 4-week time point.

Cardiac contractile function was also measured in *trans*-2-hexenal-dosed mice by echocardiography. Measurements included heart rate, end-diastolic diameter, end-systolic diameter and fractional shortening from the parasternal long axis and parasternal short axis, as well as apical four-chamber views used to obtain two-dimensional, M-node and spectral Doppler images. At the 1 and 10 mg/kg bw per week doses, mice exhibited depressed left-ventricular function (increased end-systolic dimension) and decreased fractional shortening compared with controls. These results indicate that 2-hexenal causes a significant impairment of basal left-ventricular contractile function. Mice in the 50 mg/kg bw per week dose group exhibited significant increases in heart rate compared with controls.

In addition to the echocardiography, histological examination of cardiac tissue was conducted. At all dose levels, there were no obvious morphological alterations in the heart tissue other than a few condensed nuclei, suggesting that *trans*-2-hexenal does not induce necrosis. Alpha-cardiac muscle actinin was analysed by immunohistochemistry using confocal microscopy and western blot. Mice receiving all doses exhibited no significant myofibril disarray and no changes in expression of alpha-cardiac muscle actinin. These results suggest that the structural skeletal and muscle sarcomeric proteins involved in contractile function are not altered by *trans*-2-hexenal treatment, and there was no evidence of histopathology in the cardiac tissue that could explain the altered cardiac function.

To evaluate cardiomyocyte apoptosis, a terminal deoxynucleotidyl transferase–mediated 2′-deoxyuridine 5′-triphosphate (dUTP) nick-end labelling (TUNEL) assay of fragmented nuclei was performed. TUNEL assay sections were counterstained with 4′,6′-diamidino-2-phenylindole hydrochloride (DAPI), and the percentage of cardiomyocytes with deoxyribonucleic acid (DNA) nick-end labelling was then determined by counting cells exhibiting cyan nuclei (produced by overlapping the TUNEL and DAPI signals) under a confocal fluorescent microscope. *trans*-2-Hexenal treatment at all doses increased cardiac cell apoptosis compared with controls, with significance at and above 1 mg/kg bw per week. These results indicate that exposure to *trans*-2-hexenal induces apoptosis in cardiomyocytes. The formation of protein adducts in mitochondria was measured in mitochondria isolated from untreated mouse hearts and incubated with *trans*-2-hexenal (1 µmol/l per 10 µg of mitochondrial proteins) at 37 °C for 30 minutes. Increases in the formation of hexenal–protein adducts in 2-hexenal-exposed mitochondria were observed in the western blot analysis.

To further elucidate the mechanism of *trans*-2-hexenal-induced cardiac cell apoptosis, cytosolic fractions of mouse hearts were isolated, and cytochrome *c* release and caspase-9 activation were determined by western blotting. Caspase-3 activation was evaluated by immunofluorescent staining of the mouse heart sections examined by confocal fluorescent microscopy. At all doses of *trans*-2-hexenal, cytochrome *c* release from the mitochondria into the cytosol was observed, and proteolytic processing of the cleavage of caspase-9 occurred, indicating that *trans*-2-hexenal activates the mitochondrial apoptotic pathway in cardiomyocytes. In addition, the result obtained from the caspase-3 immunofluorescent confocal microscopy indicated that *trans*-2-hexenal induced activation of caspase-3, the downstream apoptotic effector protein of caspase-9. This data set indicates that 2-hexenal impaired cardiac contractile function, resulted in the formation of protein adducts in the cardiac tissue and activated the mitochondrial cytochrome *c*–mediated apoptotic pathway in cardiomyocytes. However, histological analysis of the tissue did not reveal loss of muscle fibres, myofibril disarray or any overt pathology that would account for the decrease in function (Ping et al., 2003). The increase in the incidence of caspase 3–mediated apoptosis in the cardiomyocytes can be attributed to the increased oxidative stress and stress in general to the animals associated with bolus administration of *trans*-2-hexenal. The relevance of this observation to the safety assessment of this group of flavouring agents is unclear, as there were no deaths due to myocardial infarction or other significant pathology of the cardiovascular tissues, including the heart, among mice in the National Toxicology Program 2-year bioassay for the structurally related 2,4-hexadienal (National Toxicology Program, 2003).

3. REFERENCES

Bhatia SP, Politano VT, Api AM (2010). Genotoxicity tests conducted on a group of structurally related aldehydes. Presentation at the 49th Annual Meeting of the Society of Toxicology, 7–11 March 2010, Salt Lake City, UT, USA [abstract reported in *Toxicologist*, 114(1):216].

Cramer GM, Ford RA, Hall RL (1978). Estimation of toxic hazard—a decision tree approach. *Food and Cosmetics Toxicology*, 16(3):255–276.

Dieckhaus CM et al. (2001). Role of glutathione *S*-transferase A1-1, M1-1, and P1-1 in the detoxification of 2-phenylpropenal, a reactive felbamate metabolite. *Chemical Research in Toxicology*, 14(5):511–516.

European Flavor and Fragrance Association (2004). European inquiry on volume use. Private communication to the Flavor and Extract Manufacturers Association of the United States, Washington, DC, USA. Submitted to WHO by the International Organization of the Flavor Industry, Brussels, Belgium.

Gavin CL, Williams MC, Hallagan JB (2008). *Flavor and Extract Manufacturers Association of the United States 2005 poundage and technical effects update survey.* Washington, DC, USA, Flavor and Extract Manufacturers Association of the United States.

Honarvar N et al. (2007a). Micronucleus assay in bone marrow cells of the mouse with *trans*-2-hexenal (code 56250). Unpublished report no. 1064307 from RCC Cytotest Cell Research GmbH, Rossdorf, Germany. Submitted to WHO by the International Organization of the Flavor Industry, Brussels, Belgium.

Honarvar N et al. (2007b). Micronucleus assay in bone marrow cells of the mouse with *trans*-2-dodecenal (code 55330). Unpublished report no. 1064909 from RCC Cytotest Cell Research GmbH, Rossdorf, Germany. Submitted to WHO by the International Organization of the Flavor Industry, Brussels, Belgium.

Honarvar N et al. (2008). Micronucleus assay in bone marrow cells of the mouse with 2-nonenal. Unpublished report no. 1114009 from RCC Cytotest Cell Research GmbH, Rossdorf, Germany. Submitted to WHO by the International Organization of the Flavor Industry, Brussels, Belgium.

International Organization of the Flavor Industry (2011). Interim inquiry on volume use and added use levels for flavoring agents to be presented at the 76th JECFA meeting. Private communication to the Flavor and Extract Manufacturers Association of the United States, Washington, DC, USA. Submitted to WHO by the International Organization of the Flavor Industry, Brussels, Belgium.

Japan Flavor and Fragrance Materials Association (2005). Japanese inquiry on volume use. Private communication to the Flavor and Extract Manufacturers Association of the United States, Washington, DC, USA. Submitted to WHO by the International Organization of the Flavor Industry, Brussels, Belgium.

National Toxicology Program (2003). *NTP technical report on the toxicology and carcinogenesis studies of 2,4-hexadienal (89% trans,trans isomer, CAS No. 142-83-6; 11% cis,trans isomer) in F344/N rats and B6C3F1 mice (gavage studies)*. National Toxicology Program Technical Report Series 509, NIH Publication No. 04-4443. Research Triangle Park, NC, USA, United States Department of Health and Human Services, National Institutes of Health, National Institute of Environmental and Health Sciences, National Toxicology Program, pp. 1–290.

Nelson DL, Cox MM (2008). *Lehninger principles of biochemistry*, 3rd ed. New York, NY, USA, Worth Publishers.

Nijssen LM, van Ingen-Visscher CA, Donders JJH (2011). Volatile Compounds in Food (VCF) database, version 13.1. Zeist, the Netherlands, TNO Triskelion (http://www.vcf-online.nl/VcfHome.cfm).

Ping P et al. (2003). Cardiac toxic effects of *trans*-2-hexenal are mediated by induction of cardiomyocyte apoptotic pathways. *Cardiovascular Toxicology*, 3(4):341–351.

Sokolowski A et al. (2007a). *Salmonella typhimurium* reverse mutation assay with *trans*-2-hexenal. Unpublished report no. 1064901 from RCC Cytotest Cell Research GmbH, Rossdorf, Germany. Submitted to WHO by the International Organization of the Flavor Industry, Brussels, Belgium.

Sokolowski A et al. (2007b). *Salmonella typhimurium* reverse mutation assay with *trans*-2-dodecenal. Unpublished report no. 1064902 from RCC Cytotest Cell Research GmbH, Rossdorf, Germany. Submitted to WHO by the International Organization of the Flavor Industry, Brussels, Belgium.

Thompson CD, Kinter MT, Macdonald TL (1996). Synthesis and in vitro reactivity of 3-carbamoyl-2-phenylpropionaldehyde and 2-phenylpropenal: putative reactive metabolites of felbamate. *Chemical Research in Toxicology*, 9(8):1225–1229.

Thompson CD, Gulden PH, Macdonald TL (1997). Identification of modified atropaldehyde mercapturic acids in rat and human urine after felbamate administration. *Chemical Research in Toxicology*, 10(4):457–462.

Zhang GF et al. (2009). Catabolism of 4-hydroxyacids and 4-hydroxynonenal via 4-hydroxy-4-phosphoacyl-CoAs. *Journal of Biological Chemistry*, 284(48):33 521–33 534.

AMINO ACIDS AND RELATED SUBSTANCES (addendum)

First draft prepared by

S.M.F. Jeurissen[1], M. DiNovi[2], A.G. Renwick[3] and P. Sinhaseni[4]

[1] *Centre for Substances and Integrated Risk Assessment, National Institute for Public Health and the Environment, Bilthoven, the Netherlands*
[2] *Center for Food Safety and Applied Nutrition, Food and Drug Administration, College Park, Maryland, United States of America (USA)*
[3] *School of Medicine, University of Southampton, Southampton, England*
[4] *Community Risk Analysis Research and Development Center, Bangkok, Thailand*

1. Evaluation 133
   1.1 Introduction 133
   1.2 Assessment of dietary exposure 134
   1.3 Absorption, distribution, metabolism and elimination 138
   1.4 Application of the Procedure for the Safety Evaluation of
       Flavouring Agents 138
   1.5 Consideration of combined intakes from use as flavouring
       agents 140
   1.6 Conclusions 140
2. Relevant background information 141
   2.1 Explanation 141
   2.2 Additional considerations on dietary exposure 141
   2.3 Biological data 141
       2.3.1 Biochemical data: absorption, distribution, metabolism
             and elimination 141
       2.3.2 Toxicological studies 142
             (a) Acute toxicity 142
             (b) Short-term studies of toxicity 142
             (c) Genotoxicity 144
3. References 145

1. EVALUATION

1.1 Introduction

The Committee evaluated six additional flavouring agents belonging to the group of amino acids and related substances. The additional flavouring agents included three L-amino acids (Nos 2118–2120), two dipeptides (Nos 2121 and 2122) and one tripeptide (No. 2123). None of these flavouring agents have previously been evaluated by the Committee. The safety of the submitted substance (3R)-4-[[(1S)-1-benzyl-2-methoxy-2-oxo-ethyl]amino]-3-[3-(3-hydroxy-4-methoxyphenyl)propylamino]-4-oxo-butanoic acid hydrate (Advantame, No. 2124) was not assessed; the Committee decided that it would not be appropriate to evaluate this substance as a flavouring agent, because it is a low-calorie intense sweetener.

Three of the flavouring agents in this group (Nos 2119, 2121 and 2123) evaluated at this meeting are reported to be flavour modifiers.

The Committee considered that the use of the Procedure for the Safety Evaluation of Flavouring Agents (see Figure 1, Introduction) (Annex 1, reference *131*) was inappropriate for two members of this group—namely, L-isoleucine (No. 2118) and L-threonine (No. 2119). These substances are macronutrients and normal components of protein; as such, human exposure through food is orders of magnitude higher than the anticipated level of exposure from their use as flavouring agents. For the remaining four members of the group (Nos 2120–2123), the evaluations were conducted according to the Procedure for the Safety Evaluation of Flavouring Agents.

The Committee previously evaluated 20 other members of this group of flavouring agents at its sixty-third meeting (Annex 1, reference *173*). The Committee concluded that all 20 flavouring agents in that group were of no safety concern at estimated dietary exposures.

As noted by the Committee at its sixty-third meeting (Annex 1, reference *173*), amino acids may react with other food constituents upon heating. The mixtures thus formed are commonly referred to as "process flavours". The safety of process flavours has not been reviewed by the Committee at the sixty-third meeting or at the current meeting and may be considered at a future meeting. The evaluation of the flavouring agents belonging to the group of amino acids and related substances is therefore conducted on the basis that these flavouring agents are present in an unchanged form at the point of consumption.

In addition to Nos 2118 and 2119, which are normal components of protein, Nos 2120 and 2123 have been reported to occur in protein-rich foods, fish sauce, soya sauce, shrimp paste and scallops (Flavor and Extract Manufacturers Association of the United States, private communication, 2010).

1.2 Assessment of dietary exposure

The total annual volume of production of the six amino acids and related substances for use as flavouring agents only is approximately 49 140 kg in the USA, with no reported volume available for Europe or Japan (European Flavour and Fragrance Association, 2004; Japan Flavor and Fragrance Materials Association, 2005; Gavin, Williams & Hallagan, 2008; International Organization of the Flavor Industry, 2011). Approximately 92% of the total annual volume of production in the USA is accounted for by one substance in this group—namely, L-alanyl-L-glutamine (No. 2121).

Dietary exposures were estimated using both the single portion exposure technique (SPET) and the maximized survey-derived intake (MSDI) method. The highest estimated dietary exposure for each flavouring agent is reported in Table 1. The estimated daily dietary exposure is highest for L-alanyl-L-glutamine (No. 2121) (280 000 µg, the SPET value obtained from milk products). For the other flavouring agents, the estimated daily dietary exposures, calculated using either the SPET or the MSDI method, range from 0.02 to 60 000 µg, with the SPET yielding the highest estimates.

AMINO ACIDS AND RELATED SUBSTANCES (addendum)

Table 1. Summary of the results of the safety evaluations of amino acids and related substances used as flavouring agents[a,b,c]

Flavouring agent	No.	CAS No. and structure	Step A3[1] Does estimated dietary exposure exceed the threshold of concern?	Step A4 Is the flavouring agent or are its metabolites endogenous?	Step A5 Adequate margin of exposure for the flavouring agent or a related substance?	Related structure name (No.) and structure (if applicable)	Comments on predicted metabolism	Conclusion based on current estimated dietary exposure
Structural class I								
L-Ornithine (as the monochlorohydrate)	2120	3184-13-2	Yes, SPET: 30 000	Yes	NR	NR	Note 1	No safety concern
L-Alanyl-L-glutamine	2121	39537-23-0	Yes, SPET: 280 000	Yes	NR	NR	Note 2	No safety concern
L-Methionylglycine	2122	14486-03-4	No, SPET: 400	NR	NR	NR	Note 2	No safety concern

continued

Table 1 (continued)

Flavouring agent	No.	CAS No. and structure	Step A3[1] Does estimated dietary exposure exceed the threshold of concern?	Step A4 Is the flavouring agent or are its metabolites endogenous?	Step A5 Adequate margin of exposure for the flavouring agent or a related substance?	Related structure name (No.) and structure (if applicable)	Comments on predicted metabolism	Conclusion based on current estimated dietary exposure
Glutamyl-valyl-glycine	2123	38837-70-6	Yes, SPET: 4000	No	Yes. The NOAEL of 3130 mg/kg bw per day in a 91-day study in rats for the related compound L-alanyl-L-glutamine (No. 2121) (Oda et al., 2008) is 47 000 times the estimated daily dietary exposure to No. 2123 when used as a flavouring agent.	L-Alanyl-L-glutamine (No. 2121)	Note 2	No safety concern

Amino acids not evaluated by the Procedure

L-α-Amino acids	No.	CAS No. and structure	Conclusion
L-Isoleucine	2118	73-32-5	This substance is a macronutrient and a normal component of protein; as such, human exposure through food is orders of magnitude higher than the anticipated level of exposure from its use as a flavouring agent.

| L-Threonine | 2119 72-19-5 | This substance is a macronutrient and a normal component of protein; as such, human exposure through food is orders of magnitude higher than the anticipated level of exposure from its use as a flavouring agent. |

bw, body weight; CAS, Chemical Abstracts Service; NOAEL, no-observed-adverse-effect level; NR, not required for evaluation

[a] Twenty flavouring agents in this group were previously evaluated by the Committee at its sixty-third meeting (Annex 1, reference *173*).
[b] *Step 1*: All of the flavouring agents in this group are in structural class I.
[c] *Step 2*: All of the flavouring agents in this group can be predicted to be metabolized to innocuous products.
[d] The threshold for human dietary exposure for structural class I is 1800 µg/person per day. All dietary exposure values are expressed in µg/day. The dietary exposure value listed represents the highest estimated dietary exposure calculated using either the SPET or the MSDI method. The SPET gave the highest estimated dietary exposure in each case.

Notes:
1. As part of the urea cycle, a carbamoyl group is transferred to ornithine from citrulline.
2. Hydrolysed to constituent amino acids.

Annual volumes of production of this group of flavouring agents as well as the daily dietary exposures calculated using both the MSDI method and the SPET are summarized in Table 2.

1.3 Absorption, distribution, metabolism and elimination

Information on the absorption, distribution, metabolism and elimination of amino acids and related substances has previously been described in the monograph of the sixty-third meeting (Annex 1, reference *173*). Also, dipeptides and tripeptides are readily hydrolysed into constituent amino acids in the intestine during absorption.

1.4 Application of the Procedure for the Safety Evaluation of Flavouring Agents

Step 1. In applying the Procedure for the Safety Evaluation of Flavouring Agents to the four flavouring agents in this group of amino acids and related substances, the Committee assigned all four flavouring agents (Nos 2120–2123) to structural class I (Cramer, Ford & Hall, 1978).

Step 2. All four of the flavouring agents in this group can be predicted to be metabolized to innocuous products. The evaluation of all of these flavouring agents therefore proceeded via the A-side of the Procedure.

Step A3. The highest estimated dietary exposure to one flavouring agent in structural class I (No. 2122) is below the threshold of concern (i.e. 1800 μg/person per day for class I). According to the Procedure, this flavouring agent is not of safety concern at current estimated dietary exposure. The highest estimated dietary exposures of the three remaining flavouring agents in structural class I are above the threshold of concern. Accordingly, the evaluation of these flavouring agents proceeded to step A4.

Step A4. L-Ornithine (No. 2120) is an endogenous compound, and L-alanyl-L-glutamine (No. 2121) is metabolized to the non-essential, endogenous amino acids, L-alanine and L-glutamine. According to the Procedure, these two flavouring agents are not of safety concern at current estimated dietary exposures. Also, glutamyl-valyl-glycine (No. 2123) is metabolized to its constituent amino acids, but valine is an essential amino acid that cannot be synthesized in the human body and is therefore not an endogenous compound. Therefore, the evaluation of this flavouring agent proceeded to step A5.

Step A5. For glutamyl-valyl-glycine (No. 2123), the no-observed-adverse-effect level (NOAEL) of 3130 mg/kg of body weight (bw) per day in a 91-day study in rats for the related compound L-alanyl-L-glutamine (No. 2121) (Oda et al., 2008) provides a margin of exposure of approximately 47 000 in relation to the highest estimated dietary exposure to No. 2123 (SPET = 4000 μg/day) when used as a flavouring agent. The Committee therefore concluded that glutamyl-valyl-glycine is not of safety concern at current estimated dietary exposure.

Table 1 summarizes the evaluations of the four amino acids and related substances used as flavouring agents in this group, as well as the evaluations of the two amino acids for which the Procedure was not used.

Table 2. Annual volumes of production and daily dietary exposures for amino acids and related substances used as flavouring agents in Europe, the USA and Japan

Flavouring agent (No.)	Most recent annual volume of production (kg)[a]	Dietary exposure				Natural occurrence in foods
		MSDI[b]		SPET[c]		
		µg/day	µg/kg bw per day	µg/day	µg/kg bw per day	
L-Isoleucine (2118)				60 000	1000	+
Europe	ND	ND	ND			
USA	7	0.8	0.01			
Japan	ND	ND	ND			
L-Threonine (2119)				15 000	250	+
Europe	ND	ND	ND			
USA	3569	394	7			
Japan	ND	ND	ND			
L-Ornithine (as the monochlorohydrate) (2120)				30 000	500	+
Europe	ND	ND	ND			
USA	113	12	0.2			
Japan	ND	ND	ND			
L-Alanyl-L-glutamine (2121)				280 000	4667	−
Europe	ND	ND	ND			
USA	45 450	5021	84			
Japan	ND	ND	ND			
L-Methionylglycine (2122)				400	7	−
Europe	ND	ND	ND			
USA	1	0.06	0.001			
Japan	ND	ND	ND			
Glutamyl-valyl-glycine (2123)				4000	67	+
Europe	ND	ND	ND			
USA	0.2	0.02	0.0004			
Japan	ND	ND	ND			

Table 2 (continued)

Flavouring agent (No.)	Most recent annual volume of production (kg)[a]	Dietary exposure				Natural occurrence in foods
		MSDI[b]		SPET[c]		
		µg/day	µg/kg bw per day	µg/day	µg/kg bw per day	
Total						
Europe	ND					
USA	49 140					
Japan	ND					

bw, body weight; ND, no data reported; +, reported to occur naturally in foods (Food and Extract Manufacturers Association of the United States, private communication, 2010), but no quantitative data; −, not reported to occur naturally in foods

[a] From European Flavour and Fragrance Association (2004), Japan Flavor and Fragrance Materials Association (2005), Gavin, Williams & Hallagan (2008) and International Organization of the Flavor Industry (2011). Values greater than 0 kg but less than 0.1 kg were reported as 0.1 kg.

[b] MSDI (µg/person per day) calculated as follows:
(annual volume, kg) × (1 × 10^9 µg/kg)/(population × survey correction factor × 365 days), where population (10%, "eaters only") = 32 × 10^6 for Europe, 31 × 10^6 for the USA and 13 × 10^6 for Japan; and where survey correction factor = 0.8 for the surveys in Europe, the USA and Japan, representing the assumption that only 80% of the annual flavour volume was reported in the poundage surveys (European Flavour and Fragrance Association, 2004; Japan Flavor and Fragrance Materials Association, 2005; Gavin, Williams & Hallagan, 2008; International Organization of the Flavor Industry, 2011).
MSDI (µg/kg bw per day) calculated as follows:
(µg/person per day)/body weight, where body weight = 60 kg. Slight variations may occur from rounding.

[c] SPET (µg/person per day) calculated as follows:
(standard food portion, g/day) × (highest usual use level) (International Organization of the Flavor Industry, 2011). The dietary exposure from the single food category leading to the highest dietary exposure from one portion is taken as the SPET estimate.
SPET (µg/kg bw per day) calculated as follows:
(µg/person per day)/body weight, where body weight = 60 kg. Slight variations may occur from rounding.

1.5 Consideration of combined intakes from use as flavouring agents

The four flavouring agents evaluated using the Procedure are efficiently metabolized and eliminated, and the overall evaluation of the data indicates that combined intake would not raise any safety concerns at current estimated dietary exposures.

1.6 Conclusions

In the previous evaluation of flavouring agents in the group of amino acids and related substances, biochemical data and studies of acute toxicity, short-term

AMINO ACIDS AND RELATED SUBSTANCES (addendum)

studies of toxicity, long-term studies of toxicity and carcinogenicity, and studies of in vitro and in vivo genotoxicity and reproductive toxicity were available (Annex 1, reference 173). The toxicity data for the current evaluation (biochemical data, acute toxicity studies, short-term studies of toxicity and in vitro genotoxicity studies) supported the previous safety evaluation.

In view of the fact that No. 2118 (L-isoleucine) and No. 2119 (L-threonine) are macronutrients and normal components of protein, the use of these substances as flavouring agents would not raise any safety concerns at current estimated dietary exposures. The Committee also concluded that the use of the other four flavouring agents in this group of amino acids and related substances would not raise any safety concerns.

2. RELEVANT BACKGROUND INFORMATION

2.1 Explanation

This monograph summarizes key aspects relevant to the safety evaluation of six amino acids and related substances, which are additions to a group of 20 flavouring agents evaluated previously by the Committee at its sixty-third meeting (Annex 1, reference 173).

2.2 Additional considerations on dietary exposure

Annual volumes of production and dietary exposures estimated using both the MSDI method and the SPET for each flavouring agent are reported in Table 2.

Four of the six flavouring agents (Nos 2118–2120 and 2123) are natural components of food and have been reported to occur in protein-rich foods, fish sauce, soya sauce, shrimp paste and scallops (Flavor and Extract Manufacturers Association of the United States, private communication, 2010), but quantitative data on their natural occurrence have not been reported. L-Isoleucine (No. 2118), L-threonine (No. 2119) and L-valine (a constituent amino acid of No. 2123) cannot be synthesized to meet body needs and are considered to be essential in the human diet (Institute of Medicine, 2005; WHO/FAO/UNU, 2007). The daily requirements for adults for these amino acids are 20 mg/kg bw for L-isoleucine, 15 mg/kg bw for L-threonine and 26 mg/kg bw for L-valine (WHO/FAO/UNU, 2007). These daily requirements are significantly higher than the daily dietary exposures from their use as flavouring agents.

2.3 Biological data

2.3.1 Biochemical data: absorption, distribution, metabolism and elimination

There are numerous studies on the absorption, distribution, metabolism and excretion of amino acids in the scientific literature, only some of which are described in the monograph of the sixty-third meeting (Annex 1, reference 173). Some additional studies on dipeptides and tripeptides are described below.

Dipeptides are absorbed more rapidly than amino acids in the small intestine (Newey & Smyth, 1960; Adibi, 1971; Webb, 1990; Gilbert, Wong & Webb, 2008). In an in vivo study with healthy volunteers, amino acid absorption rates in different intestinal segments were significantly higher from test solutions containing the same amount of amino acids in dipeptide form (e.g. glycylglycine) than in free form (e.g. glycine) (Adibi, 1971). The absorption of dipeptides and tripeptides occurs through an independent transport system for peptides (intestinal peptide transporter, or "PepT1") (Gilbert, Wong & Webb, 2008).

Dipeptides and tripeptides are readily degraded into their constituent amino acids upon intestinal absorption (Minami, Morse & Adibi, 1992; Herzog et al., 1996; Klassen et al., 2000; Rogero et al., 2006; Bishay, 2009a,b). Levels of alanine and glutamine in blood have been found to rise within 15–30 minutes of oral administration of L-alanyl-L-glutamine (No. 2121) to both rats (Rogero et al., 2006; only glutamine levels were measured) and humans (Klassen et al., 2000).

Glutamyl-valyl-glycine (No. 2123) (16.5 µmol/l) was readily hydrolysed during incubation with the homogenate and the microsomal fraction of human small intestinal mucosa at 37 °C. During the incubations with the homogenate, approximately 20%, 2.6% and 2.0% of the initial dose were recovered after, respectively, 30, 60 and 90 minutes. Valyl-glycine, which was formed by the degradation of glutamyl-valyl-glycine, was also hydrolysed, and after 90 minutes of incubation, the concentration of valyl-glycine was almost the same as at the start of the reaction (the level existing in the homogenate of small intestinal mucosa) (Bishay, 2009a). During the incubations with the microsomal fraction, 21% and 2.5% of the initial dose were recovered after, respectively, 5 and 10 minutes. After 15 minutes, the concentration of glutamyl-valyl-glycine was below the limit of quantification (Bishay, 2009b).

2.3.2 Toxicological studies

(a) Acute toxicity

An oral median lethal dose (LD_{50} value) has been reported for one of the six flavouring agents in this group. For L-alanyl-L-glutamine (No. 2121), an oral LD_{50} value of greater than 2000 mg/kg bw has been reported in rats (Oda et al., 2008). This result supports the findings in the previous evaluation that the oral acute toxicity of these flavouring agents is low.

(b) Short-term studies of toxicity

The short-term studies of toxicity with amino acids and related substances are summarized in Table 3 and described below.

(i) L-Alanyl-L-glutamine (No. 2121)

In a 14-day range-finding study from the public literature, groups of five male and five female Sprague-Dawley SPF rats were administered L-alanyl-L-glutamine at dietary concentrations of 0%, 1%, 3% and 5% (weight per weight [w/w]) (equal to doses of 0, 970, 2930 and 4880 mg/kg bw per day for males and 0, 970, 2880 and 4910 mg/kg bw per day for females) (Oda et al., 2008). The general condition

Table 3. Results of short-term studies of toxicity with amino acids and related substances used as flavouring agents

No.	Flavouring agent	Species; sex	No. of test groups[a] / no. per group[b]	Route	Duration (days)	NOAEL (mg/kg bw per day)	Reference
2121	L-Alanyl-L-glutamine	Rat; M, F	3/20	Diet	91	3130[c]	Oda et al. (2008)

F, female; M, male

[a] Total number of test groups does not include control animals.
[b] Total number per test group includes both male and female animals.
[c] Highest dose tested.

of the rats was usually observed twice daily. Body weights and feed consumption rates were recorded on days 1, 4, 8, 11 and 15. Routine haematological and clinical chemistry parameters were determined. Necropsy included an external examination and observation of the major organs in the thoracic and abdominal cavities. Weights of adrenals, spleen, heart, lung, liver, kidney, testis and ovary were recorded, and samples of these tissues were preserved for possible histopathological analysis.

Only incidental effects were observed, including a slightly higher red blood cell count in males of the high-dose group, an increase in haemoglobin in males of the mid-dose group, a slightly increased alanine aminotransferase activity in the mid-dose females and a slight increase in relative kidney weight in females of the low-dose group. These effects were not considered to be treatment related (Oda et al., 2008).

In the subsequent 13-week study (Oda et al., 2008), groups of 10 male and 10 female Sprague-Dawley SPF rats were administered L-alanyl-L-glutamine at dietary concentrations of 0%, 1%, 3% and 5% (w/w) (equal to doses of 0, 630, 1860 and 3130 mg/kg bw per day in males and 0, 730, 2160 and 3600 mg/kg bw per day in females). An additional group of 10 males and 10 females received 5% L-alanyl-L-glutamine that was produced differently from the test compound in diet (equal to doses of 3150 mg/kg bw per day in males and 3630 mg/kg bw per day in females). The highest dose was set based on the results of the 14-day range-finding study. The study was reported to be conducted under good laboratory practice (GLP).

The general condition of the rats was observed twice daily. Body weights and feed consumption were recorded on dosing days following overnight fasting. Body weights were also recorded on the day of necropsy. Ophthalmic examinations were performed in all animals before the start of the treatment and in five male and five female rats per group in week 13. Urine analysis was performed in week 13. At necropsy, blood was collected from all rats for determination of routine haematological and clinical chemistry parameters. Animals were killed by exsanguination. Necropsy included an external examination and opening of the cephalic, thoracic and abdominal cavities for observation of all major organs. The absolute and relative weights of major organs and tissues were determined. All organs and tissues of the controls and the animals in the high-dose group were examined microscopically. In addition, sections of testis and epididymides were examined in the low- and mid-dose groups.

The treatment did not result in deaths or clinical symptoms or in changes in body weight, feed consumption or feed efficiency. Ophthalmoscopy and urine analysis did not reveal any effects. Haematological and clinical chemistry analyses showed a significant increase in mean corpuscular haemoglobin concentration in males exposed to the differently produced test compound, a significantly higher aspartate aminotransferase activity in the low-dose males and an increased total cholesterol concentration in the high-dose males. Although statistically significant, these changes were small and not considered to be toxicologically relevant.

The macroscopic examinations revealed no treatment-related changes. Thymus weight was increased and adrenal gland weight was decreased in males of the low- and high-dose groups. An increased absolute weight of the heart was reported in the high-dose females. These incidental changes were not considered to be toxicologically relevant. In 3 of the 10 males of the high-dose group, one or more of the following effects were observed: seminiferous tubular atrophy, degeneration of spermatogenic cells, vacuolation of the Sertoli cells in the testis, with associated hypospermia, or intraductal cell debris in the epididymis. However, the effects were graded as minimal to mild, were within the range of historical control values for the age and strain of rat used at the testing laboratory and were not seen in rats administered 5% L-alanyl-L-glutamine that was produced differently from the test compound in the diet. Therefore, the authors considered that these changes were not treatment related.

The NOAEL was 3130 mg/kg bw per day, the highest dose tested (Oda et al., 2008).

(c) Genotoxicity

(i) In vitro

Studies of genotoxicity in vitro have been reported for one of the flavouring agents in this group (L-alanyl-L-glutamine, No. 2121). The results of these studies are summarized below and in Table 4.

No evidence of mutagenicity was observed in the Ames assay when L-alanyl-L-glutamine (No. 2121) was incubated with *Salmonella typhimurium* strains TA98, TA100, TA1535 and TA1537 and *Escherichia coli* strain WP2*uvrA*⁻, with or without metabolic activation, at concentrations up to 5000 µg/plate. The study was reported to be conducted under GLP (Oda et al., 2008).

L-Alanyl-L-glutamine (No. 2121) was negative in a chromosomal aberration assay in Chinese hamster lung cells at concentrations up to 2180 µg/ml, with and without metabolic activation. The study was reported to be conducted under GLP (Oda et al., 2008).

(ii) Conclusions

The available data for the current evaluation are in line with the conclusion in the previous evaluation that the amino acids and related substances tested were non-genotoxic in a variety of test systems (Annex 1, reference *173*).

Table 4. Results of in vitro studies of genotoxicity with amino acids and related substances used as flavouring agents

No.	Flavouring agent	End-point	Test system	Concentration	Results	Reference
2121	L-Alanyl-L-glutamine	Reverse mutation	*Salmonella typhimurium* strains TA98, TA100, TA1535 and TA1537, *Escherichia coli* WP2*uvrA*⁻	50–5000 μg/plate, ±S9[a]	Negative	Oda et al. (2008)
2121	L-Alanyl-L-glutamine	Chromosomal aberration	Chinese hamster lung cells	1090, 1635 or 2180 μg/ml, ±S9[b]	Negative	Oda et al. (2008)

S9, 9000 × g supernatant fraction of rat liver homogenate

[a] Two independent experiments were performed using the plate incorporation method.
[b] Two independent experiments were performed. Cells were analysed either 18 hours after 6 hours of treatment (experiment 1) or immediately after 24 or 48 hours of treatment (experiment 2).

3. REFERENCES

Adibi SA (1971). Intestinal transport of dipeptides in man: relative importance of hydrolysis and intact absorption. *Journal of Clinical Investigation*, 50:2266–2275.

Bishay I (2009a). The in vitro study on degradation of glutamyl-valyl-glycine (5 ppm) in the homogenate of human small intestinal mucosa. Unpublished report. Submitted to WHO by the International Organization of the Flavor Industry, Brussels, Belgium.

Bishay I (2009b). The in vitro study on the degradation of glutamyl-valyl-glycine (5 ppm) in the microsomal fraction from human small intestinal mucosa. Unpublished report. Submitted to WHO by the International Organization of the Flavor Industry, Brussels, Belgium.

Cramer GM, Ford RA, Hall RL (1978). Estimation of toxic hazard—a decision tree approach. *Food and Cosmetics Toxicology*, 16:255–276.

European Flavour and Fragrance Association (2004). European inquiry on volume use. Private communication to the Flavor and Extract Manufacturers Association of the United States, Washington, DC, USA. Submitted to WHO by the International Organization of the Flavor Industry, Brussels, Belgium.

Gavin CL, Williams MC, Hallagan JB (2008). *Flavor and Extract Manufacturers Association of the United States 2005 poundage and technical effects update survey*. Washington, DC, USA, Flavor and Extract Manufacturers Association of the United States.

Gilbert ER, Wong EA, Webb KE Jr (2008). Board-invited review: Peptide absorption and utilization: implications for animal nutrition and health. *Journal of Animal Science*, 86:2135–2155.

Herzog B et al. (1996). In vitro peptidase activity of rat mucosa cell fractions against glutamine-containing dipeptides. *Nutritional Biochemistry*, 7:135–141.

Institute of Medicine (2005). *Dietary reference intakes for energy, carbohydrate, fiber, fat, fatty acids, cholesterol, protein, and amino acids*. Washington, DC, USA, National Academies Press.

International Organization of the Flavor Industry (2011). Interim inquiry on volume use and added use levels for flavoring agents to be presented at the 76th JECFA meeting. Private communication to the Flavor and Extract Manufacturers Association of the United States,

Washington, DC, USA. Submitted to WHO by the International Organization of the Flavor Industry, Brussels, Belgium.

Japan Flavor and Fragrance Materials Association (2005). Japanese inquiry on volume use. Private communication to the Flavor and Extract Manufacturers Association of the United States, Washington, DC, USA. Submitted to WHO by the International Organization of the Flavor Industry, Brussels, Belgium.

Klassen P et al. (2000). The pharmacokinetic responses of humans to 20 g of alanyl-glutamine dipeptide differ with the dosing protocol but not with gastric acidity or in patients with acute Dengue fever. *Human Nutrition and Metabolism*, 130:177–182.

Minami H, Morse EL, Adibi SA (1992). Characteristics and mechanism of glutamine dipeptide absorption in human intestine. *Gastroenterology*, 103:3–11.

Newey H, Smyth DH (1960). Intracellular hydrolysis of dipeptides during intestinal absorption. *Journal of Physiology*, 152:367–380.

Oda S et al. (2008). Safety studies of L-alanyl-L-glutamine (L-AG). *Regulatory Toxicology and Pharmacology*, 50:226–238.

Rogero MM et al. (2006). Effect of alanyl-glutamine supplementation on plasma and tissue glutamine concentrations in rats submitted to exhaustive exercise. *Nutrition*, 22:564–571.

Webb KE Jr (1990). Intestinal absorption of protein hydrolysis products: a review. *Journal of Animal Science*, 68:3011–3022.

WHO/FAO/UNU (2007). *Protein and amino acid requirements in human nutrition. Report of a Joint WHO/FAO/UNU Expert Consultation, Geneva, 9–16 April 2002.* Geneva, Switzerland, World Health Organization (WHO Technical Report Series, No. 935; http://whqlibdoc.who.int/trs/WHO_TRS_935_eng.pdf).

EPOXIDES (addendum)

First draft prepared by

I.G. Sipes[1], M. DiNovi[2] and A. Mattia[2]

[1] Department of Pharmacology, College of Medicine, University of Arizona, Tucson, Arizona, United States of America (USA)
[2] Center for Food Safety and Applied Nutrition, Food and Drug Administration, College Park, Maryland, USA

1. Evaluation 147
   1.1 Introduction 147
   1.2 Assessment of dietary exposure 148
   1.3 Absorption, distribution, metabolism and elimination 148
   1.4 Application of the Procedure for the Safety Evaluation of
       Flavouring Agents 148
   1.5 Consideration of combined intakes from use as flavouring
       agents 154
   1.6 Consideration of secondary components 154
   1.7 Conclusions 155
2. Relevant background information 155
   2.1 Explanation 155
   2.2 Additional considerations on dietary exposure 155
   2.3 Biological data 155
       2.3.1 Biochemical data: hydrolysis, absorption, distribution,
             metabolism and elimination 155
       2.3.2 Toxicological studies 156
             (a) Acute toxicity 156
             (b) Short-term studies of toxicity 156
             (c) Genotoxicity 156
3. References 159

1. EVALUATION

1.1 Introduction

The Committee evaluated a group of seven epoxides that includes two phenylglycidate derivatives (Nos 2143 and 2144), two terpene epoxides (Nos 2145 and 2146) and three aliphatic epoxides (Nos 2147–2149). The evaluations were conducted according to the Procedure for the Safety Evaluation of Flavouring Agents (see Figure 1, Introduction) (Annex 1, reference *131*). None of these flavouring agents have previously been evaluated.

The Committee previously evaluated nine other members of this group of flavouring agents at the sixty-fifth meeting (Annex 1, reference *178*). The Committee concluded that the nine flavouring agents in that group were of no safety concern at estimated dietary exposures.

One of the seven flavouring agents (No. 2145) has been reported to occur naturally in food and has been detected in angelica oil, black currants, cardamom, dill, ginger, orange oil and pepper (Nijssen, van Ingen-Visscher & Donders, 2011).

1.2 Assessment of dietary exposure

The total annual volume of production of the seven epoxides is approximately 19 kg in Japan, with no reported volumes of production from the USA or Europe (European Flavour and Fragrance Association, 2004; Japan Flavor and Fragrance Materials Association, 2005; Gavin, Williams & Hallagan, 2008; International Organization of the Flavor Industry, 2011). Two flavouring agents in this group, methyl β-phenylglycidate (No. 2144) and ethyl α-ethyl-β-methyl-β-phenylglycidate (No. 2143), with annual volumes of production of 8 kg and 6 kg, respectively, account for approximately 74% of the total annual volume of production in Japan.

Dietary exposures were estimated for each flavouring agent using both the maximized survey-derived intake (MSDI) method and the single portion exposure technique (SPET), with the highest values reported in Table 1. The estimated dietary exposures are highest for ethyl α-ethyl-β-methyl-β-phenylglycidate (No. 2143) and methyl β-phenylglycidate (No. 2144) (SPET = 60 μg/day for each flavouring agent). This SPET value was obtained from gelatines and non-alcoholic beverages for No. 2143 and from non-alcoholic beverages for No. 2144. For the other flavouring agents, the estimated daily dietary exposures range from 0.03 to 40 μg, with the SPET yielding the highest estimates.

Annual volumes of production of this group of flavouring agents as well as the daily dietary exposures calculated using both the MSDI method and the SPET are summarized in Table 2.

1.3 Absorption, distribution, metabolism and elimination

The hydrolysis, absorption, distribution, metabolism and elimination of flavouring agents belonging to the group of epoxides have previously been described in the report of the sixty-fifth meeting (Annex 1, reference *178*).

Epoxides are three-membered rings containing an oxygen atom. The inherent ring strain and polarity of the C–O bond in the epoxide ring are factors that promote cleavage of the three-membered ring in the presence of suitable nucleophiles. Epoxide hydrolases, enzymes with wide tissue distribution, catalyse epoxide ring cleavage by water to yield *trans*-diols. The diols are then excreted primarily in the urine unchanged or as the glucuronic acid or sulfate conjugates. Alternatively, epoxides may be conjugated with glutathione mediated by glutathione S-transferases to yield the corresponding mercapturic acid conjugates, which are also excreted in the urine.

1.4 Application of the Procedure for the Safety Evaluation of Flavouring Agents

Step 1. In applying the Procedure for the Safety Evaluation of Flavouring Agents to the seven flavouring agents in this group of epoxides, the Committee assigned all seven (Nos 2143–2149) to structural class III (Cramer, Ford & Hall, 1978).

Table 1. Summary of the results of the safety evaluations of epoxides used as flavouring agents[a,b,c]

Flavouring agent	No.	CAS No. and structure	Step B3[d] Does estimated dietary exposure exceed the threshold of concern?	Step B4[e] Adequate margin of exposure for the flavouring agent or a related substance?	Step B5 Do the conditions of use result in an estimated dietary exposure greater than 1.5 µg/day?	Comments on predicted metabolism	Related structure name (No.) and structure (if applicable)	Conclusion based on current estimated dietary exposure
Structural class III								
Ethyl α-ethyl-β-methyl-β-phenylglycidate	2143	19464-94-9	No, SPET: 60	Yes. The NOEL of 35 mg/kg bw per day for the structurally related ethyl methylphenylglycidate (No. 1577) in a 2-year study in rats (Dunnington et al., 1981) is 35 000 times the estimated dietary exposure to No. 2143 when used as a flavouring agent.	NR	Note 1	Ethyl methylphenylglycidate (No. 1577)	No safety concern
Methyl β-phenylglycidate	2144	37161-74-3	No, SPET: 60	Yes. The NOEL of 35 mg/kg bw per day for the structurally related ethyl methylphenylglycidate (No. 1577) in a 2-year study in rats (Dunnington et al., 1981) is 35 000 times the estimated dietary exposure to No. 2144 when used as a flavouring agent.	NR	Note 2	Ethyl methylphenylglycidate (No. 1577)	No safety concern

continued

Table 1 (continued)

Flavouring agent	No.	CAS No. and structure	Step B3[f] Does estimated dietary exposure exceed the threshold of concern?	Step B4[e] Adequate margin of exposure for the flavouring agent or a related substance?	Step B5 Do the conditions of use result in an estimated dietary exposure greater than 1.5 µg/day?	Comments on predicted metabolism	Related structure name (No.) and structure (if applicable)	Conclusion based on current estimated dietary exposure
d-8-p-Menthene-1,2-epoxide	2145	1195-92-2	No, SPET: 15	Yes. The NOEL of 48 mg/kg bw per day for the structurally related piperitenone oxide (No. 1574) in a 28-day study in rats (Bauter, 2011) is 190 000 times the estimated dietary exposure to No. 2145 when used as a flavouring agent.	NR	Note 3	Piperitenone oxide (No. 1574)	No safety concern
l-8-p-Menthene-1,2-epoxide	2146	203719-53-3	No, SPET: 40	Yes. The NOEL of 48 mg/kg bw per day for the structurally related piperitenone oxide (No. 1574) in a 28-day study in rats (Bauter, 2011) is 72 000 times the estimated dietary exposure to No. 2146 when used as a flavouring agent.	NR	Note 3	Piperitenone oxide (No. 1574)	No safety concern

EPOXIDES (addendum)

Name	No.	CAS				
2,3-Epoxyoctanal	2147	42134-50-9	No, SPET: 30	No	Yes	Note 3	Additional data required to complete evaluation
2,3-Epoxyheptanal	2148	58936-30-4	No, SPET: 30	No	Yes	Note 3	Additional data required to complete evaluation
2,3-Epoxydecanal	2149	102369-06-2	No, SPET: 30	No	Yes	Note 3	Additional data required to complete evaluation

bw, body weight; CAS, Chemical Abstracts Service; NOEL, no-observed-effect level

[a] Nine flavouring agents in this group were previously evaluated by the Committee (Annex 1, reference *178*).
[b] *Step 1*: The seven flavouring agents in this group (Nos 2143–2149) are in structural class III.
[c] *Step 2*: The seven epoxides (Nos 2143–2149) are not expected to be metabolized to innocuous products.
[d] The threshold for human dietary exposure for structural class III is 90 μg/person per day. All dietary exposure values are expressed in μg/day. The dietary exposure value listed represents the highest estimated dietary exposure calculated using either the SPET or the MSDI method. The SPET gave the highest estimated dietary exposure in each case.
[e] The margins of exposure were calculated based on the estimated dietary exposures calculated using the SPET.

Notes:
1. The ester group is hydrolysed by carboxyl esterases, followed by loss of carbon dioxide and rearrangement to 2-phenyl-3-pentanone.
2. The ester group is hydrolysed by carboxyl esterases, followed by loss of carbon dioxide and rearrangement to phenylacetaldehyde.
3. The epoxide is hydrolysed via epoxide hydrolase to form a vicinal *trans*-diol, which forms a glucuronic acid conjugate and is eliminated in the urine, and/or the epoxide is directly conjugated with glutathione via glutathione S-transferase and is eliminated in the urine.

Table 2. Annual volumes of production and daily dietary exposures for epoxides used as flavouring agents in Europe, the USA and Japan

Flavouring agent (No.)	Most recent annual volume of production (kg)[a]	Dietary exposure				Natural occurrence in foods
		MSDI[b]		SPET[c]		
		µg/ day	µg/kg bw per day	µg/ day	µg/kg bw per day	
Ethyl α-ethyl-β-methyl-β-phenylglycidate (2143)				60	1	–
Europe	ND	ND	ND			
USA	ND	ND	ND			
Japan	6	2	0.03			
Methyl β-phenylglycidate (2144)				60	1	–
Europe	ND	ND	ND			
USA	ND	ND	ND			
Japan	8	2	0.04			
d-8-p-Menthene-1,2-epoxide (2145)				15	0.3	+
Europe	ND	ND	ND			
USA	ND	ND	ND			
Japan	2	1	0.01			
l-8-p-Menthene-1,2-epoxide (2146)				40	0.7	–
Europe	ND	ND	ND			
USA	ND	ND	ND			
Japan	2	1	0.01			
2,3-Epoxyoctanal (2147)				30	0.5	–
Europe	ND	ND	ND			
USA	ND	ND	ND			
Japan	0.3	0.09	0.001			
2,3-Epoxyheptanal (2148)				30	0.5	–
Europe	ND	ND	ND			
USA	ND	ND	ND			
Japan	0.1	0.03	0.0005			
2,3-Epoxydecanal (2149)				30	0.5	–
Europe	ND	ND	ND			
USA	ND	ND	ND			

EPOXIDES (addendum)

Table 2 (continued)

Flavouring agent (No.)	Most recent annual volume of production (kg)[a]	Dietary exposure MSDI[b] μg/day	MSDI[b] μg/kg bw per day	SPET[c] μg/day	SPET[c] μg/kg bw per day	Natural occurrence in foods
Japan	0.1	0.03	0.0005			
Total						
Europe	ND					
USA	ND					
Japan	19					

ND, no data reported; +, reported to occur naturally in foods (Nijssen, van Ingen-Visscher & Donders, 2011), but no quantitative data; –, not reported to occur naturally in foods

[a] From European Flavour and Fragrance Association (2004), Japan Flavor and Fragrance Materials Association (2005), Gavin, Williams & Hallagan (2008) and International Organization of the Flavor Industry (2011). Values greater than 0 kg but less than 0.1 kg were reported as 0.1 kg.

[b] MSDI (μg/person per day) calculated as follows:
(annual volume, kg) × (1 × 10^9 μg/kg)/(population × survey correction factor × 365 days), where population (10%, "eaters only") = 32 × 10^6 for Europe, 31 × 10^6 for the USA and 13 × 10^6 for Japan; and where survey correction factor = 0.8 for the surveys in Europe, the USA and Japan, representing the assumption that only 80% of the annual flavour volume was reported in the poundage surveys (European Flavour and Fragrance Association, 2004; Japan Flavor and Fragrance Materials Association, 2005; Gavin, Williams & Hallagan, 2008; International Organization of the Flavor Industry, 2011).
MSDI (μg/kg bw per day) calculated as follows:
(μg/person per day)/body weight, where body weight = 60 kg. Slight variations may occur from rounding.

[c] SPET (μg/person per day) calculated as follows:
(standard food portion, g/day) × (highest usual use level) (International Organization of the Flavor Industry, 2011). The dietary exposure from the single food category leading to the highest dietary exposure from one portion is taken as the SPET estimate.
SPET (μg/kg bw per day) calculated as follows:
(μg/person per day)/body weight, where body weight = 60 kg. Slight variations may occur from rounding.

Step 2. None of the flavouring agents in this group can be predicted to be metabolized to innocuous products. The evaluation of these flavouring agents therefore proceeded via the B-side of the Procedure.

Step B3. The highest dietary exposures to the seven flavouring agents in this group are below the threshold of concern (i.e. 90 μg/person per day for class III). Accordingly, the evaluation of all seven flavouring agents in the group proceeded to step B4.

Step B4. For ethyl α-ethyl-β-methyl-β-phenylglycidate (No. 2143), the no-observed-effect level (NOEL) of 35 mg/kg of body weight (bw) per day for the structurally related flavouring agent ethyl methylphenylglycidate (No. 1577) in a

2-year study in rats (Dunnington et al., 1981) provides a margin of exposure of 35 000 in relation to the dietary exposure to No. 2143 (SPET = 60 µg/day) when used as a flavouring agent.

For methyl β-phenylglycidate (No. 2144), the NOEL of 35 mg/kg bw per day for the structurally related flavouring agent ethyl methylphenylglycidate (No. 1577) in a 2-year study in rats (Dunnington et al., 1981) provides a margin of exposure of 35 000 in relation to the dietary exposure to No. 2144 (SPET = 60 µg/day) when used as a flavouring agent.

For d-8-*p*-menthene-1,2-epoxide (No. 2145), the NOEL of 48 mg/kg bw per day for the structurally related flavouring agent piperitenone oxide (No. 1574) in a 28-day study in rats (Bauter, 2011) provides a margin of exposure of 190 000 in relation to the dietary exposure to No. 2145 (SPET = 15 µg/day) when used as a flavouring agent.

For l-8-*p*-menthene-1,2-epoxide (No. 2146), the NOEL of 48 mg/kg bw per day for the structurally related flavouring agent piperitenone oxide (No. 1574) in a 28-day study in rats (Bauter, 2011) provides a margin of exposure of 72 000 in relation to the dietary exposure to No. 2146 (SPET = 40 µg/day) when used as a flavouring agent.

For 2,3-epoxyoctanal (No. 2147), 2,3-epoxyheptanal (No. 2148) and 2,3-epoxydecanal (No. 2149), no toxicological data are available on the flavouring agents or structurally related substances with which to calculate margins of exposure. Therefore, the evaluation of these flavouring agents proceeded to step B5.

Step B5. The conditions of use for 2,3-epoxyoctanal (No. 2147), 2,3-epoxyheptanal (No. 2148) and 2,3-epoxydecanal (No. 2149) result in dietary exposures greater than 1.5 µg/day. Therefore, the Committee concluded that additional data would be necessary to complete the evaluation of these flavouring agents.

Table 1 summarizes the evaluations of the seven additional flavouring agents in this group of epoxides (Nos 2143–2149).

1.5 Consideration of combined intakes from use as flavouring agents

A number of the flavouring agents in this group (Nos 1576–1578, 2143 and 2144) are alkyl phenylglycidates or closely related compounds. If these were to be consumed at the same time, the combined intake would exceed the class III threshold, primarily due to Nos 1576 and 1577; their combined intake was considered in detail at the sixty-fifth meeting of the Committee (Annex 1, reference *178*). The other flavouring agents in this group have diverse structures, with various potential sites of metabolism, and are not likely to be metabolized to common products. The Committee concluded that the combined intake of these flavouring agents was not a safety concern.

1.6 Consideration of secondary components

Three flavouring agents in this group (Nos 2144, 2148 and 2149) have minimum assay values of less than 95% (see Annex 5). The secondary component of methyl β-phenylglycidate (No. 2144) is ethyl β-phenylglycidate (No. 1576); the secondary component of 2,3-epoxyheptanal (No. 2148) is *trans*-2-heptenal (No. 1360); and the secondary component of 2,3-epoxydecanal (No. 2149) is *trans*-2-decenal (No. 1349). These secondary components are considered not to present a safety concern at current estimated dietary exposures.

1.7 Conclusions

In the previous evaluation of flavouring agents in this group of epoxides, studies of acute toxicity, short-term studies of toxicity, long-term studies of toxicity and carcinogenicity, and studies of genotoxicity and reproductive toxicity were available. The results of a short-term study of the toxicity of No. 1574 and in vitro genotoxicity studies on Nos 1574 and 2145 considered at this meeting support the previous safety evaluation (Annex 1, reference *178*).

The Committee concluded that four flavouring agents (Nos 2143–2146), which are additions to the group of epoxides evaluated previously, would not give rise to safety concerns at the current estimated dietary exposures.

The Committee concluded that additional toxicity data on the flavouring agents or structurally related substances would be necessary to complete the evaluations of Nos 2147–2149.

2. RELEVANT BACKGROUND INFORMATION

2.1 Explanation

This monograph summarizes the key data relevant to the safety evaluation of a group of seven epoxides that includes two phenylglycidate derivatives (Nos 2143 and 2144), two terpene epoxides (Nos 2145 and 2146) and three aliphatic epoxides (Nos 2147–2149), which are additions to the group of epoxides evaluated previously (Annex 1, reference *178*).

2.2 Additional considerations on dietary exposure

One of the seven flavouring agents (No. 2145) has been reported to occur naturally in food and has been detected in angelica oil, black currants, cardamom, dill, ginger, orange oil and pepper (Nijssen, van Ingen-Visscher & Donders, 2011).

2.3 Biological data

2.3.1 Biochemical data: hydrolysis, absorption, distribution, metabolism and elimination

No relevant information additional to that available and described in the monograph of the sixty-fifth meeting (Annex 1, reference *177*) was available on the absorption, distribution, metabolism and elimination of flavouring agents belonging to the group of epoxides.

2.3.2 Toxicological studies

(a) Acute toxicity

No relevant information additional to that available and described in the monograph of the sixty-fifth meeting (Annex 1, reference *177*) was available on the acute toxicity of flavouring agents belonging to the group of epoxides.

(b) Short-term studies of toxicity

Results of a short-term study of toxicity are available for piperitenone oxide (No. 1574). They are summarized in Table 3 and described below.

(i) Piperitenone oxide (No. 1574)

In a 28-day dietary repeated-dose toxicity study, Hsd:SD rats (five of each sex per dietary concentration) were provided a diet designed to deliver 0, 120, 360 or 720 mg of *Mentha longifolia* oil containing 85% piperitenone oxide (No. 1574) per kilogram (Bauter, 2011). The author of the study determined that these concentrations of *Mentha longifolia* oil correspond to doses of *Mentha longifolia* of 0, 10.0, 28.2 and 56.6 mg/kg bw per day in males and 0, 10.2, 29.9 and 61.7 mg/kg bw per day in females. All animals had ad libitum access to feed. The rats were observed for routine signs of clinical toxicity. Body weight gain was recorded weekly. Blood was sampled on day 27 of the study and again on day 30 prior to necropsy. Gross examinations were performed for all animals. Selected organs and tissues were preserved for microscopic examination.

All animals survived to the termination of the study, with no clinical signs of toxicity observed. Body weight gains, feed consumption and efficiency and ophthalmological measurements were comparable between test and control groups. There were no significant or biologically relevant changes in blood chemistry, haematology or urine analysis parameters measured when test groups were compared with control groups.

Based on the lack of any toxicologically relevant findings, the author determined the NOAEL for *Mentha longifolia* oil in the diet of the rat to be 720 mg/kg, or *Mentha longifolia* doses of 56.6 and 61.7 mg/kg bw per day for males and females, respectively. These levels correspond to a NOAEL for piperitenone oxide of 48 and 52 mg/kg bw per day for males and females, respectively, the highest dose tested.

(c) Genotoxicity

Genotoxicity studies with epoxides used as flavouring agents are summarized in Table 4 and described below.

(i) In vitro

d-8-*p*-Menthene-1,2-epoxide (No. 2145) was tested for mutagenicity to bacteria in *Salmonella typhimurium* strains TA98, TA100, TA1535 and TA1537,

Table 3. Results of short-term studies of toxicity of epoxides used as flavouring agents

No.	Flavouring agent	Species; sex	No. of test groups[a] / no. per group[b]	Duration (days)	NOAEL (mg/kg bw per day)	Reference
1574	Piperitenone oxide	Rat; M, F	3/10	28	M: 48 F: 52	Bauter (2011)

F, female; M, male; NOAEL, no-observed-adverse-effect level

[a] Total number of test groups does not include control animals.
[b] Total number per test group includes both male and female animals.

either alone or with an exogenous induced male rat liver bioactivation system, using the plate incorporation assay. The test substance was toxic at 1782 µg/plate in *S. typhimurium*. At five concentrations up to 1782 µg/plate, the test substance produced no increase in revertant mutants in any strain, whereas appropriate control substances established the responsiveness of the tester strains (Basler, von der Hude & Seelbach, 1989; von der Hude, Seelbach & Basler, 1990).

d-8-*p*-Menthene-1,2-epoxide (No. 2145) was tested for mutagenicity to bacteria in *Escherichia coli* PQ37, either alone or with an exogenous induced male rat liver bioactivation system, using the plate incorporation assay. At four concentrations up to 588 µg/ml, the test substance produced no increase in mutants, whereas appropriate control substances established the responsiveness of the tester strain (Basler, von der Hude & Seelbach, 1989; von der Hude, Seelbach & Basler, 1990).

d-8-*p*-Menthene-1,2-epoxide (No. 2145) was tested for induction of unscheduled DNA synthesis in rat hepatocytes. The test substance was toxic at a concentration of 178 µg/ml. At three concentrations up to178 µg/ml, the test substance produced no unscheduled DNA synthesis, whereas appropriate control substances established the responsiveness of the rat hepatocytes (von der Hude et al., 1989; von der Hude, Mateblowski & Basler, 1990).

Mentha longifolia oil containing 85% piperitenone oxide (No. 1574) was tested for mutagenicity to bacteria in *S. typhimurium* strains TA98, TA100, TA1535 and TA1537 and *E. coli* WP2uvrA, either alone or with an exogenous induced male rat liver bioactivation system, using both the plate incorporation and preincubation assays. The test substance was toxic at 500 µg/plate in *S. typhimurium* and at 1500 µg/plate in *E. coli*. At all concentrations up to 5000 µg/plate, the test substance produced no increase in revertant mutants in any strain, whereas appropriate control substances established the responsiveness of the tester strains (Thompson & Bowles, 2011).

(ii)   Conclusions for genotoxicity

The available data for the current evaluation are in line with the conclusion in the previous evaluation of the group of epoxides that flavouring agents in this group are not expected to exhibit any mutagenic or genotoxic potential.

Table 4. Genotoxicity studies with epoxides used as flavouring agents

No.	Flavouring agent	End-point	Test object	Concentration	Results	Reference
In vitro						
2145	d-8-p-Menthene-1,2-epoxide	Reverse mutation	Salmonella typhimurium TA98, TA100, TA1535, TA1537	18, 53, 178, 588 and 1782 μg/plate (0.1, 0.3, 1.0, 3.3 and 10.0 μmol/plate)[a]	Negative[b,c]	von der Hude, Seelbach & Basler (1990)
2145	d-8-p-Menthene-1,2-epoxide	Reverse mutation	S. typhimurium, strains unspecified	Concentrations unspecified	Negative[b]	Basler, von der Hude & Seelbach (1989)
2145	d-8-p-Menthene-1,2-epoxide	SOS chromotest	Escherichia coli PQ37	Concentrations unspecified	Negative[b]	Basler, von der Hude & Seelbach (1989)
2145	d-8-p-Menthene-1,2-epoxide	SOS chromotest	E. coli PQ37	18, 53, 178 and 588 μg/ml (0.1, 0.3, 1.0 and 3.3 mmol/l)[a]	Negative	von der Hude, Seelbach & Basler (1990)
2145	d-8-p-Menthene-1,2-epoxide	Sister chromatid exchange	V79 cells	Concentrations unspecified	Negative[d]	von der Hude et al. (1989)
2145	d-8-p-Menthene-1,2-epoxide	Unscheduled DNA synthesis	Rat hepatocytes	Concentrations unspecified	Negative	von der Hude et al. (1989)
2145	d-8-p-Menthene-1,2-epoxide	Unscheduled DNA synthesis	Rat hepatocytes	2, 18 and 178 μg/ml (0.01, 0.1 and 1.0 mmol/l)[a]	Negative[b,c]	von der Hude, Mateblowski & Basler (1990)
1574	Piperitenone oxide	Reverse mutation	S. typhimurium TA98, TA100, TA1535, TA1537	5, 15, 50, 150, 500, 1500, 3000 and 5000 μg/plate	Negative[b]	Thompson & Bowles (2011)
1574	Piperitenone oxide	Reverse mutation	E. coli WP2uvrA	5, 15, 50, 150, 500, 1500, 3000 and 5000 μg/plate	Negative[b]	Thompson & Bowles (2011)

DNA, deoxyribonucleic acid

[a] Calculated using a relative molecular mass of 178.18.
[b] With and without exogenous metabolic bioactivation.
[c] The test material was cytotoxic to the cell line at the highest concentration tested.
[d] Performed in the absence of exogenous metabolic activation.

3. REFERENCES

Basler A, von der Hude W, Seelbach A (1989). Genotoxicity of epoxides. I. Investigations with the SOS chromotest and *Salmonella*/mammalian microsome test. *Mutagenesis*, 4(4):313–314 (abstract).

Bauter MR (2011). ML-2: a 28-day dietary study in rats. Unpublished report no. 31278 from Product Safety Labs, East Brunswick, NJ, USA. Submitted to WHO by the International Organization of the Flavor Industry, Brussels, Belgium.

Cramer GM, Ford RA, Hall RL (1978). Estimation of toxic hazard—a decision tree approach. *Food and Cosmetics Toxicology*, 16(3):255–276.

Dunnington D et al. (1981). Long-term toxicity study of ethyl methylphenylglycidate (strawberry aldehyde) in the rat. *Food and Cosmetics Toxicology*, 19(6):691–699.

European Flavour and Fragrance Association (2004). European inquiry on volume use. Private communication to the Flavor and Extract Manufacturers Association of the United States, Washington, DC, USA. Submitted to WHO by the International Organization of the Flavor Industry, Brussels, Belgium.

Gavin CL, Williams MC, Hallagan JB (2008). *Flavor and Extract Manufacturers Association of the United States 2005 poundage and technical effects update survey.* Washington, DC, USA, Flavor and Extract Manufacturers Association of the United States.

International Organization of the Flavor Industry (2011). Interim inquiry on volume use and added use levels for flavoring agents to be presented at the 76th JECFA meeting. Private communication to the Flavor and Extract Manufacturers Association of the United States, Washington, DC, USA. Submitted to WHO by the International Organization of the Flavor Industry, Brussels, Belgium.

Japan Flavor and Fragrance Materials Association (2005). Japanese inquiry on volume use. Private communication to the Flavor and Extract Manufacturers Association of the United States, Washington, DC, USA. Submitted to WHO by the International Organization of the Flavor Industry, Brussels, Belgium.

Nijssen LM, van Ingen-Visscher CA, Donders JJH (2011). Volatile Compounds in Food (VCF) database, version 13.1. Zeist, the Netherlands, TNO Triskelion (http://www.vcf-online.nl/VcfHome.cfm).

Thompson PW, Bowles A (2011). ML-2: reverse mutation assay "Ames test" using *Salmonella typhimurium* and *Escherichia coli*. Unpublished report no. 41005038 from Harlan Laboratories Ltd, Derbyshire, England. Submitted to WHO by the International Organization of the Flavor Industry, Brussels, Belgium.

von der Hude W, Mateblowski R, Basler A (1990). Induction of DNA-repair synthesis in primary rat hepatocytes by epoxides. *Mutation Research*, 245(3):145–150.

von der Hude W, Seelbach A, Basler A (1990). Epoxides: comparison of the induction of SOS repair in *Escherichia coli* PQ37 and the bacterial mutagenicity in the Ames test. *Mutation Research*, 231(2):205–218.

von der Hude W et al. (1989). Genotoxicity of epoxides. II. In vitro investigations with the sister-chromatid exchange (SCE) test and the unscheduled DNA synthesis (UDS) test. *Mutagenesis*, 4(4):323–324 (abstract).

FURFURYL ALCOHOL AND RELATED SUBSTANCES (addendum)

First draft prepared by

B. Fields[1], M. DiNovi[2], J.-C. Leblanc[3], A. Renwick[4] and U. Mueller[1]

[1] Food Standards Australia New Zealand, Canberra, ACT, Australia
[2] Center for Food Safety and Applied Nutrition, Food and Drug Administration, College Park, Maryland, United States of America (USA)
[3] Agence nationale de sécurité sanitaire de l'alimentation, de l'environnement et du travail (ANSES), Maisons-Alfort, France
[4] School of Medicine, University of Southampton, Southampton, England

1. Evaluation 161
    1.1 Introduction 161
    1.2 Metabolism 163
    1.3 Genotoxicity 163
    1.4 Conclusion 164
2. Relevant background information 164
    2.1 Biochemical data: absorption, distribution, metabolism and elimination 164
    2.2 Toxicological studies 164
        2.2.1 Acute toxicity 164
        2.2.2 Short-term studies of toxicity 164
            (a) Furfural (No. 450) 164
            (b) Furfuryl alcohol (No. 451) 165
        2.2.3 Genotoxicity 166
            (a) In vitro 166
            (b) In vivo 169
3. References 170

1. EVALUATION

1.1 Introduction

At the present meeting, the Committee considered four additional flavouring agents (Nos 2099–2102) belonging to the group of furfuryl alcohol and related substances: one furfuryl alcohol (No. 2099), one furfural acetal (No. 2100) and two furfuryl esters (Nos 2101 and 2102) (Table 1). Three flavouring agents (Nos 2103–2105) were originally assigned to the current group, but it was considered appropriate for these three to be evaluated in the group of aliphatic hydrocarbons, alcohols, aldehydes, ketones, carboxylic acids and related esters, sulfides, disulfides and ethers containing furan substitution (see relevant monograph in this volume).

Furfural (No. 450) was evaluated previously by the Committee at its thirty-ninth and fifty-first meetings (Annex 1, references *101* and *137*). An acceptable daily intake (ADI) was not established at either meeting because of concern about the findings of tumours in male mice given furfural by gavage and the fact that no no-observed-effect level (NOEL) was identified for hepatotoxicity in male rats.

Table 1. Chemical structures of the four additional flavouring agents belonging to the group of furfuryl alcohol and related substances

Flavouring agent	No.	CAS No. and structure
5-Methylfurfuryl alcohol	2099	3857-25-8
Furfural propyleneglycol acetal	2100	4359-54-0
Furfuryl formate	2101	13493-97-5
Furfuryl decanoate	2102	39252-05-6

In the mouse study, the combined incidence of adenomas and carcinomas was increased in males at the highest dose tested. In order to address the concern regarding the formation of liver tumours in mice, the Committee at its fifty-first meeting requested the results of studies of deoxyribonucleic acid (DNA) binding or adduct formation in vivo to clarify whether furfural interacts with DNA in the liver of mice (Annex 1, reference *137*). While no specific studies of DNA binding were submitted, the results of an assay for unscheduled DNA synthesis (UDS) in mice in vivo was evaluated by the Committee at its fifty-fifth meeting (Annex 1, reference *149*). This study, in which single doses of up to 350 mg/kg of body weight (bw) were given, was considered to be particularly relevant, as it addressed potential DNA repair in the cells in which tumours arose—namely, hepatocytes. The negative results obtained in this assay were considered by the Committee to provide evidence that the liver tumours observed in the long-term study in mice were unlikely to have occurred through a genotoxic mechanism. The Committee concluded that the concerns raised previously with respect to the liver tumours in mice were adequately addressed by this study and that a study of DNA binding was unnecessary.

For furfuryl alcohol (No. 451), carcinogenicity studies in mice and rats using the inhalation route were considered by the Committee at its fifty-fifth meeting (Annex 1, reference *150*). Under the conditions of these 2-year inhalation studies, male mice exposed to the highest tested concentration had a significantly increased incidence of renal tubular degeneration and a significantly increased combined incidence of renal tubular adenomas and carcinomas. The Committee considered that although these studies are of limited value for assessing the potential toxicity of oral exposure to furfuryl alcohol, the findings may provide some indication of potential long-term toxicity.

The Committee evaluated 15 members of this group of flavouring agents at its fifty-fifth meeting and concluded that all 15 were of no safety concern at estimated dietary exposures (Annex 1, reference *149*).

A group ADI of 0–0.5 mg/kg bw was established by the Committee at its fifty-fifth meeting for 12 members of the group—namely, furfural (No. 450), furfuryl alcohol (No. 451), furfuryl acetate (No. 739), furfuryl propionate (No. 740), furfuryl pentanoate (No. 741), furfuryl octanoate (No. 742), furfuryl 3-methylbutanoate (No. 743), methyl 2-furoate (No. 746), propyl 2-furoate (No. 747), amyl 2-furoate (No. 748), hexyl 2-furoate (No. 749) and octyl 2-furoate (No. 750). Furfural, furfuryl alcohol and these 10 derivatives are metabolized to the same metabolite, 2-furoic acid (Annex 1, reference *149*). The ADI was established on the basis of a NOEL of 53 mg/kg bw per day in a 13-week rat study with furfural and using a safety factor of 100.

At the current meeting, the Committee included new in vitro and in vivo genotoxicity studies on several members of this group in its re-evaluation of the group of furfuryl alcohol and related substances. Genotoxicity data were available for one of the four additional flavouring agents in this group (5-methylfurfuryl alcohol; No. 2099). Additional genotoxicity studies were also available on two flavouring agents previously evaluated by the Committee—namely, furfuryl alcohol (No. 451) and furfuryl acetate (No. 739). In addition, new published genotoxicity data on furfural (No. 450) were available; however, these data were evaluated by the Committee at the fifty-fifth meeting (Annex 1, reference *149*) as separate unpublished in vitro and in vivo studies.

1.2 Metabolism

All four additional members of this group of flavouring agents would be expected to be metabolized to furfuryl alcohol or a structural analogue that would undergo further oxidation to the corresponding 2-furoic acid.

1.3 Genotoxicity

Positive genotoxicity findings were observed in several in vitro and in vivo studies with three flavouring agents in this group. In reverse mutation assays with *Salmonella typhimurium* TA100-derived strains engineered to express various mouse and human sulfotransferases (SULT), furfuryl alcohol (No. 451) and 5-methylfurfuryl alcohol (No. 2099) exhibited mutagenic activity. Furfuryl acetate

(No. 739) was tested in the conventional TA100 strain only and exhibited mutagenic activity. These studies suggested that furfuryl alcohol is converted by intracellular sulfate conjugation to 2-sulfo-oxymethylfuran, an electrophile reacting with DNA. Subsequent experiments resulted in the detection of nucleoside adducts of 2'-deoxyadenosine and 2'-deoxyguanosine in porcine liver DNA incubated with freshly prepared 2-sulfo-oxymethylfuran. These adducts were also observed in the DNA of liver, lung and kidney of mice administered furfuryl alcohol (No. 451) at a dose of 390 mg/kg bw per day via drinking-water for 28 days. Another study indicated that furfuryl alcohol (No. 451) can induce sister chromatid exchange (SCE) in human lymphocytes in vitro and in vivo.

1.4 Conclusion

New in vitro and in vivo studies raise concerns regarding the potential genotoxicity of furfuryl alcohol and derivatives that can be metabolized to furfuryl alcohol (e.g. furfuryl esters). The Committee concluded that this group of flavouring agents could not be evaluated according to the Procedure for the Safety Evaluation of Flavouring Agents because of the unresolved concerns regarding genotoxicity. In addition, the group ADI previously established by the Committee will need to be reconsidered at a future meeting.

2. RELEVANT BACKGROUND INFORMATION

2.1 Biochemical data: absorption, distribution, metabolism and elimination

General information on the metabolism of furfuryl alcohol and related substances was previously provided in the report of the fifty-fifth meeting (Annex 1, reference *149*). No information relevant to the flavouring agents considered at the current meeting was available.

2.2 Toxicological studies

2.2.1 Acute toxicity

No acute toxicity data are available on the four additional flavouring agents in this group.

2.2.2 Short-term studies of toxicity

No data are available from short-term studies of the toxicity of the four additional flavouring agents in this group. Additional short-term studies on the toxicity of two flavouring agents previously evaluated by the Committee—namely, furfural (No. 450) and furfuryl alcohol (No. 451)—are described below and summarized in Table 2.

(a) *Furfural (No. 450)*

In a published study, furfural was administered to F344 rats (five of each sex per group) by gavage at 0, 6, 12, 24, 48, 96 or 192 mg/kg bw per day for

Table 2. Results of short-term studies of the toxicity of furfuryl alcohol and related substances

No.	Flavouring agent	Species; sex	No. of test groups[a] / no. per group[b]	Route	Duration (days)	NOAEL (mg/kg bw per day)	Reference
450	Furfural	Rat; M, F	6/10	Gavage	28	96	Arts et al. (2004)
451	Furfuryl alcohol	Mouse; M, F	15/4	Gavage	5, 10, 20, 30, 60, 90	NI	Sujatha (2008a)

F, female; M, male; NI, not identified; NOAEL, no-observed-adverse-effect level
[a] Total number of test groups does not include control animals.
[b] Total number per test group includes both male and female animals.

28 days. It was not stated whether the study was conducted according to good laboratory practice or whether a specific test guideline (e.g. Organisation for Economic Co-operation and Development) was followed. Animals were observed daily for clinical signs of toxicity, body weights were measured once or twice a week and feed consumption was measured weekly. At the end of the study, haematology and blood chemistry parameters were measured, and a complete necropsy and histopathological examination were conducted.

One female and one male in the highest-dose group were found dead on day 1 and day 4 of the study, respectively. Based on these deaths, the highest dose of 192 mg/kg bw per day was lowered to 144 mg/kg bw per day for both male and female rats. An additional female death occurred on day 10, and one female was euthanized in extremis on day 11 due to lethargy and breathing difficulties. The highest dose level was therefore further reduced to 120 mg/kg bw per day for both sexes. On the last day of the study, another female was found dead; however, the possible cause of death was not discussed. No other treatment-related clinical signs, effects on body weight gain or feed consumption, changes in haematology or clinical chemistry, lesions or microscopic changes upon necropsy were observed in the study. The rats that did not survive to completion of the study were examined and had a red infiltration of the lungs and a gaseous substance in the stomach, but did not display treatment-related histopathological changes. Increased mean kidney weight, by 21% for absolute weight ($P < 0.01$) and by 12% for body weight–relative weight ($P < 0.01$), and increased mean absolute liver weight (29%; $P < 0.01$) were observed in surviving females in the highest-dose group.

Because of the death at the final high dose of 120 mg/kg bw per day, the no-observed-adverse-effect level (NOAEL) was considered to be 96 mg/kg bw per day (Arts et al., 2004).

(b)   Furfuryl alcohol (No. 451)

In a published study that investigated a limited range of toxicological end-points, Swiss albino mice (four per group per period; sex not stated) received daily oral gavage doses of furfuryl alcohol of 0, 0.5, 1 or 2 mg/day for 5, 10, 20, 30, 60 or

90 days. Using the reported mean body weight of the mice at the start of the study (25 g), the initial group doses are calculated to be 0, 20, 40 and 80 mg/kg bw per day. Assuming a final mean body weight of 40 g, the group doses at the end of the 90 days are calculated to be 0, 12.5, 25 and 50 mg/kg bw per day. The mice were sacrificed 24 hours after the last exposure for histopathological evaluation of the liver and kidney. Total tissue protein content and estimates of activities of aspartate aminotransferase (AST) and alanine aminotransferase (ALT) were determined for liver and kidney tissue.

Mice receiving 0.5 and 1 mg/day were reported to exhibit dyspnoea and loss of hair from the neck region. Information on the onset and duration of these clinical signs was not provided. Dark red liver was reported in all mice receiving 2 mg/day for 10–60 days. Mice receiving 1 and 2 mg/day showed signs of hepatotoxicity in the form of pycnosis, vacuolation and focal necrosis after 60 and 90 days. Kidneys were not affected by 30 and 90 days of treatment, but a dose of 2 mg/day for 60 days resulted in damage to the tubular epithelium. Total protein content in liver was increased ($P < 0.05$) after 10–30 days of exposure to 1 and 2 mg/day, but was decreased ($P < 0.05$) after 60 days. Kidney protein content was not affected. At 1 and 2 mg/day, increased activities of AST and ALT ($P < 0.05$) were observed in the liver after 20, 30 and 60 days of exposure. At 1 and 2 mg/day, increased activity of AST ($P < 0.05$) was observed in the kidney after 60 days of exposure. For 90 days of treatment, information on tissue levels of total protein, AST and ALT was not provided.

A NOAEL was not set for this study because of the limited range of endpoints investigated, the uncertainty in the doses received on a body weight basis and the limited level of reporting. The Committee considered that the results of this study were not suitable for risk assessment purposes (Sujatha, 2008a).

2.2.3 Genotoxicity

Genotoxicity data are available on one of the four additional flavouring agents in this group (5-methylfurfuryl alcohol; No. 2099). Additional genotoxicity studies are available on three flavouring agents previously evaluated by the Committee—namely, furfural (No. 450), furfuryl alcohol (No. 451) and furfuryl acetate (No. 739). These studies are described below and summarized in Table 3.

(a) In vitro

The ability of furfural (No. 450) to induce UDS in human liver slices was studied. Precision-cut human liver slices from four donors were cultured for 24 hours in medium containing [^{3}H]thymidine and 0–10 mmol of furfural per litre. UDS was evaluated by quantitative assessment of [^{3}H]thymidine uptake in hepatocyte nuclei and cytoplasm. Small, statistically significant increases in the net silver grain count (i.e. nuclear grain count minus cytoplasmic grain count) were observed in hepatocytes with 2–10 mmol of furfural per litre. However, these increases in net count were due to concentration-dependent furfural-induced cytotoxicity resulting in decreased mean cytoplasmic grain counts, whereas nuclear grain counts were unaffected. Positive control compounds induced large increases in nuclear grain count (Lake et al., 2001).

Table 3. Results of genotoxicity studies with furfuryl alcohol and related substances

No.	Flavouring agent	End-point	Test object	Concentration/dose	Results	Reference
In vitro						
450	Furfural	UDS	Human liver slices	0.005–10 mmol/l	Negative	Lake et al. (2001)[a]
451	Furfuryl alcohol	SCE	Human lymphocytes	240, 480 and 960 μg/ml	Positive[b,c] Positive[b,d] Negative[b,e]	Sujatha (2008b)
451	Furfuryl alcohol	Chromosomal aberration	Human lymphocytes	240, 480 and 960 μg/ml	Negative[b,f]	Sujatha (2008b)
451	Furfuryl alcohol	Reverse mutation	*Salmonella typhimurium* TA100 and TA100-derived strains expressing human SULT 1A1	0.1–100 μg/plate	Negative[b] (TA100) Positive[b] (TA100-SULT)	Monien et al. (2011)
451	Furfuryl alcohol	Reverse mutation	*S. typhimurium* TA100 and TA100-derived strains engineered to express human, rat and mouse SULT	2–980 μg/plate	Negative[b] (TA100) Positive[b] (TA100-SULT)	Glatt et al. (2012)
739	Furfuryl acetate	Reverse mutation	*S. typhimurium* TA100	0.14–280 μg/plate[g]	Positive[b]	Glatt et al. (2012)
2099	5-Methylfurfuryl alcohol	Reverse mutation	*S. typhimurium* TA100 and TA100-derived strains engineered to express human, rat and mouse SULT	6–220 μg/plate	Negative[b] (TA100) Positive[b] (TA100-SULT)	Glatt et al. (2012)
In vivo						
450	Furfural	UDS	B6C3F1 mice	50–320 mg/kg bw (single gavage dose)	Negative	Lake et al. (2001)[a]
450	Furfural	UDS	F344 rats	5, 17 or 50 mg/kg bw (single gavage dose)	Negative	Lake et al. (2001)[a]

continued

Table 3 (continued)

No.	Flavouring agent	End-point	Test object	Concentration/dose	Results	Reference
451	Furfuryl alcohol	SCE	Swiss albino mice	20, 40 and 80 mg/kg bw (single gavage dose)	Positive	Sujatha (2007)
451	Furfuryl alcohol	SCE	Swiss albino mice	20, 40 and 80 mg/kg bw per day for 5 days (gavage)	Positive	Sujatha (2007)
451	Furfuryl alcohol	DNA adduct formation	FVB/N mice	390 mg/kg bw per day for 28 days (via drinking-water)	Positive	Monien et al. (2011)

[a] The studies described in Lake et al. (2001) were evaluated by the Committee at the fifty-fifth meeting (Annex 1, reference 149) as separate unpublished in vitro and in vivo studies.
[b] Without metabolic activation.
[c] For 24-hour exposure at 480 µg/ml only.
[d] For 48-hour exposure at all concentrations.
[e] For 72-hour exposure at all concentrations.
[f] For 24-, 48- and 72-hour exposure at all concentrations.
[g] Cytotoxic at the next highest concentration tested (1400 µg/plate).

The potential for furfuryl alcohol (No. 451) to induce chromosomal aberrations and SCE in human lymphocytes was investigated. Furfuryl alcohol concentrations of 0, 240, 480 and 960 µg/ml were evaluated in both assays. After 24, 48 and 72 hours of exposure, slides were prepared and analysed for variations in cell division (mitotic index) and chromosomal aberrations. There were no statistically significant effects on mitotic index or chromosomal aberrations under any of the tested conditions. Furfuryl alcohol exposure resulted in statistically significant increases in SCE at 480 µg/ml, but not at 960 µg/ml, with 24-hour exposure. For 48-hour exposure, all concentrations resulted in statistically significant increases in SCE, which were concentration dependent. No effect was observed with 72-hour exposure. A metabolic activation system (e.g. S9 mix) and positive control compounds were not used in these assays (Sujatha, 2008b).

In reverse mutation assays with *Salmonella typhimurium* strain TA100 and TA100-derived strains engineered to express human SULT 1A1*1 and 1A1*1Y, furfuryl alcohol (No. 451) was inactive in TA100 but mutagenic in TA100 strains expressing human SULT 1A1. All assays were conducted in the absence of S9 metabolic activation. Positive control compounds were not used in these assays (Monien et al., 2011).

In reverse mutation assays with *S. typhimurium* strain TA100 and TA100-derived strains expressing human or rodent SULT, furfuryl alcohol (No. 451) and 5-methylfurfuryl alcohol (No. 2099) were inactive in TA100 but mutagenic in TA100 strains expressing human SULT 1A1, 1A2 and 1C2 and murine SULT 1a1. Furfuryl acetate (No. 739) was tested with the conventional TA100 strain only and was mutagenic when tested in water in the concentration range of 100–280 µg/plate. When tested in the presence of magnesium sulfate (82 mmol/l), putatively leading to the formation of furfuryl sulfate, a slight reduction in mutagenicity was observed. The presence of potassium chloride (123 mmol/l), putatively leading to the formation of furfuryl chloride, strongly enhanced bacteriotoxicity, with a concomitant decrease in the maximal mutagenic response. All assays were conducted in the absence of S9 metabolic activation. Positive control compounds were not used in these assays (Glatt et al., 2012).

(b) In vivo

The ability of furfural (No. 450) to induce UDS in isolated hepatocytes of male and female B6C3F1 mice and male F344 rats was studied. Furfural was administered as a single oral gavage dose at 0, 50, 175 or 320 mg/kg bw to mice (three of each sex per group) and 0, 5, 17 or 50 mg/kg bw to male rats (three per group). Hepatocytes were isolated by liver perfusion either 2–4 hours or 12–16 hours after treatment, cultured in medium containing [^{3}H]thymidine for 4 hours and assessed for UDS by grain counting of autoradiographs. Furfural treatment did not produce any statistically significant increase or any dose-related effects on UDS in mouse or rat hepatocytes. Positive control compounds induced UDS as expected (Lake et al., 2001).

The ability of furfuryl alcohol (No. 451) to induce SCE in mouse bone marrow was evaluated. Groups of female Swiss albino mice received 0, 20, 40 or 80 mg of furfuryl alcohol per kilogram of body weight by oral gavage as a single dose or once a day for 5 consecutive days. 5-Bromodeoxyuridine (5-BrdU) was intraperitoneally injected at hourly intervals for 6 hours at a dose of 0.04 mg/g bw. For both single and repeated dosing, mice were exposed to 5-BrdU for 26 hours and were sacrificed 12, 24 and 48 hours (two animals per dose per period) following the last feeding. Bone marrow slides were prepared and stained for evaluation of SCE. Furfuryl alcohol treatment resulted in a dose-dependent induction of SCE in the single-dose experiments, which were statistically significant at doses of 40 and 80 mg/kg bw for the 12- and 48-hour sacrifice times and at all doses for the 24-hour sacrifice. In the repeated-dose experiments, treatment with furfuryl alcohol resulted in dose-dependent increases in SCE at all doses and time points, which were statistically significant at all dose levels for the 12- and 24-hour sacrifices and at the two highest doses at the 48-hour sacrifice. The magnitudes of the increase were greatest at the 24-hour sacrifice for both single and repeated dosing. A positive control compound was not used in this study (Sujatha, 2007).

FVB/N mice (five of each sex) received furfuryl alcohol (No. 451) via drinking-water for 28 days, resulting in an achieved dose of approximately 390 mg/kg bw per day. Control mice received drinking-water alone. DNA was isolated from liver, lungs, kidneys and colon mucosa and analysed for the adducts N^2-[(furan-2-yl)methyl]-2′-deoxyguanosine (N^2-MFdG) and N^6-[(furan-2-yl)methyl]-2′-deoxyadenosine (N^6-MFdA). Adduct levels were similar in liver, kidney and lung, with 0.8–5 molecules of N^2-MFdG per 10^8 nucleosides and 0.03–0.3 molecules of N^6-MFdA per 10^8 nucleosides. In contrast, N^2-MFdG and N^6-MFdA were not detected in colon mucosa. In liver, kidney and lung, N^2-MFdG levels were generally 10- to 20-fold higher than those of N^6-MFdA. The latter were close to, or in some samples even below, the limit of detection (0.04 molecules of N^6-MFdA per 10^8 nucleosides). In contrast, the levels of N^2-MFdG were 8- to 50-fold higher than the limit of detection (0.1 molecules of N^2-MFdG per 10^8 nucleosides). There were no detectable DNA adducts in tissue samples from control animals. The authors proposed that adduct formation results from conversion of furfuryl alcohol by intracellular sulfate conjugation to 2-sulfo-oxymethylfuran, which subsequently undergoes reaction with DNA (Monien et al., 2011).

3. REFERENCES

Arts JHE et al. (2004). Subacute (28-day) toxicity of furfural in Fischer 344 rats: a comparison of the oral and inhalation route. *Food and Chemical Toxicology*, 42(9):1389–1399.

Glatt H et al. (2012). Hydroxymethyl-substituted furans: mutagenicity in *Salmonella typhimurium* strains engineered for expression of various human and rodent sulphotransferases. *Mutagenesis*, 27(1):41–48.

Lake BG et al. (2001). Lack of effect of furfural on unscheduled DNA synthesis in the in vivo rat and mouse hepatocyte DNA repair assays and in precision-cut human liver slices. *Food and Chemical Toxicology*, 39(10):999–1011.

Monien BH et al. (2011). Metabolic activation of furfuryl alcohol: formation of 2-methylfuranyl DNA adducts in *Salmonella typhimurium* strains expressing human sulfotransferase 1A1 and in FVB/N mice. *Carcinogenesis*, 32(10):1533–1539.

Sujatha PS (2007). Genotoxic evaluation of furfuryl alcohol and 2-furyl methyl ketone by sister chromatid exchange (SCE) analysis. *Journal of Health Science*, 53(1):124–127.

Sujatha PS (2008a). Monitoring cytotoxic potentials of furfuryl alcohol and 2-furyl methyl ketone in mice. *Food and Chemical Toxicology*, 46(1):286–292.

Sujatha PS (2008b). Response of human lymphocytes in vitro to two dietary furans, furfuryl alcohol and 2-furyl methyl ketone. *Journal of Health Science*, 54(1):66–71.

LINEAR AND BRANCHED-CHAIN ALIPHATIC, UNSATURATED, UNCONJUGATED ALCOHOLS, ALDEHYDES, ACIDS AND RELATED ESTERS (addendum)

First draft prepared by

I.G. Sipes[1], J. Alexander[2] and M. DiNovi[3]

[1] Department of Pharmacology, College of Medicine, University of Arizona, Tucson, Arizona, United States of America (USA)
[2] Norwegian Institute of Public Health, Oslo, Norway
[3] Center for Food Safety and Applied Nutrition, Food and Drug Administration, College Park, Maryland, USA

1. Evaluation 171
   1.1 Introduction 171
   1.2 Assessment of dietary exposure 172
   1.3 Absorption, distribution, metabolism and elimination 172
   1.4 Application of the Procedure for the Safety Evaluation of Flavouring Agents 179
   1.5 Consideration of combined intakes from use as flavouring agents 179
   1.6 Conclusions 180
2. Relevant background information 180
   2.1 Explanation 180
   2.2 Additional considerations on dietary exposure 180
   2.3 Biological data 180
      2.3.1 Biochemical data: hydrolysis, absorption, distribution, metabolism and elimination 180
      2.3.2 Toxicological studies 181
         (a) Acute toxicity 181
         (b) Short-term studies of toxicity 181
         (c) Genotoxicity 182
3. References 183

1. EVALUATION

1.1 Introduction

The Committee evaluated nine flavouring agents (Nos 2177–2185) in the group of linear and branched-chain aliphatic, unsaturated, unconjugated alcohols, aldehydes, acids and related esters using the Procedure for the Safety Evaluation of Flavouring Agents (see Figure 1, Introduction) (Annex 1, reference *131*). Five of the nine flavouring agents are esters (Nos 2179, 2180 and 2182–2184). The remaining four flavouring agents are linear unsaturated alcohols (Nos 2177 and 2178), an unsaturated acid (No. 2181) and an aldehyde (No. 2185). None of these flavouring agents have previously been evaluated.

The Committee evaluated 42 other members of this group of flavouring agents at the fifty-first meeting (Annex 1, reference *137*). Forty-one of the 42 flavouring agents in that group were concluded to be of no safety concern at estimated dietary exposures. The evaluation of one flavouring agent, ethyl 2-methyl-3,4-pentadienoate (No. 353), was deferred pending review of a 90-day rat feeding study. It was evaluated at the sixty-eighth meeting of the Committee (Annex 1, reference *187*) and concluded to be of no safety concern at the estimated dietary exposure.

The Committee evaluated 20 other members of this group of flavouring agents at the sixty-first meeting (Annex 1, reference *166*). All of the flavouring agents in that group were concluded to be of no safety concern at estimated dietary exposures.

The nine flavouring agents (Nos 2177–2185) in this group are natural components of foods. They have been detected in cucumber, maize, melon, peas, potato, pulasam, apple, banana, guava, raspberry, blackberry, boysenberry, tomato, wine, pork, grape, starfruit, tea, pepino fruit and prickly pear (Estévez et al., 2003; Nijssen, van Ingen-Visscher & Donders, 2011).

1.2 Assessment of dietary exposure

The total annual volumes of production of the nine linear and branched-chain aliphatic, unsaturated, unconjugated alcohols, aldehydes, acids and related esters are approximately 56 kg in Europe, 18 kg in the USA and 6 kg in Japan (European Flavour and Fragrance Association, 2004; Japan Flavor and Fragrance Materials Association, 2005; Gavin, Williams & Hallagan, 2008; International Organization of the Flavor Industry, 2011). Approximately 98% of the total annual volume of production in Europe is accounted for by one flavouring agent in this group, *trans*-3-hexenyl acetate (No. 2180). Two flavouring agents, *trans*-3-nonen-1-ol (No. 2178) and *cis*,*cis*-3,6-nonadienyl acetate (No. 2179), with annual volumes of production of 10 kg and 7 kg, respectively, account for 94% of the volume in the USA. In Japan, all of the volume of production is accounted for by *cis*-3-nonen-1-ol (No. 2177) and *trans*-3-hexenyl acetate (No. 2180), with 1 kg and 5 kg, respectively.

Dietary exposures were estimated for each flavouring agent using both the single portion exposure technique (SPET) and the maximized survey-derived intake (MSDI) method, with the highest values reported in Table 1. The estimated daily dietary exposure is highest for (*Z*)-5-octenyl acetate (No. 2184) (2400 µg, the SPET value obtained from instant tea and coffee). For the other flavouring agents, the estimated daily dietary exposures range from 0.01 to 600 µg, with the SPET yielding the highest estimates.

Annual volumes of production of this group of flavouring agents as well as the daily dietary exposures calculated using both the MSDI method and the SPET are summarized in Table 2.

1.3 Absorption, distribution, metabolism and elimination

Information on the hydrolysis, absorption, distribution, metabolism and elimination of flavouring agents belonging to this group of linear and branched-chain aliphatic, unsaturated, unconjugated alcohols, aldehydes, acids and related esters has previously been described in the reports of the fifty-first and sixty-first meetings (Annex 1, references *137* and *166*).

Table 1. Summary of the results of the safety evaluations of linear and branched-chain aliphatic, unsaturated, unconjugated alcohols, aldehydes, acids and related esters used as flavouring agents[a,b,c]

Flavouring agent	No.	CAS No. and structure	Step A3[1] Does estimated dietary exposure exceed the threshold of concern?	Step A4 Is the flavouring agent or are its metabolites endogenous?	Step A5[e] Adequate margin of exposure for the flavouring agent or a related substance?	Comments on predicted metabolism	Related structure name (No.) and structure (if applicable)	Conclusion based on current estimated dietary exposure
Structural class I								
cis-3-Nonen-1-ol	2177	10340-23-5	No, SPET: 30	NR	NR	Note 1		No safety concern
trans-3-Nonen-1-ol	2178	10339-61-4	No, SPET: 80	NR	NR	Note 1		No safety concern
cis,cis-3,6-Nonadienyl acetate	2179	83334-93-4	No, SPET: 2	NR	NR	Note 2		No safety concern

continued

Table 1 (continued)

Flavouring agent	No.	CAS No. and structure	Step A3[1] Does estimated dietary exposure exceed the threshold of concern?	Step A4 Is the flavouring agent or are its metabolites endogenous?	Step A5[2] Adequate margin of exposure for the flavouring agent or a related substance?	Comments on predicted metabolism	Related structure name (No.) and structure (if applicable)	Conclusion based on current estimated dietary exposure
trans-3-Hexenyl acetate	2180	3681-82-1	No, SPET: 600	NR	NR	Note 2		No safety concern
cis-3-Hexenoic acid	2181	1775-43-5	No, SPET: 600	NR	NR	Note 3		No safety concern
cis-3-Nonenyl acetate	2182	13049-88-2	No, SPET: 2	NR	NR	Note 2		No safety concern
cis-6-Nonenyl acetate	2183	76238-22-7	No, SPET: 2	NR	NR	Note 2		No safety concern

LINEAR/BRANCHED-CHAIN ALIPHATIC, UNSATURATED, UNCONJUGATED ALCOHOLS

(Z)-5-Octenyl acetate	2184	71978-00-2	[structure]	Yes, SPET: 2400	No	Yes. The NOEL of 120 mg/kg bw per day in a 98-day study in rats for the structurally related cis-3-hexenol (No. 315) (Gaunt et al., 1969) is 3000 times the estimated dietary exposure to No. 2184 when used as a flavouring agent.	Note 2	cis-3-Hexenol (No. 315) [structure]	No safety concern
(E)-4-Undecenal	2185	68820-35-9	[structure]	No, SPET: 15	NR	NR	Note 4		No safety concern

CAS, Chemical Abstracts Service; NR, not required for evaluation because dietary exposure to the flavouring agent was determined to be of no safety concern at step A3 of the Procedure

[a] Sixty-two flavouring agents in this group were previously evaluated by the Committee (Annex 1, references 137 and 166).
[b] Step 1: All nine additional flavouring agents in this group (Nos 2177–2185) are in structural class I (Cramer, Ford & Hall, 1978).
[c] Step 2: All of the flavouring agents in this group can be predicted to be metabolized to innocuous products.
[d] The threshold for human dietary exposure for structural class I is 1800 μg/person per day. All dietary exposure values are expressed in μg/day. The dietary exposure value listed represents the highest estimated dietary exposure calculated using either the SPET or the MSDI method. The highest estimated dietary exposures were calculated by the SPET in all cases.
[e] The margins of exposure were calculated based on the estimated dietary exposures calculated using the SPET.

Notes:
1. The primary alcohol is oxidized to the corresponding aldehyde and carboxylic acid, which is completely metabolized in the fatty acid and tricarboxylic acid pathways to carbon dioxide and water.
2. The ester is expected to undergo hydrolysis to the corresponding primary alcohol and carboxylic acid. The alcohol is oxidized to the corresponding aldehyde and carboxylic acid, which is completely metabolized in the fatty acid and tricarboxylic acid pathways to carbon dioxide and water.
3. The carboxylic acid is completely metabolized in the fatty acid and tricarboxylic acid pathways.
4. The aldehyde is oxidized to the corresponding carboxylic acid, which is subsequently oxidized via the fatty acid pathway and the tricarboxylic acid cycle.

Table 2. Annual volumes of production and daily dietary exposures for linear and branched-chain aliphatic, unsaturated, unconjugated alcohols, aldehydes, acids and related esters used as flavouring ingredients in Europe, the USA and Japan

Flavouring agent (No.)	Most recent annual volume of production (kg)[a]	Dietary exposure				Annual volume of consumption via natural occurrence in foods (kg)[d]	Consumption ratio[e]
		MSDI[b]		SPET[c]			
		µg/ day	µg/kg bw per day	µg/ day	µg/kg bw per day		
***cis*-3-Nonen-1-ol (2177)**				30	0.5	+	NA
Europe	0.1	0.01	0.0002				
USA	ND	ND	ND				
Japan	1	0.1	0.002				
***trans*-3-Nonen-1-ol (2178)**				80	1.33	+	NA
Europe	ND	ND	ND				
USA	10	1	0.02				
Japan	ND	ND	ND				
***cis,cis*-3,6-Nonadienyl acetate (2179)**				2	0.03	127	18
Europe	ND	ND	ND				
USA	7	1	0.01				
Japan	ND	ND	ND				
***trans*-3-Hexenyl acetate (2180)**				600	10	135	NA
Europe	55	6	0.1				
USA	ND	ND	ND				
Japan	5	1	0.02				

LINEAR/BRANCHED-CHAIN ALIPHATIC, UNSATURATED, UNCONJUGATED ALCOHOLS

cis-3-Hexenoic acid (2181)	600	10	117	117
Europe	ND	ND		
USA	1	0.1	ND	0.002
Japan	ND	ND		
cis-3-Nonenyl acetate (2182)	2	0.03	15	150
Europe	ND	ND		
USA	0.1	0.01	ND	0.0002
Japan	ND	ND		
cis-6-Nonenyl acetate (2183)	2	0.03	50	500
Europe	0.4	0.04	0.001	
USA	0.1	0.01	ND	0.0002
Japan	ND	ND		
(Z)-5-Octenyl acetate (2184)	2400	40	136	1360
Europe	ND	ND		
USA	0.1	0.01	ND	0.0002
Japan	ND	ND		
(E)-4-Undecenal (2185)	15	0.25	+[f]	NA
Europe	ND	ND		
USA	0.1	0.01	ND	0.0002
Japan	ND	ND		

continued

Table 2 (continued)

Flavouring agent (No.)	Most recent annual volume of production (kg)[a]	Dietary exposure				Annual volume of consumption via natural occurrence in foods (kg)[d]	Consumption ratio[e]
		MSDI[b]		SPET[c]			
		µg/day	µg/kg bw per day	µg/day	µg/kg bw per day		
Total							
Europe	56						
USA	18						
Japan	6						

NA, not applicable; ND, no data reported; +, reported to occur naturally in foods (Nijssen, van Ingen-Visscher & Donders, 2011), but no quantitative data; −, not reported to occur naturally in foods

[a] From European Flavour and Fragrance Association (2004), Japan Flavor and Fragrance Materials Association (2005), Gavin, Williams & Hallagan (2008) and International Organization of the Flavor Industry (2011). Values greater than 0 kg but less than 0.1 kg were reported as 0.1 kg.
[b] MSDI (µg/person per day) calculated as follows:
(annual volume, kg) × (1 × 10^9 µg/kg)/(population × survey correction factor × 365 days), where population (10%, "eaters only") = 32 × 10^6 for Europe, 31 × 10^6 for the USA and 13 × 10^6 for Japan; and where survey correction factor = 0.8 for the surveys in Europe, the USA and Japan, representing the assumption that only 80% of the annual flavour volume was reported in the poundage surveys (European Flavour and Fragrance Association, 2004; Japan Flavor and Fragrance Materials Association, 2005; Gavin, Williams & Hallagan, 2008; International Organization of the Flavor Industry, 2011).
MSDI (µg/kg bw per day) calculated as follows:
(µg/person per day)/body weight, where body weight = 60 kg. Slight variations may occur from rounding.
[c] SPET (µg/person per day) calculated as follows:
(standard food portion, g/day) × (highest usual use level) (International Organization of the Flavor Industry, 2011). The dietary exposure from the single food category leading to the highest dietary exposure from one portion is taken as the SPET estimate.
SPET (µg/kg bw per day) calculated as follows:
(µg/person per day)/body weight, where body weight = 60 kg. Slight variations may occur from rounding.
[d] Quantitative data for the USA reported by Stofberg & Grundschober (1987).
[e] The consumption ratio is calculated as follows:
(annual consumption via natural occurrence in foods in the USA, kg)/(most recent reported volume of production as a flavouring agent in the USA, kg).
[f] Natural occurrence in food reported in Estévez et al. (2003).

The aliphatic esters in this group are predicted to be hydrolysed to the corresponding unsaturated aliphatic alcohol and carboxylic acid in the gastrointestinal tract (Gangolli & Shilling, 1968; Junge & Heymann, 1979; Heymann, 1980). Once formed, the linear and branched-chain unsaturated primary alcohols will be rapidly absorbed (Dawson, Holdsworth & Webb, 1964; Gaillard & Derache, 1965) and oxidized to their corresponding aldehydes and acids. Aliphatic aldehydes and acids are readily absorbed. Once absorbed, the aldehydes are oxidized to their corresponding unsaturated carboxylic acids.

1.4 Application of the Procedure for the Safety Evaluation of Flavouring Agents

Step 1. In applying the Procedure for the Safety Evaluation of Flavouring Agents to the above-mentioned flavouring agents, the Committee assigned all nine flavouring agents (Nos 2177–2185) to structural class I (Cramer, Ford & Hall, 1978).

Step 2. All the flavouring agents in this group are expected to be metabolized to innocuous products. The evaluation of all flavouring agents in this group therefore proceeded via the A-side of the Procedure.

Step A3. The estimated dietary exposures to eight of the flavouring agents in this group are below the threshold of concern (i.e. 1800 µg/person per day for class I). According to the Procedure, the safety of these eight flavouring agents raises no concern at their current estimated dietary exposures. The highest estimated dietary exposure for one flavouring agent (No. 2184) in this group is above the threshold of concern (i.e. 1800 µg/person per day for class I). Accordingly, the evaluation of this flavouring agent proceeded to step A4.

Step A4. The flavouring agent No. 2184 and its metabolites are not endogenous, and therefore its evaluation proceeded to step A5.

Step A5. For (*Z*)-5-octenyl acetate (No. 2184), the no-observed-effect level (NOEL) of 120 mg/kg of body weight (bw) per day for the structurally related *cis*-3-hexenol (No. 315) obtained in a 98-day study in rats (Gaunt et al., 1969) provides a margin of exposure of 3000 in relation to the estimated dietary exposure to No. 2184 (SPET = 2400 µg/day) when used as a flavouring agent.

The Committee concluded that none of the nine additional flavouring agents (Nos 2177–2185) belonging to the group of linear and branched-chain aliphatic, unsaturated, unconjugated alcohols, aldehydes, acids and related esters would pose a safety concern at current estimated dietary exposures.

Table 1 summarizes the evaluations of these additional flavouring agents.

1.5 Consideration of combined intakes from use as flavouring agents

The nine additional flavouring agents in this group of linear and branched-chain aliphatic, unsaturated, unconjugated alcohols, aldehydes, acids and related esters have very low MSDIs (0.01–1 µg/day). Consideration of combined intake is not deemed necessary, because these additional flavouring agents would not contribute significantly to the combined intake of this flavouring group.

1.6 Conclusions

In the previous evaluations of flavouring agents in this group of linear and branched-chain aliphatic, unsaturated, unconjugated alcohols, aldehydes, acids and related esters, studies of acute toxicity, short-term studies of toxicity, long-term studies of toxicity and carcinogenicity, and studies of genotoxicity and reproductive toxicity were available (Annex 1, references *137* and *166*). The results of a subchronic toxicity study on No. 345 and genotoxicity studies on Nos 336 and 348 considered at this meeting support the previous safety evaluations.

The Committee concluded that the nine flavouring agents evaluated at this meeting, which are additions to the group of linear and branched-chain aliphatic, unsaturated, unconjugated alcohols, aldehydes, acids and related esters evaluated previously, would not present safety concerns at the current estimated dietary exposures.

2. RELEVANT BACKGROUND INFORMATION

2.1 Explanation

This monograph summarizes the key data relevant to the safety evaluation of five esters (Nos 2179, 2180 and 2182–2184), two linear unsaturated alcohols (Nos 2177 and 2178), one unsaturated acid (No. 2181) and one aldehyde (No. 2185), which are additions to the group of linear and branched-chain aliphatic, unsaturated, unconjugated alcohols, aldehydes, acids and related esters evaluated previously.

2.2 Additional considerations on dietary exposure

The nine flavouring agents (Nos 2177–2185) in this group are natural components of foods. They have been detected in cucumber, maize, melon, peas, potato, pulasam, apple, banana, guava, raspberry, blackberry, boysenberry, tomato, wine, pork, grape, starfruit, tea, pepino fruit and prickly pear (Estévez et al., 2003; Nijssen, van Ingen-Visscher & Donders, 2011) (see Table 2).

The estimated dietary exposures for each flavouring agent, calculated using either the MSDI method or the SPET, are reported in Table 1. The estimated daily dietary exposure is greatest for (Z)-5-octenyl acetate (No. 2184) (2400 µg, the SPET value obtained from instant tea and coffee). For the other flavouring agents, the estimated daily dietary exposures range from 0.01 to 600 µg, with the SPET yielding the highest estimates.

2.3 Biological data

2.3.1 Biochemical data: hydrolysis, absorption, distribution, metabolism and elimination

No relevant information additional to that available and described in the monographs of the fifty-first and sixty-first meetings (Annex 1, references *137* and *166*) was available on the hydrolysis, absorption, metabolism or elimination of flavouring agents belonging to the group of linear and branched-chain aliphatic, unsaturated, unconjugated alcohols, aldehydes, acids and related esters.

2.3.2 Toxicological studies

(a) Acute toxicity

No relevant information additional to that available and described in the monographs of the fifty-first and sixty-first meetings (Annex 1, references *137* and *166*) was available on the acute toxicity of flavouring agents belonging to the group of linear and branched-chain aliphatic, unsaturated, unconjugated alcohols, aldehydes, acids and related esters.

(b) Short-term studies of toxicity

Relevant information additional to that available and described in the monographs of the fifty-first and sixty-first meetings (Annex 1, references *137* and *166*) was available from short-term studies on the toxicity of flavouring agents belonging to the group of linear and branched-chain aliphatic, unsaturated, unconjugated alcohols, aldehydes, acids and related esters (Table 3).

(i) Ethyl oleate (No. 345)

In a 91-day feeding study, Sprague-Dawley rats (20 of each sex per group) were provided diets delivering 0%, 3.3%, 6.7% or 10% ethyl oleate (No. 345). These dietary concentrations correspond to estimated daily doses of 0, 1650, 3350 and 5000 mg/kg bw (USFDA, 1993). The diets for all groups were adjusted with up to 10% high–oleic acid safflower oil and an additional 5% of soya bean oil to ensure matched calorie content and 15% final fat content. The rats were observed twice daily for mortality and moribund behaviour. The animals were observed weekly for clinical signs of toxicity and motor activity assessments. Body weight and feed consumption were monitored and recorded weekly. Beginning at week 11 and continuing for the next 21 consecutive days, all females were subject to daily vaginal smears to evaluate the estrous cycle. At week 13, ophthalmic examinations were performed for all animals. To assess clinical chemistry parameters and plasma oleate concentrations, blood was drawn on days 30, 60 and 91 of the study. Urine was collected overnight prior to blood collection. On days 29 and 59, the faeces were collected from 10 rats of each sex per group, and on day 89, faeces from all animals were collected overnight for faecal fat evaluation. For males, spermatocyte assessment was performed. All surviving rats were sacrificed and necropsied following the end of the study. At sacrifice, the following organs were weighed: adrenal gland, brain, epididymis, heart, kidney, liver, ovary, pituitary gland, spleen, testis, thymus, thyroid with parathyroid and uterus. A wide selection of tissue types from the control and high-dose groups were preserved in 10% neutral buffered formalin, and slides were prepared. The testes were preserved in Bouin's fixative.

Three unscheduled deaths among male rats (one at 3.3% and two at 6.7% ethyl oleate) were reported prior to the end of the study. There were no pathological findings to relate the deaths to the test diet. Absolute terminal body weight and body

Table 3. Results of short-term studies on the toxicity of linear and branched-chain aliphatic, unsaturated, unconjugated alcohols, aldehydes, acids and related esters

No.	Flavouring agent	Species; sex	No. of test groups[a]/ no. per group[b]	Route	Duration (days)	NOAEL (mg/kg bw per day)	Reference
345	Ethyl oleate	Rat; M, F	3/40	Diet	91	5000[c]	Bookstaff et al. (2004)

F, female; M, male; NOAEL, no-observed-adverse-effect level

[a] Total number of test groups does not include control animals.
[b] Total number per test group includes both male and female animals.
[c] Highest dose tested. The actual NOAEL may be higher.

weight gains for the 6.7% and 10% female groups were significantly reduced compared with controls. Ophthalmic examinations revealed no differences between test animals and controls. Haematological and clinical chemistry results revealed minimally lower calcium and mildly lower inorganic phosphate levels for males fed the 10% ethyl oleate diet. All other parameters measured were unremarkable in both sexes at all dose levels. The majority of the rats showed plasma levels of ethyl oleate of less than 0.02 μg/ml, which was the limit of detection. Urine analysis was comparable across all test groups and the control group. There was a concentration-dependent increase in the faecal fat content; faecal fat content was notably higher among males than among females of equal exposure levels. There were no test material–related effects on total faecal weights or visual observations when test groups were compared with controls. Terminal organ weights revealed a slight increase in relative brain weights for females in the 6.7% and 10% ethyl oleate groups. The authors of the study attributed this increase to the decrease in body weight for these two groups of females. The only test material–related histopathological finding was an increased incidence of hepatocellular vacuolation typical of fat accumulation in controls and the highest-dose group for both sexes. No treatment-related effects were observed for the reproductive parameters: sperm motility, sperm count, sperm morphology and estrous cycle.

Based on a clear lack of adverse effects, the authors concluded that the NOAEL for ethyl oleate in the diet of rats is 5000 mg/kg bw per day, the highest dose tested (Bookstaff et al., 2004).

(c) *Genotoxicity*

Studies of the genotoxicity of linear and branched-chain aliphatic, unsaturated, unconjugated alcohols, aldehydes, acids and related esters used as flavouring agents are summarized in Table 4 and described below.

(i) *In vitro*

In vitro genotoxicity studies have not been reported for the nine flavouring agents evaluated in this group. Supplementary studies have been published for flavouring agents previously evaluated in this group (Annex 1, references *43, 107* and *166*).

Table 4. Studies of the genotoxicity of linear and branched-chain aliphatic, unsaturated, unconjugated alcohols, aldehydes, acids and related esters used as flavouring agents

No.	Flavouring agent	End-point	Test object	Concentration	Results	Reference
In vitro						
336	cis-3-Hexenyl cis-3-hexenoate	Reverse mutation	Salmonella typhimurium TA98, TA100, TA102, TA1535, TA1537	3–5000 µg/plate	Negative[a]	Sokolowski (2000)
349	2,6-Dimethyl-5-heptenal	Reverse mutation	S. typhimurium TA98, TA100, TA1535, TA1537	3–5000 µg/plate	Negative[a]	Sokolowski (2006)
349	2,6-Dimethyl-5-heptenal	Reverse mutation	Escherichia coli WP2uvrA	3–5000 µg/plate	Negative[a]	Sokolowski (2006)

[a] With and without metabolic activation.

cis-3-Hexenyl cis-3-hexenoate (No. 336) was evaluated for its ability to cause mutations in Salmonella typhimurium TA98, TA100, TA102, TA1535 and TA1537, with and without metabolic activation, at 3, 10, 33, 100, 1000, 2500 and 5000 µg/plate. Duplicate assays with plate incorporation and preincubation were performed. No chemical-related mutagenic responses were reported for any of the strains. Under the conditions of the assays, which included positive and negative controls, cis-3-hexenyl cis-3-hexenoate was considered to be non-mutagenic (Sokolowski, 2000).

2,6-Dimethyl-5-heptenal (No. 349) was evaluated for the ability to cause mutations in S. typhimurium strains TA98, TA100, TA1535 and TA1537 and in Escherichia coli WP2uvrA with and without metabolic activation at 3, 10, 33, 100, 1000, 2500 and 5000 µg/plate. Duplicate assays with plate incorporation and preincubation were performed. No chemical-related mutagenic responses were reported for any of the strains. Under the conditions of the assay, which included positive and negative controls, 2,6-dimethyl-5-heptenal was considered to be non-mutagenic (Sokolowski, 2006).

3. REFERENCES

Bookstaff RC et al. (2004). The safety of ethyl oleate is supported by a 91-day feeding study in rats. *Regulatory Toxicology and Pharmacology*, 39(2):202–213.

Cramer GM, Ford RA, Hall RL (1978). Estimation of toxic hazard—a decision tree approach. *Food and Cosmetics Toxicology*, 16:255–276.

Dawson AM, Holdsworth CD, Webb J (1964). Absorption of short chain fatty acids in man. *Proceedings of the Society for Experimental Biology and Medicine*, 117:97–100.

Estévez M et al. (2003). Analysis of volatiles in meat from Iberian pigs and lean pigs after refrigeration and cooking by using SPME-GC-MS. *Journal of Agricultural and Food Chemistry*, 39:1839–1847.

European Flavour and Fragrance Association (2004). European inquiry on volume use. Private communication to the Flavor and Extract Manufacturers Association of the United States, Washington, DC, USA. Submitted to WHO by the International Organization of the Flavor Industry, Brussels, Belgium.

Gaillard D, Derache R (1965). Metabolism of different alcohols, present in alcoholic beverages, in the rat. *Travaux de la Société de pharmacie de Montpellier*, 25(1):51–62.

Gangolli SD, Shilling WH (1968). Hydrolysis of esters by artificial gastric and pancreatic juices. Unpublished report no. 11 submitted to the Flavor and Extract Manufacturers Association of the United States, Washington, DC, USA. Submitted to WHO by the International Organization of the Flavor Industry, Brussels, Belgium.

Gaunt IF et al. (1969). Acute (rat and mouse) and short-term (rat) toxicity studies on *cis*-3-hexen-1-ol. *Food and Cosmetics Toxicology*, 7(5):451–459.

Gavin CL, Williams MC, Hallagan JB (2008). *Flavor and Extract Manufacturers Association of the United States 2005 poundage and technical effects update survey.* Washington, DC, USA, Flavor and Extract Manufacturers Association of the United States.

Heymann E (1980). Chapter 16: Carboxylesterases and amidases. In: Jakobi WB, ed. *Enzymatic basis of detoxification. Vol. II.* New York, NY, USA, Academic Press, pp. 291–323.

International Organization of the Flavor Industry (2011). Interim inquiry on volume use and added use levels for flavoring agents to be presented at the 76th JECFA meeting. Private communication to the Flavor and Extract Manufacturers Association of the United States, Washington, DC, USA. Submitted to WHO by the International Organization of the Flavor Industry, Brussels, Belgium.

Japan Flavor and Fragrance Materials Association (2005). Japanese inquiry on volume use. Private communication to the Flavor and Extract Manufacturers Association of the United States, Washington, DC, USA. Submitted to WHO by the International Organization of the Flavor Industry, Brussels, Belgium.

Junge W, Heymann E (1979). Characterization of the isoenzymes of pig-liver esterase. 2. Kinetic studies. *European Journal of Biochemistry*, 95(3):519–525.

Nijssen LM, van Ingen-Visscher CA, Donders JJH (2011). Volatile Compounds in Food (VCF) database, version 13.1. Zeist, the Netherlands, TNO Triskelion (http://www.vcf-online.nl/VcfHome.cfm).

Sokolowski A (2000). *Salmonella typhimurium* reverse mutation assay with hexenyl-3-*cis* hexenoate. Unpublished report no. 659903 from RCC Cytotest Cell Research GmbH, Rossdorf, Germany. Submitted to WHO by the International Organization of the Flavor Industry, Brussels, Belgium.

Sokolowski A (2006). *Salmonella typhimurium* and *Escherichia coli* reverse mutation assay with melonal. Unpublished report no. 910500 from RCC Cytotest Cell Research GmbH, Rossdorf, Germany. Submitted to WHO by the International Organization of the Flavor Industry, Brussels, Belgium.

Stofberg J, Grundschober F (1987). Consumption ratio and food predominance of flavoring materials. *Perfumer & Flavorist*, 12:27–68.

USFDA (1993). Priority-based assessment of food additives (PAFA) database. United States Food and Drug Administration, Center for Food Safety and Applied Nutrition, p. 58.

MISCELLANEOUS NITROGEN-CONTAINING SUBSTANCES (addendum)

First draft prepared by

S.M.F. Jeurissen[1], M. DiNovi[2], A. Mattia[2] and A.G. Renwick[3]

[1] Centre for Substances and Integrated Risk Assessment, National Institute for Public Health and the Environment, Bilthoven, the Netherlands
[2] Center for Food Safety and Applied Nutrition, Food and Drug Administration, College Park, Maryland, United States of America (USA)
[3] School of Medicine, University of Southampton, Southampton, England

1. Evaluation 185
   1.1 Introduction 185
   1.2 Assessment of dietary exposure 186
   1.3 Absorption, distribution, metabolism and elimination 186
   1.4 Application of the Procedure for the Safety Evaluation of
       Flavouring Agents 186
   1.5 Consideration of combined intakes from use as flavouring
       agents 190
   1.6 Conclusions 191
2. Relevant background information 191
   2.1 Explanation 191
   2.2 Additional considerations on dietary exposure 191
   2.3 Biological data 191
       2.3.1 Biochemical data: absorption, distribution, metabolism
             and elimination 191
             (a) Imidazolidines 191
       2.3.2 Toxicological studies 193
             (a) Acute toxicity 193
             (b) Short-term studies of toxicity 193
             (c) Genotoxicity 196
3. References 199

1. EVALUATION

1.1 Introduction

The Committee evaluated two additional flavouring agents belonging to the group of miscellaneous nitrogen-containing substances. The additional flavouring agents were imidazolidines (Nos 2161 and 2162) that were initially submitted as additional flavouring agents to the group of pyridine, pyrrole and quinoline derivatives. The evaluations were conducted according to the Procedure for the Safety Evaluation of Flavouring Agents (see Figure 1, Introduction) (Annex 1, reference *131*). Neither of these flavouring agents has previously been evaluated by the Committee. Both flavouring agents in this group that were evaluated at this meeting are reported to be flavour modifiers.

The Committee previously evaluated 16 other members of this group of flavouring agents at its sixty-fifth meeting (Annex 1, reference *178*). The Committee concluded that the use of these 16 flavouring agents would not present a safety concern at estimated dietary exposures. For 10 flavouring agents, the evaluation was conditional, because the estimated exposures were based on anticipated annual volumes of production. At its sixty-ninth meeting (Annex 1, reference *190*), the actual volumes of production for these flavouring agents were provided, and the Committee confirmed that these 10 flavouring agents were of no safety concern based on estimated dietary exposures. The Committee also evaluated 14 other members of this group of flavouring agents at its sixty-ninth meeting and concluded that all 14 flavouring agents in that group were of no safety concern at estimated dietary exposures.

1.2 Assessment of dietary exposure

The total annual volume of production of the two miscellaneous nitrogen-containing substances is 2.2 kg in the USA, with no reported data from Europe or Japan (European Flavour and Fragrance Association, 2004; Japan Flavor and Fragrance Materials Association, 2005; Gavin, Williams & Hallagan, 2008; International Organization of the Flavor Industry, 2011). 3-(1-((3,5-Dimethylisoxazol-4-yl)methyl)-1H-pyrazol-4-yl)-1-(3-hydroxybenzyl)-imidazolidine-2,4-dione (No. 2161), with 2 kg, accounts for approximately 90% of the total annual volume of production.

Dietary exposure estimates were made using the maximized survey-derived intake (MSDI) method and the single portion exposure technique (SPET), with the highest values (calculated using the SPET) reported in Table 1. The highest estimated daily dietary exposure is 4000 μg for each flavouring agent (the SPET value obtained from gelatines and puddings and from reconstituted vegetables).

Annual volumes of production of this group of flavouring agents as well as the daily dietary exposures calculated using both the MSDI method and the SPET are summarized in Table 2.

1.3 Absorption, distribution, metabolism and elimination

Information on the absorption, distribution, metabolism and elimination of the flavouring agents belonging to the group of miscellaneous nitrogen-containing substances has previously been described in the monographs of the sixty-fifth and sixty-ninth meetings (Annex 1, references *179* and *191*).

The imidazolidines under evaluation are rapidly absorbed, metabolized and excreted. They undergo hydroxylation of the benzyl ring as well as of the dimethylisoxazole group and conjugation with sulfates or glucuronides.

1.4 Application of the Procedure for the Safety Evaluation of Flavouring Agents

Step 1. In applying the Procedure for the Safety Evaluation of Flavouring Agents to the two flavouring agents in this group of miscellaneous nitrogen-containing substances, the Committee assigned both flavouring agents to structural class III (Cramer, Ford & Hall, 1978).

Table 1. Summary of the results of the safety evaluations of miscellaneous nitrogen-containing substances used as flavouring agents[a,b,c]

Flavouring agent	No.	CAS No. and structure	Step B3[1] Does estimated dietary exposure exceed the threshold of concern?	Follow-on from step B3[e] Are additional data available for flavouring agent with an estimated dietary exposure exceeding the threshold of concern?	Comments on predicted metabolism	Conclusion based on current estimated dietary exposure
Structural class III						
3-(1-((3,5-Dimethylisoxazol-4-yl)methyl)-1H-pyrazol-4-yl)-1-(3-hydroxybenzyl)-imidazolidine-2,4-dione	2161	1119831-25-2	Yes, SPET: 4000	Yes. The NOAEL of 100 mg/kg bw per day from a 91-day study in rats (Chase, 2010a) is 1500 times the estimated daily dietary exposure to No. 2161 when used as a flavouring agent.	Note 1	No safety concern
3-(1-((3,5-Dimethylisoxazol-4-yl)methyl)-1H-pyrazol-4-yl)-1-(3-hydroxybenzyl)-5,5-dimethylimidazolidine-2,4-dione	2162	1217341-48-4	Yes, SPET: 4000	Yes. The NOAEL of 100 mg/kg bw per day from a 28-day study in rats (Chase, 2010b) is 1500 times the estimated daily dietary exposure to No. 2162 when used as a flavouring agent.	Note 1	No safety concern

bw, body weight; CAS, Chemical Abstracts Service; NOAEL, no-observed-adverse-effect level

[a] Thirty flavouring agents belonging to the group of miscellaneous nitrogen-containing substances were previously evaluated by the Committee at its sixty-fifth and sixty-ninth meetings (Annex 1, references 178 and 190).
[b] Step 1: Both flavouring agents in this group are in structural class III.
[c] Step 2: Neither of the flavouring agents in this group can be predicted to be metabolized to innocuous products.

d The threshold for human dietary exposure for structural class III is 90 µg/person per day. All dietary exposure values are expressed in µg/day. The dietary exposure value listed represents the highest estimated dietary exposure calculated using either the SPET or the MSDI method. The highest estimated dietary exposures were calculated by the SPET in both cases.
e The margins of exposure were calculated based on the estimated dietary exposures calculated using the SPET.

Note:
1. Ring hydroxylation and conjugation and elimination in the urine.

Table 2. Annual volumes of production and daily dietary exposures for miscellaneous nitrogen-containing substances used as flavouring agents in Europe, the USA and Japan

Flavouring agent (No.)	Most recent annual volume of production (kg)[a]	Dietary exposure				Natural occurrence in foods
		MSDI[b]		SPET[c]		
		µg/day	µg/kg bw per day	µg/day	µg/kg bw per day	
3-(1-((3,5-Dimethylisoxazol-4-yl)methyl)-1H-pyrazol-4-yl)-1-(3-hydroxybenzyl)-imidazolidine-2,4-dione (2161)				4000	67	–
Europe	ND	ND	ND			
USA	2	0.2	0.004			
Japan	ND	ND	ND			
3-(1-((3,5-Dimethylisoxazol-4-yl)methyl)-1H-pyrazol-4-yl)-1-(3-hydroxybenzyl)-5,5-dimethylimidazolidine-2,4-dione (2162)				4000	67	–
Europe	ND	ND	ND			
USA	0.2	0.02	0.0004			
Japan	ND	ND	ND			
Total						
Europe	ND					
USA	2.2					
Japan	ND					

bw, body weight; ND, no data reported; –, not reported to occur naturally in foods

[a] From European Flavour and Fragrance Association (2004), Japan Flavor and Fragrance Materials Association (2005), Gavin, Williams & Hallagan (2008) and International Organization of the Flavor Industry (2011). Values greater than 0 kg but less than 0.1 kg were reported as 0.1 kg.

[b] MSDI (µg/person per day) calculated as follows:
(annual volume, kg) × (1 × 10^9 µg/kg)/(population × survey correction factor × 365 days), where population (10%, "eaters only") = 32 × 10^6 for Europe, 31 × 10^6 for the USA and 13 × 10^6 for Japan; and where survey correction factor = 0.8 for the surveys in Europe, the USA and Japan, representing the assumption that only 80% of the annual flavour volume was reported in the poundage surveys (European Flavour and Fragrance Association, 2004; Japan Flavor and Fragrance Materials Association, 2005; Gavin, Williams & Hallagan, 2008; International Organization of the Flavor Industry, 2011).
MSDI (µg/kg bw per day) calculated as follows:
(µg/person per day)/body weight, where body weight = 60 kg. Slight variations may occur from rounding.

[c] SPET (µg/person per day) calculated as follows:
(standard food portion, g/day) × (highest usual use level) (International Organization of the Flavor Industry, 2011). The dietary exposure from the single food category leading to the highest dietary exposure from one portion is taken as the SPET estimate.
SPET (µg/kg bw per day) calculated as follows:
(µg/person per day)/body weight, where body weight = 60 kg. Slight variations may occur from rounding.

Step 2. The two flavouring agents in this group cannot be predicted to be metabolized to innocuous products. Therefore, the evaluation of these flavouring agents proceeded via the B-side of the Procedure.

Step B3. For both flavouring agents, the highest estimated dietary exposure is 4000 μg/day (calculated using the SPET) and above the threshold of concern (i.e. 90 μg/person per day for class III). Therefore, additional data are necessary for the evaluation of these flavouring agents.

Consideration of flavouring agents with high exposure evaluated via the B-side of the decision-tree:

In accordance with the Procedure, additional data were evaluated for 3-(1-((3,5-dimethylisoxazol-4-yl)methyl)-1H-pyrazol-4-yl)-1-(3-hydroxybenzyl)-imidazolidine-2,4-dione (No. 2161) and 3-(1-((3,5-dimethylisoxazol-4-yl)methyl)-1H-pyrazol-4-yl)-1-(3-hydroxybenzyl)-5,5-dimethylimidazolidine-2,4-dione (No. 2162), as their estimated dietary exposures exceeded the threshold of concern for structural class III (90 μg/person per day).

For No. 2161, data on kinetics, oral 28- and 91-day studies of toxicity in rats and in vitro and in vivo studies of genotoxicity are available. For No. 2162, data on kinetics, an oral 28-day study of toxicity in rats and in vitro studies of genotoxicity are available.

Nos 2161 and 2162 undergo hydroxylation of the benzyl ring as well as of the dimethylisoxazole group and conjugation with sulfate or glucuronic acid (Liu, 2010a,b; Zhu, 2010). No. 2162 is rapidly excreted. Twenty-four hours after oral administration of 10–100 mg/kg of body weight (bw) and 4–8 hours after intravenous administration of 1 mg/kg bw, blood levels of No. 2162 were below the limit of quantification (Taoudi, 2010).

No. 2161 was negative in a bacterial reverse mutation test with and without S9, an in vitro chromosomal aberration test in human lymphocytes and an in vivo micronucleus test in mice (Barfield, 2010; May, 2010a; Woods, 2010). No. 2162 was negative in two bacterial reverse mutation assays with and without S9 (May, 2009, 2010b).

The no-observed-adverse-effect level (NOAEL) of 100 mg/kg bw per day in a 91-day study in rats (Chase, 2010a) provides a margin of exposure of 1500 for No. 2161 (SPET = 4000 μg/day) when used as a flavouring agent. The NOAEL of 100 mg/kg bw per day in a 28-day study in rats (Chase, 2010b) provides a margin of exposure of 1500 for No. 2162 (SPET = 4000 μg/day) when used as a flavouring agent. The Committee therefore concluded that these two flavouring agents are not of safety concern at current estimated dietary exposures.

Table 1 summarizes the evaluations of the two miscellaneous nitrogen-containing substances (Nos 2161 and 2162) in this group.

1.5 Consideration of combined intakes from use as flavouring agents

Both additional flavouring agents in this group of miscellaneous nitrogen-containing substances have MSDI values less than 20% of the threshold of concern for structural class III. Consideration of combined intakes is therefore not deemed necessary.

1.6 Conclusions

In the previous evaluation of flavouring agents in the group of miscellaneous nitrogen-containing substances, studies of acute toxicity, short-term studies of toxicity, long-term studies of toxicity and carcinogenicity, and studies of genotoxicity were available (Annex 1, references *179* and *191*). The majority of these data were on alkyl isothiocyanates. The additional toxicity data available for this evaluation on the two imidazolidines do not raise safety concerns.

The Committee concluded that the two flavouring agents evaluated at this meeting (Nos 2161 and 2162), which are additions to the group of miscellaneous nitrogen-containing substances evaluated previously, would not give rise to safety concerns at current estimated dietary exposures.

2. RELEVANT BACKGROUND INFORMATION

2.1 Explanation

This monograph summarizes key aspects relevant to the safety evaluation of two miscellaneous nitrogen-containing substances, which are additions to a group of 30 flavouring agents evaluated previously by the Committee at its sixty-fifth and sixty-ninth meetings (Annex 1, references *178* and *190*).

2.2 Additional considerations on dietary exposure

Annual volumes of production and dietary exposures estimated using both the MSDI method and the SPET for each flavouring agent are reported in Table 2.

Both flavouring agents evaluated at the current meeting do not occur naturally in food (Table 2).

2.3 Biological data

2.3.1 Biochemical data: absorption, distribution, metabolism and elimination

Information on the hydrolysis, absorption, distribution, metabolism and elimination of flavouring agents belonging to the group of miscellaneous nitrogen-containing substances has been described in the monographs of the sixty-fifth and sixty-ninth meetings (Annex 1, references *178* and *190*). Additional data on the imidazolidines are summarized below.

(a) Imidazolidines

3-(1-((3,5-Dimethylisoxazol-4-yl)methyl)-1H-pyrazol-4-yl)-1-(3-hydroxybenzyl)-imidazolidine-2,4-dione (No. 2161) (1, 10 and 50 µmol/l) was incubated with pooled rat and pooled human liver microsomes (0.5 mg/ml) for up to 120 minutes. Samples were analysed using liquid chromatography with mass spectrometry. Rat liver microsomes formed one metabolite that was monohydroxylated at an undetermined position of the dimethylisoxazole group. Human liver microsomes formed the 3,4-dihydroxybenzyl and 2,5-dihydroxybenzyl metabolites and at least two other identifiable metabolites that were monohydroxylated at undetermined positions of the dimethylisoxazole group (Zhu, 2010).

Groups of four male Sprague-Dawley rats were given a single oral dose of 3-(1-((3,5-dimethylisoxazol-4-yl)methyl)-1H-pyrazol-4-yl)-1-(3-hydroxybenzyl)-imidazolidine-2,4-dione (No. 2161) at 100 mg/kg bw, and blood samples were collected at 0.5, 2 and 4 hours post-dosing. The maximum concentration of the parent compound in rat plasma was observed at the 0.5-hour time point in all test animals (mean plasma concentration was 105.4 ng/ml). The mean concentrations of a glucuronide derivative and a sulfate derivative of the parent compound were in the range of 32–166 ng/ml and 334–555 ng/ml, respectively, during the 4-hour sampling period. The mean concentration of a monohydroxylated 2,3-dihydroxybenzyl metabolite was in the range of 12.6–18.9 ng/ml. Additionally, metabolites in which the isoxazole group of the parent compound has undergone oxidation and sulfate conjugates of the monohydroxylated metabolites were observed, but not quantified. Samples were also analysed for the presence of 3,4-dihydroxybenzyl and 2,5-dihydroxybenzyl metabolites, but these were either absent or below the limit of quantification (Liu, 2010a).

As part of a 90-day study of toxicity, plasma concentrations of 3-(1-((3,5-dimethylisoxazol-4-yl)methyl)-1H-pyrazol-4-yl)-1-(3-hydroxybenzyl)-imidazolidine-2,4-dione (No. 2161) were determined in blood samples collected from Crl:CD(SD) rats given No. 2161 at dietary concentrations equal to doses of 0, 10, 30 or 100 mg/kg bw per day (Chase, 2010a). The samples were taken from three male and three female rats from each dose group at six time points over a 24-hour period on day 7, at week 6 and at week 13 and were analysed using liquid chromatography with tandem mass spectrometry. At week 13, maximum plasma concentrations (C_{max}) were 54.7 ng/ml and 107 ng/ml for males and females of the high-dose group, respectively, and areas under the mean plasma concentration–time curve estimated over a 24-hour interval (AUC_{24h}) were 925 ng·h/ml and 1670 ng·h/ml for males and females of the high-dose group, respectively. The C_{max} and AUC_{24h} were characterized by non-linear (dose-dependent) kinetics. Increasing the dose above 30 mg/kg bw per day was considered likely to result in a lower systemic exposure in males, but a higher systemic exposure in females, than would be predicted from a linear relationship, but the evidence of non-proportionality did not attain statistical significance. The study also provided evidence that, in general, there were no differences in the systemic exposures of male and female rats to No. 2161 at the low and middle doses, but that at the high dose (100 mg/kg bw per day), the systemic exposure of females was higher than that of males. There were no time-related differences in systemic exposure in either sex (Chase, 2010a).

To investigate the in vivo pharmacokinetics and metabolism of 3-(1-((3,5-dimethylisoxazol-4-yl)methyl)-1H-pyrazol-4-yl)-1-(3-hydroxybenzyl)-5,5-dimethylimidazolidine-2,4-dione (No. 2162), male Sprague-Dawley rats were administered a single dose of 30 mg/kg bw via gavage. Blood samples were collected at five time points (0.5, 2, 4, 8 and 24 hours) after dose administration. The maximum concentration of the parent compound in rat plasma (C_{max}) was 5.2

± 7.1 ng/ml. The time to C_{max} (T_{max}) was estimated to occur between 2 and 8 hours. The primary metabolites identified were glucuronide and sulfate derivatives of the parent compound, as well as a sulfate derivative of a monohydroxylated metabolite. The metabolic profile of this compound shows a similar biotransformation pathway to that of No. 2161, although plasma concentrations of the parent compound are significantly less than those seen with No. 2161 (Liu, 2010b).

Groups of three male rats were administered 3-(1-((3,5-dimethylisoxazol-4-yl)methyl)-1H-pyrazol-4-yl)-1-(3-hydroxybenzyl)-5,5-dimethylimidazolidine-2,4-dione (No. 2162) by oral gavage or intravenously at single doses of 10, 30 or 100 mg/kg bw (oral gavage) or 1 mg/kg bw (intravenously). Blood samples were collected periodically for 48 hours, and analyses by liquid chromatography with tandem mass spectrometry were performed to determine concentrations of No. 2162. Twenty-four hours after oral administration and 4–8 hours after intravenous administration, blood levels of No. 2162 were below the limit of quantification (Taoudi, 2010).

2.3.2 Toxicological studies

(a) Acute toxicity

An oral median lethal dose (LD_{50} value) has been published for one of the two flavouring agents in this group. For 3-(1-((3,5-dimethylisoxazol-4-yl)methyl)-1H-pyrazol-4-yl)-1-(3-hydroxybenzyl)imidazolidine-2,4-dione (No. 2161), an oral LD_{50} value of greater than 50 mg/kg bw (the highest dose tested) was reported in female rats (Arulnesan, 2008).

(b) Short-term studies of toxicity

Short-term studies of the toxicity of miscellaneous nitrogen-containing substances used as flavouring agents are summarized in Table 3 and described below.

(i) 3-(1-((3,5-Dimethylisoxazol-4-yl)methyl)-1H-pyrazol-4-yl)-1-(3-hydroxybenzyl)-imidazolidine-2,4-dione (No. 2161)

In a 28-day range-finding study, groups of five male and five female Crl:CD(SD) rats were administered 3-(1-((3,5-dimethylisoxazol-4-yl)methyl)-1H-pyrazol-4-yl)-1-(3-hydroxybenzyl)-imidazolidine-2,4-dione at dietary concentrations equal to doses of 10, 30 and 100 mg/kg bw per day for males and 10, 30 and 110 mg/kg bw per day for females. The study generally followed good laboratory practice (GLP) principles. Observations included clinical condition, body weight, feed consumption, haematology, blood chemistry, organ weight, detailed macropathology and histopathological examinations of the liver.

Effects observed included slight reductions of haematocrit (−6%), haemoglobin concentration (−6%) and red blood cell count (−4%) in females and reduced plasma calcium concentrations in males (2.63 mmol/l in all treatment groups compared with 2.75 mmol/l in the controls) and in high-dose females (2.60 mmol/l compared with 2.71 mmol/l in the controls). Also, in female rats, alanine aminotransferase, aspartate aminotransferase and phosphorus levels were reduced (up to −20%, −26% and −17%, respectively) and glucose levels were increased (up

Table 3. Results of short-term studies of the oral toxicity of miscellaneous nitrogen-containing substances used as flavouring agents

No.	Flavouring agent	Species; sex	No. of test groups[a] / no per group[b]	Route	Duration (days)	NOAEL (mg/ kg bw per day)	Reference
2161	3-(1-((3,5-Dimethylisoxazol-4-yl)methyl)-1H-pyrazol-4-yl)-1-(3-hydroxybenzyl)-imidazolidine-2,4-dione	Rats; M, F	3/10	Diet	28	100[c]	Chase (2009)
2161	3-(1-((3,5-Dimethylisoxazol-4-yl)methyl)-1H-pyrazol-4-yl)-1-(3-hydroxybenzyl)-imidazolidine-2,4-dione	Rats; M, F	3/40	Diet	91	100[c]	Chase (2010a)
2162	3-(1-((3,5-Dimethylisoxazol-4-yl)methyl)-1H-pyrazol-4-yl)-1-(3-hydroxybenzyl)-5,5-dimethylimidazolidine-2,4-dione	Rats; M, F	3/20	Diet	28	100[c]	Chase (2010b)

F, female; M, male

[a] Total number of test groups does not include control animals.
[b] Total number per test group includes both male and female animals.
[c] Highest dose tested.

to 28%) at all dose levels. All haematological and biochemical findings were minor, observed in one sex only and/or within historical control ranges. Thyroid weight was slightly increased in males (0.019, 0.016 and 0.019 g in the respective treatment groups compared with 0.015 g in controls), but no dose–response relationship was observed. Macroscopic and microscopic (liver only) examinations did not reveal treatment-related effects.

The NOAEL was 100 mg/kg bw per day, the highest dose tested (Chase, 2009).

In the subsequent 91-day study of toxicity, groups of 20 male and 20 female Crl:CD(SD) rats were given 3-(1-((3,5-dimethylisoxazol-4-yl)methyl)-1H-pyrazol-4-yl)-1-(3-hydroxybenzyl)-imidazolidine-2,4-dione at dietary concentrations equal to doses of 0, 10, 30 and 100 mg/kg bw per day. The highest dose was based on the 28-day range-finding study. The study was performed according to Organisation for Economic Co-operation and Development (OECD) Test Guideline 408 (Repeated Dose 90-Day Oral Toxicity Study in Rodents). Clinical condition, detailed physical examination and arena observations, sensory reactivity, grip strength, motor activity, body weight, feed consumption, ophthalmic examination, haematology, blood chemistry, urine analysis, toxicokinetics, organ weight, macropathology and histopathological investigations were recorded.

No treatment-related effects were observed on mortality, clinical symptoms, sensory reactivity, grip strength, body weight, feed consumption, ophthalmoscopy or urine analysis. Motor activity scores were slightly but significantly increased for high-dose males. Monocyte counts were significantly increased in high-dose males at all time points, whereas in females, monocyte counts were significantly reduced in the mid- and high-dose groups on day 14 and in all dose groups in week 7. Haematocrit and haemoglobin levels were significantly increased in high-dose males and platelet counts were significantly lower in males of all doses in week 13. These effects were not observed in females. Glucose concentrations were increased in male rats in week 7 (significant only in high-dose males) and week 13 (no dose–response relationship). These changes were minor, and the majority of the individual values were within the background range. As a consequence, these differences were not considered to be toxicologically significant. Haematology and clinical chemistry revealed a few additional incidental findings, including changes in prothrombin time, neutrophil count, and urea, potassium, chloride and phosphorus levels. However, these were observed in one sex only and at intermediate time points (after 14 days and/or in week 7 only) and were not considered toxicologically relevant. No significant changes in organ weights were observed, and macroscopic and microscopic examinations did not reveal treatment-related effects.

The NOAEL of this study was 100 mg/kg bw per day, the highest dose tested (Chase, 2010a).

(ii) *3-(1-((3,5-Dimethylisoxazol-4-yl)methyl)-1H-pyrazol-4-yl)-1-(3-hydroxybenzyl)-5,5-dimethylimidazolidine-2,4-dione (No. 2162)*

Groups of 10 male and 10 female Crl:CD(SD) rats were administered 3-(1-((3,5-dimethylisoxazol-4-yl)methyl)-1H-pyrazol-4-yl)-1-(3-hydroxybenzyl)-5,5-dimethylimidazolidine-2,4-dione at dietary concentrations equal to doses of 10, 30 and 100 mg/kg bw per day for males and 10, 30 and 110 mg/kg bw per day for females for 28 days. The study was performed according to OECD Test Guideline 407. Observations included clinical condition, detailed physical examinations and arena observations, sensory reactivity, grip strength, motor activity, body weight, feed consumption, ophthalmic examination, haematology, blood chemistry, urine analysis, organ weights and macroscopic and microscopic investigations.

No premature deaths and no clinical signs related to treatment occurred. Functional observational tests, urine analysis and ophthalmoscopy did not reveal treatment-related effects. Body weight gain was significantly reduced between day 1 and day 8 in males of all treatment groups and between day 15 and day 22 in high-dose females. Feed consumption was significantly reduced in high-dose males between day 1 and day 3. These incidental findings were minor and were variable during the study. Therefore, they were not considered to be of toxicological relevance. Haematology revealed a minimal, but statistically significant, prolongation in prothrombin times in high-dose male rats (15.1 seconds compared with 14.7 seconds in controls). As the difference was minor and not seen in female rats and as all individual values were within the background range, it was considered to represent normal biological variation. Biochemical examination of blood plasma revealed reduced phosphorus concentrations in males (up to −6%) and females (up to −10%) in all dose groups, which were statistically significant in the mid- and high-dose groups. The majority of the individual values were also slightly below historical control values. Calcium levels were reduced, although not statistically significantly, in male rats (up to −2%) and female rats (up to −6%) in all dose groups. In addition, chloride levels were significantly increased in female rats (up to +3%) in all dose groups. Most of the individual values were slightly outside the background range. Alkaline phosphatase (−18%) and aspartate aminotransferase (−11%) activities were slightly lower in males of the high-dose group, and glucose level was elevated (+17%) in males of the mid- and high-dose groups. These findings were not observed in females. In males, slightly but significantly reduced adrenal weights were observed in the high-dose group (0.049 g compared with 0.055 g in controls). Macroscopic and histopathological examinations did not reveal any treatment-related abnormalities. A slight decrease in the incidence of corticomedullary mineralization was seen in the kidneys of females of the high-dose group. The author indicated that as this is a common background finding in female rats of this age, which occurs at a variable incidence, this alteration was considered to have arisen spontaneously and to not be caused by treatment.

The Committee concluded that the biochemical effects observed are not considered to be toxicologically relevant given the absence of any related microscopic findings. The NOAEL was 100 mg/kg bw per day, the highest dose tested (Chase, 2010b).

(c)   Genotoxicity

Studies of genotoxicity in vitro and in vivo have been reported for both miscellaneous nitrogen-containing substances in this group. The results of these studies are summarized in Table 4 and described below.

Table 4. Studies of genotoxicity in vitro and in vivo with miscellaneous nitrogen-containing substances used as flavouring agents

No.	Flavouring agent	End-point	Test system	Concentration	Results	Reference
In vitro						
2161	3-(1-(((3,5-Dimethylisoxazol-4-yl)methyl)-1H-pyrazol-4-yl)-1-(3-hydroxybenzyl))imidazolidine-2,4-dione	Reverse mutation	*Salmonella typhimurium* TA98, TA100	62–5000 µg/plate, ±S9	Negative[a]	Zhang (2008)
2161	3-(1-(((3,5-Dimethylisoxazol-4-yl)methyl)-1H-pyrazol-4-yl)-1-(3-hydroxybenzyl))imidazolidine-2,4-dione	Reverse mutation	*S. typhimurium* TA98, TA100, TA1535, TA1537; *Escherichia coli* WP2uvrA (pKM101)	5–5000 µg/plate, ±S9	Negative[b]	May (2010a)
2161	3-(1-(((3,5-Dimethylisoxazol-4-yl)methyl)-1H-pyrazol-4-yl)-1-(3-hydroxybenzyl))imidazolidine-2,4-dione	Chromosomal aberration	Human lymphocytes	1st experiment: 297, 494 and 824 µg/ml, ±S9 2nd experiment: 200, 400 and 600 µg/ml, −S9 1100, 1200 and 1300 µg/ml, +S9	Negative[c]	Woods (2010)
2162	3-(1-((3,5-Dimethylisoxazol-4-yl)methyl)-1H-pyrazol-4-yl)-1-(3-hydroxybenzyl)-5,5-dimethylimidazolidine-2,4-dione	Reverse mutation	*S. typhimurium* TA98, TA100	5–5000 µg/plate, ±S9	Negative[a]	May (2009)
2162	3-(1-((3,5-Dimethylisoxazol-4-yl)methyl)-1H-pyrazol-4-yl)-1-(3-hydroxybenzyl)-5,5-dimethylimidazolidine-2,4-dione	Reverse mutation	*S. typhimurium* TA98, TA100, TA1535, TA1537; *E. coli* WP2uvrA (pKM101)	5–5000 µg/plate, ±S9	Negative[b]	May (2010b)

continued

Table 4 (continued)

No.	Flavouring agent	End-point	Test system	Concentration	Results	Reference
In vivo						
2161	3-(1-((3,5-Dimethylisoxazol-4-yl)methyl)-1H-pyrazol-4-yl)-1-(3-hydroxybenzyl)imidazolidine-2,4-dione	Micronucleus induction	Mice; M	500, 1000 and 2000 mg/kg bw per day[d]	Negative	Barfield (2010)

M, male; S9, 9000 × g supernatant fraction of rat liver homogenate

[a] Plate incorporation method. Minimal bacterial toxicity was reported at 5000 µg/plate.
[b] Two independent experiments using the preincubation method and the plate incorporation method, respectively.
[c] Cells were analysed 18 hours after 3 hours of treatment, except for cells in the second experiment with metabolic activation, which were analysed after 21 hours of treatment.
[d] The test compound was administered via gavage for 2 subsequent days. Bone marrow samples were collected 24 hours after administration of the second dose. No signs of toxicity were observed at the highest dose level.

(i)   In vitro

No evidence of mutagenicity was observed in bacterial reverse mutation assays when 3-(1-((3,5-dimethylisoxazol-4-yl)methyl)-1H-pyrazol-4-yl)-1-(3-hydroxybenzyl)-imidazolidine-2,4-dione (No. 2161) and 3-(1-((3,5-dimethylisoxazol-4-yl)methyl)-1H-pyrazol-4-yl)-1-(3-hydroxybenzyl)-5,5-dimethylimidazolidine-2,4-dione (No.2162) were incubated with *Salmonella typhimurium* strains TA98, TA100, TA1535 and TA1537 and *Escherichia coli* WP2*uvrA*, with or without metabolic activation, at concentrations up to 5000 µg/plate (Zhang, 2008; May, 2009, 2010a,b). These assays were performed according to OECD Test Guideline 471 (Bacterial Reverse Mutation Test).

Negative results were obtained in a chromosomal aberration assay in human lymphocyte cells with 3-(1-((3,5-dimethylisoxazol-4-yl)methyl)-1H-pyrazol-4-yl)-1-(3-hydroxybenzyl)imidazolidine-2,4-dione (No. 2161) (Woods, 2010). This study was performed according to OECD Test Guideline No. 473 (In Vitro Mammalian Chromosome Aberration Test).

(ii)   In vivo

No genotoxic potential was demonstrated in a mouse micronucleus assay with 3-(1-((3,5-dimethylisoxazol-4-yl)methyl)-1H-pyrazol-4-yl)-1-(3-hydroxybenzyl)-imidazolidine-2,4-dione (No. 2161) (Barfield, 2010). The study was performed according to OECD Test Guideline 474 (Mammalian Erythrocyte Micronucleus Test).

(iii)   Conclusion

In the previous evaluations of this group, numerous studies on isothiocyanates were evaluated. Based on these studies, it was concluded by the Committee at its sixty-ninth meeting that at low levels, it is likely that isothiocyanates exhibit limited genotoxic activity (Annex 1, reference *191*). The data available for the current evaluation do not raise concerns with respect to genotoxicity for the two imidazolidines currently under evaluation.

3.   REFERENCES

Arulnesan N (2008). Acute oral toxicity study with S6821 in rats. Unpublished report no. 206071 from Nucro-Technics, Toronto, Ontario, Canada. Submitted to WHO by the International Organization of the Flavor Industry, Brussels, Belgium.

Barfield W (2010). S6821: Mouse in vivo micronucleus test. Unpublished report no. TEI0019 from Huntingdon Life Sciences, Cambridgeshire, England. Submitted to WHO by the International Organization of the Flavor Industry, Brussels, Belgium.

Chase KR (2009). S6821: Toxicity study by dietary administration to CD rats for 4 weeks. Unpublished report no. TEI0005 from Huntingdon Life Sciences, Cambridgeshire, England. Submitted to WHO by the International Organization of the Flavor Industry, Brussels, Belgium.

Chase KR (2010a). S6821: Toxicity study by dietary administration to CD rats for 13 weeks. Unpublished report no. TEI0016 from Huntingdon Life Sciences, Cambridgeshire, England. Submitted to WHO by the International Organization of the Flavor Industry, Brussels, Belgium.

Chase KR (2010b). S7958: Toxicity study by dietary administration to CD rats for 4 weeks. Unpublished report no. TEI0027 from Huntingdon Life Sciences, Cambridgeshire,

England. Submitted to WHO by the International Organization of the Flavor Industry, Brussels, Belgium.

Cramer GM, Ford RA, Hall RL (1978). Estimation of toxic hazard—a decision tree approach. *Food and Cosmetics Toxicology*, 16:255–276.

European Flavour and Fragrance Association (2004). European inquiry on volume use. Private communication to the Flavor and Extract Manufacturers Association of the United States, Washington, DC, USA. Submitted to WHO by the International Organization of the Flavor Industry, Brussels, Belgium.

Gavin CL, Williams MC, Hallagan JB (2008). *Flavor and Extract Manufacturers Association of the United States 2005 poundage and technical effects update survey.* Washington, DC, USA, Flavor and Extract Manufacturers Association of the United States.

International Organization of the Flavor Industry (2011). Interim inquiry on volume use and added use levels for flavoring agents to be presented at the 76th JECFA meeting. Private communication to the Flavor and Extract Manufacturers Association of the United States, Washington, DC, USA. Submitted to WHO by the International Organization of the Flavor Industry, Brussels, Belgium.

Japan Flavor and Fragrance Materials Association (2005). Japanese inquiry on volume use. Private communication to the Flavor and Extract Manufacturers Association of the United States, Washington, DC, USA. Submitted to WHO by the International Organization of the Flavor Industry, Brussels, Belgium.

Liu H (2010a). In vivo metabolism of S6821 following a single oral dose to male Sprague-Dawley rats. Unpublished report. Submitted to WHO by the International Organization of the Flavor Industry, Brussels, Belgium.

Liu H (2010b). Pharmacokinetics and metabolites of S7958 following a single oral dose to male Sprague-Dawley rats. Unpublished report. Submitted to WHO by the International Organization of the Flavor Industry, Brussels, Belgium.

May K (2009). S7958: Bacterial reverse mutation screening test. Unpublished report no. TEI0012 from Huntingdon Life Sciences, Cambridgeshire, England. Submitted to WHO by the International Organization of the Flavor Industry, Brussels, Belgium.

May K (2010a). S6821: Bacterial reverse mutation test. Unpublished report no. TEI0017 from Huntingdon Life Sciences, Cambridgeshire, England. Submitted to WHO by the International Organization of the Flavor Industry, Brussels, Belgium.

May K (2010b). S7958: Bacterial reverse mutation test. Unpublished report no. TEI0028 from Huntingdon Life Sciences, Cambridgeshire, England. Submitted to WHO by the International Organization of the Flavor Industry, Brussels, Belgium.

Taoudi M (2010). S7958: Pharmacokinetic study in rats. Unpublished report no. TEI0029 from Huntingdon Life Sciences, Cambridgeshire, England. Submitted to WHO by the International Organization of the Flavor Industry, Brussels, Belgium.

Woods I (2010). S6821: In vitro mammalian chromosome aberration test in human lymphocytes. Unpublished report no. TEI0018 from Huntingdon Life Sciences, Cambridgeshire, England. Submitted to WHO by the International Organization of the Flavor Industry, Brussels, Belgium.

Zhang B (2008). TA98 and TA100 reverse mutation test of S6821. Unpublished report no. 205777 from Nucro-Technics, Toronto, Ontario, Canada. Submitted to WHO by the International Organization of the Flavor Industry, Brussels, Belgium.

Zhu W (2010). S6821: comparative in vitro metabolism using rat and human liver microsomes. Unpublished report no. TEI0020 from Huntingdon Life Sciences, Cambridgeshire, England. Submitted to WHO by the International Organization of the Flavor Industry, Brussels, Belgium.

PHENOL AND PHENOL DERIVATIVES (addendum)

First draft prepared by

S.M.F. Jeurissen[1], J. Alexander[2], M. DiNovi[3] and A.G. Renwick[4]

[1] Centre for Substances and Integrated Risk Assessment, National Institute for Public Health and the Environment, Bilthoven, the Netherlands
[2] Norwegian Institute of Public Health, Nydalen, Norway
[3] Center for Food Safety and Applied Nutrition, Food and Drug Administration, College Park, Maryland, United States of America (USA)
[4] School of Medicine, University of Southampton, Southampton, England

1. Evaluation 201
   1.1 Introduction 201
   1.2 Assessment of dietary exposure 202
   1.3 Absorption, distribution, metabolism and elimination 202
   1.4 Application of the Procedure for the Safety Evaluation of
       Flavouring Agents 206
   1.5 Consideration of combined intakes from use as flavouring
       agents 206
   1.6 Conclusions 207
2. Relevant background information 207
   2.1 Explanation 207
   2.2 Additional considerations on dietary exposure 207
   2.3 Biological data 207
       2.3.1 Biochemical data: absorption, distribution, metabolism and
             elimination 207
       2.3.2 Toxicological studies 209
             (a) Acute toxicity 209
             (b) Short-term studies of toxicity 209
             (c) Long-term studies of toxicity and carcinogenicity ... 213
             (d) Genotoxicity 215
             (e) Reproductive/developmental toxicity 215
3. References 219

1. EVALUATION

1.1 Introduction

The Committee evaluated three additional flavouring agents belonging to the group of phenol and phenol derivatives that was evaluated previously. The additional flavouring agents included a flavone (No. 2170), a dihydrochalcone (No. 2171) and a flavanone (No. 2172). The safety of the two submitted substances rebaudioside C (No. 2168) and rebaudioside A (No. 2169) was not assessed; the Committee decided that it would not be appropriate to evaluate these substances as flavouring agents, as they had already been evaluated as food additives (sweeteners). The evaluations were conducted according to the Procedure for the Safety Evaluation of Flavouring Agents (see Figure 1, Introduction) (Annex 1, reference *131*). None of these flavouring agents have previously been evaluated by the Committee. All three flavouring agents evaluated at the current meeting are reported to be flavour modifiers.

The Committee previously evaluated 48 other members of this group of flavouring agents at its fifty-fifth meeting (Annex 1, reference *149*). The Committee concluded that all 48 flavouring agents in that group were of no safety concern at estimated dietary exposures.

The Committee also evaluated 13 other members of this group of flavouring agents at its seventy-third meeting (Annex 1, reference *202*). The Committee concluded that all 13 flavouring agents in that group were of no safety concern at estimated dietary exposures.

One of the three flavouring agents in this group (No. 2172) is a natural component of food and has been detected in grapefruit (United States Department of Agriculture, 2006).

1.2 Assessment of dietary exposure

The total annual volume of production of the three flavouring agents belonging to the group of phenol and phenol derivatives is 14.2 kg in the USA, with no reported data from Europe or Japan (European Flavour and Fragrance Association, 2004; Japan Flavor and Fragrance Materials Association, 2005; Gavin, Williams & Hallagan, 2008; International Organization of the Flavor Industry, 2011). Approximately 99% of the total annual volume of production in the USA is accounted for by trilobatin (No. 2171).

Dietary exposures were estimated using both the single portion exposure technique (SPET) and the maximized survey-derived intake (MSDI) method, with the highest values reported in Table 1. The estimated daily dietary exposure is highest for trilobatin (No. 2171) (50 000 µg, the SPET value obtained from milk products). For the other flavouring agents, the estimated daily dietary exposures range from 0.01 to 15 000 µg, with the SPET yielding the highest estimates.

Annual volumes of production of this group of flavouring agents as well as the daily dietary exposures calculated using both the MSDI method and the SPET are summarized in Table 2.

1.3 Absorption, distribution, metabolism and elimination

Information on the absorption, distribution, metabolism and elimination of the flavouring agents belonging to the group of phenol and phenol derivatives has previously been described in the monographs of the fifty-fifth and seventy-third meetings (Annex 1, references *150* and *203*). Additional information on the absorption, distribution, metabolism and elimination of polyphenols was available for this meeting.

Glycoside conjugates of polyphenols are hydrolysed on the brush border of small intestine epithelial cells or within the epithelial cells. Polyphenols are rapidly but incompletely absorbed after oral administration. Metabolism occurs in the gastrointestinal tract and after absorption. Polyphenols are metabolized through hydrolysis, sulfation, glucuronidation and/or methylation. Urinary excretion is rapid

Table 1. Summary of the results of the safety evaluations of phenol and phenol derivatives used as flavouring agents[a,b,c]

Flavouring agent	No.	CAS No. and structure	Step A3[1] Does estimated dietary exposure exceed the threshold of concern?	Step A4 Is the flavouring agent or are its metabolites endogenous?	Step A5[b] Adequate NOAEL for flavouring agent or related substance?	Comments on predicted metabolism	Related structure name (No.) and structure (if applicable)	Conclusion based on current estimated dietary exposure
Structural class III								
3′,7-Dihydroxy-4′-methoxyflavan	2170	76426-35-2	Yes, SPET: 15 000	No	Yes. The NOAEL of 760 mg/kg bw per day for structurally related neohesperidin dihydrochalcone in a 90-day study in rats (Lina, Dreef-van der Meulen & Leegwater, 1990) is 3000 times the estimated dietary exposure to No. 2170 when used as a flavouring agent.	Note 1	Neohesperidin dihydrochalcone	No safety concern
Trilobatin	2171	4192-90-9	Yes, SPET: 50 000	No	Yes. The NOAEL of 760 mg/kg bw per day for structurally related neohesperidin dihydrochalcone in a 90-day study in rats (Lina, Dreef-van der Meulen & Leegwater, 1990) is 910 times the estimated dietary exposure to No. 2171 when used as a flavouring agent.	Notes 1 and 2	Neohesperidin dihydrochalcone	No safety concern

continued

Table 1 (continued)

Flavouring agent	No.	CAS No. and structure	Step A3[d] Does estimated dietary exposure exceed the threshold of concern?	Step A4 Is the flavouring agent or are its metabolites endogenous?	Step A5[e] Adequate NOAEL for flavouring agent or related substance?	Comments on predicted metabolism	Related structure name (No.) and structure (if applicable)	Conclusion based on current estimated dietary exposure
(±)-Eriodictyol	2172	4049-38-1	Yes, SPET: 6000	No	Yes. The NOAEL of 760 mg/kg bw per day for structurally related neohesperidin dihydrochalcone in a 90-day study in rats (Lina, Dreef-van der Meulen & Leegwater, 1990) is 7600 times the estimated dietary exposure to No. 2172 when used as a flavouring agent.	Note 1	Neohesperidin dihydrochalcone	No safety concern

bw, body weight; CAS, Chemical Abstracts Service; NOAEL, no-observed-adverse-effect level

[a] In total, 61 flavouring agents in this group were previously evaluated by the Committee (Annex 1, references *149* and *202*).
[b] *Step 1*: All three flavouring agents in this group are in structural class III.
[c] *Step 2*: All three flavouring agents in this group can be predicted to be metabolized to innocuous products.
[d] The threshold for human dietary exposure for structural class III is 90 µg/person per day. All dietary exposure values are expressed in µg/day. The dietary exposure value listed represents the highest estimated dietary exposure calculated using either the SPET or the MSDI method. The SPET gave the highest estimated dietary exposure in each case.
[e] The margins of exposure were calculated based on the estimated dietary exposure calculated by the SPET.

Notes:
1. Aglycones are methylated and form sulfates or glucuronic acid conjugates prior to elimination.
2. Glycosides are expected to undergo hydrolysis to the aglycone.

Table 2. Annual volumes of production and daily dietary exposures for phenol and phenol derivatives used as flavouring agents in Europe, the USA and Japan

Flavouring agent (No.)	Most recent annual volume of production (kg)[a]	Dietary exposure				Natural occurrence in foods
		MSDI[b]		SPET[c]		
		µg/day	µg/kg bw per day	µg/day	µg/kg bw per day	
3',7-Dihydroxy-4'-methoxyflavan (2170)				15 000	250	–
Europe	ND	ND	ND			
USA	0.1	0.01	0.0002			
Japan	ND	ND	ND			
Trilobatin (2171)				50 000	833	–
Europe	ND	ND	ND			
USA	14	2	0.03			
Japan	ND	ND	ND			
(±)-Eriodictyol (2172)				6000	100	+
Europe	ND	ND	ND			
USA	0.1	0.01	0.0002			
Japan	ND	ND	ND			
Total						
Europe	ND					
USA	14.2					
Japan	ND					

bw, body weight; ND, no data reported; +, reported to occur naturally in foods (United States Department of Agriculture, 2006), but no quantitative data; –, not reported to occur naturally in foods

[a] From European Flavour and Fragrance Association (2004), Japan Flavor and Fragrance Materials Association (2005), Gavin, Williams & Hallagan (2008) and International Organization of the Flavor Industry (2011). Values greater than 0 kg but less than 0.1 kg were reported as 0.1 kg.

[b] MSDI (µg/person per day) calculated as follows:
(annual volume, kg) × (1 × 10^9 µg/kg)/(population × survey correction factor × 365 days), where population (10%, "eaters only") = 32 × 10^6 for Europe, 31 × 10^6 for the USA and 13 × 10^6 for Japan; and where survey correction factor = 0.8 for the surveys in Europe, the USA and Japan, representing the assumption that only 80% of the annual flavour volume was reported in the poundage surveys (European Flavour and Fragrance Association, 2004; Japan Flavor and Fragrance Materials Association, 2005; Gavin, Williams & Hallagan, 2008; International Organization of the Flavor Industry, 2011).
MSDI (µg/kg bw per day) calculated as follows:
(µg/person per day)/body weight, where body weight = 60 kg. Slight variations may occur from rounding.

[c] SPET (µg/person per day) calculated as follows:

(standard food portion, g/day) × (highest usual use level) (International Organization of the Flavor Industry, 2011). The dietary exposure from the single food category leading to the highest dietary exposure from one portion is taken as the SPET estimate.
SPET (μg/kg bw per day) calculated as follows:
(μg/person per day)/body weight, where body weight = 60 kg. Slight variations may occur from rounding.

to relatively slow, and biliary excretion also occurs. Metabolites not absorbed in the small intestine may undergo further metabolism in the large intestine. The microflora cleave conjugated moieties, with the resultant aglycones undergoing ring fission, leading to phenolic acid and cinnamic acid derivatives. These metabolites can be absorbed and ultimately excreted in the urine.

1.4 Application of the Procedure for the Safety Evaluation of Flavouring Agents

Step 1. In applying the Procedure for the Safety Evaluation of Flavouring Agents to the three flavouring agents in this group of phenol and phenol derivatives, the Committee assigned all three flavouring agents (Nos 2170–2172) to structural class III (Cramer, Ford & Hall, 1978).

Step 2. All three flavouring agents in this group can be predicted to be metabolized to innocuous products. The evaluation of all of these flavouring agents therefore proceeded via the A-side of the Procedure.

Step A3. The highest estimated dietary exposures to all three flavouring agents are above the threshold of concern (i.e. 90 μg/person per day for class III). Accordingly, the evaluation of all three flavouring agents proceeded to step A4.

Step A4. None of the three flavouring agents or their metabolites are endogenous substances. Accordingly, the evaluation of all three flavouring agents proceeded to step A5.

Step A5. The no-observed-adverse-effect level (NOAEL) of 760[1] mg/kg of body weight (bw) per day for the structurally related substance neohesperidin dihydrochalcone from a 90-day study in rats (Lina, Dreef-van der Meulen & Leegwater, 1990) provides adequate margins of exposure of 3000, 910 and 7600 for 3′,7-dihydroxy-4′-methoxyflavan (No. 2170; SPET = 15 000 μg/day), trilobatin (No. 2171; SPET = 50 000 μg/day) and (±)-eriodictyol (No. 2172; SPET = 6000 μg/day), respectively, when used as flavouring agents. The Committee therefore concluded that these flavouring agents would not pose a safety concern at current estimated dietary exposures.

Table 1 summarizes the evaluations of the three flavouring agents belonging to the group of phenol and phenol derivatives (Nos 2070–2072).

1.5 Consideration of combined intakes from use as flavouring agents

The three additional flavouring agents in this group of phenol and phenol derivatives have low MSDIs (0.01–2 μg/day). The Committee concluded that

---

[1] Previously rounded to 750 mg/kg bw per day (Annex 1, reference 202).

consideration of combined intakes is not necessary, because the additional flavouring agents would not contribute significantly to the combined intake of this flavouring group.

1.6 Conclusions

In the previous evaluation of flavouring agents in the group of phenol and phenol derivatives, studies of acute toxicity, short-term and long-term studies of toxicity (18 days to 2 years), and studies of carcinogenicity, genotoxicity, and reproductive and developmental toxicity were available (Annex 1, references *150* and *203*).

For the present evaluation, additional biochemical data and in vitro studies of genotoxicity were available for two flavouring agents in this group (Nos 2170 and 2172), three flavouring agents previously evaluated in this group (Nos 706–708) and one related substance (neohesperidin dihydrochalcone). For neohesperidin dihydrochalcone, studies of acute toxicity, short-term studies of toxicity, long-term studies of toxicity and carcinogenicity and studies of reproductive and developmental toxicity were also available. The studies available for the present evaluation support the previous safety evaluations.

The Committee concluded that these three flavouring agents, which are additions to the group of phenol and phenol derivatives evaluated previously, would not give rise to safety concerns at current estimated dietary exposures.

2. RELEVANT BACKGROUND INFORMATION

2.1 Explanation

This monograph summarizes key aspects relevant to the safety evaluation of three phenol derivatives, which are additions to a group of 61 flavouring agents evaluated previously by the Committee at its fifty-fifth and seventy-third meetings (Annex 1, references *149* and *202*).

2.2 Additional considerations on dietary exposure

Annual production volumes and daily dietary exposures for each flavouring agent are reported in Table 2.

One of the three flavouring agents in this group (No. 2172) is a natural component of food and has been detected in grapefruit (United States Department of Agriculture, 2006) (Table 2).

2.3 BIOLOGICAL DATA

2.3.1 Biochemical data: absorption, distribution, metabolism and elimination

Information on the absorption, distribution, metabolism and elimination of the flavouring agents belonging to the group of phenol and phenol derivatives and of structurally related agents has been reported in the monographs of the fifty-fifth and seventy-third meetings (Annex 1, references *150* and *203*). In addition, numerous studies on the absorption, distribution, metabolism and/or elimination of polyphenols are available from public literature, only some of which were submitted. Those most relevant to the flavouring agents under evaluation are described below.

Polyphenols are ubiquitous in plant-derived foods. They include anthocyanins, dihydrochalcones, flavan-3-ols, flavanones, flavones, flavonols and isoflavones, as well as tannins, phenolic acids, hydroxycinnamates and stilbenes. Naturally occurring dietary flavonoids are ingested, with the exception of the flavan-3-ols, as glycoside conjugates. Typically, the glycoside moiety is hydrolysed in the brush border of small intestine epithelial cells via lactase phlorizin hydrolase, which has substrate specificity for flavonoid-*O*-β-D-glucosides, and the aglycone is absorbed through the epithelium primarily through passive diffusion. An alternative site of hydrolysis is cytosolic β-glucosidase within the epithelial cells. For cytosolic β-glucosidase-initiated hydrolysis to occur, the polar glycosides must be transported into the epithelial cells. It has been proposed that the active sodium-dependent glucose transporter, SGLT1, may be involved (Day et al., 2000; Gee et al., 2000; Donovan et al., 2006).

Prior to entering the bloodstream, the aglycones may undergo sulfation by sulfotransferases, glucuronidation by uridine diphosphate–glucuronosyltransferases and methylation by catechol-*O*-methyltransferase within the cells of the intestinal wall (Donovan et al., 2006). Upon entry into the bloodstream, additional phase II metabolic processes, as well as deconjugation reactions, may occur in the liver.

Metabolites not absorbed in the small intestine may undergo further metabolism in the large intestine. The microflora cleave conjugated moieties, with the resultant aglycones undergoing ring fission, leading to phenolic acid and cinnamic acid derivatives. These metabolites can be absorbed and ultimately excreted in the urine (Donovan et al., 2006).

In a review on the bioavailability of polyphenols, a single 50 mg dose of aglycone equivalent from 18 major polyphenols was calculated to correspond with plasma concentrations of total metabolites ranging from 0 to 4 µmol/l, and a relative urinary excretion ranged from 0.3% to 43% of the dose. The plasma kinetics also differed among polyphenol classes, with the peak plasma concentration (C_{max}) being reached after 1.5 or 5.5 hours, depending on the site of intestinal absorption. The polyphenols that are most well absorbed in humans are isoflavones and gallic acid, followed by catechins, flavanones and quercetin glucosides. The least well absorbed polyphenols are the proanthocyanidins, the galloylated tea catechins and the anthocyanins (Manach et al., 2005).

To determine whether stereoselectivity plays a role in the pharmacokinetics of (±)-eriodictyol (No. 2172), six male Sprague-Dawley rats were administered 20 mg of racemic eriodictyol per kilogram of body weight intravenously, and body fluid samples were analysed using chiral separation methods. Blood samples were collected after 0, 1, 10 and 30 minutes and 1, 2, 4, 6, 12, 24, 48, 72, 96 and 120 hours. Urine was collected 0, 2, 6, 12, 24, 48, 72, 96 and 120 hours post-injection. Estimated half-lives for the two stereoisomers were 4 hours based on serum data and 48 hours based on urinary data. A low proportion of a dose of

(±)-eriodictyol is excreted in the urine (5% for the *R*-enantiomer and 7% for the *S*-enantiomer). The large volume of distribution, 4.9 l/kg for either isomer, indicates that the eriodictyol leaves the blood compartment and is present largely in the tissues. It was concluded that the eriodictyol enantiomers are not metabolized in a significantly different manner from one another (Yáñez et al., 2007).

Neohesperidin dihydrochalcone labelled with ^{14}C was administered to rats at doses of 1, 10 or 100 mg/kg bw by oral gavage. Approximately 80% of the administered radiolabelled substance was excreted with the urine within 24 hours. The remaining radioactivity was excreted with the faeces or was recovered from the intestinal contents. Only traces of radioactivity were found in various tissues after 24 hours, and the recovery of radioactivity in the respiratory carbon dioxide was 0.1% or less in 24 hours. The chemical identity of the radioactivity excreted with the urine and faeces was not determined (Gumbmann et al., 1978).

Neohesperidin dihydrochalcone was fed to rats at a dietary concentration of 0.5%, and the urine was examined for metabolic products. It was not specified when the urinary samples were taken. The aglycone of neohesperidin dihydrochalcone was the major metabolite, and lesser amounts of the parent compound were detected. *m*-Hydroxyphenylpropionic acid was not detected in urine (Booth, Robbins & Gagne, 1965). However, incubation of neohesperidin dihydrochalcone with rabbit faecal matter for 30 minutes under anaerobic conditions led to the formation of *m*-hydroxyphenylpropionic acid, but the major metabolite was dihydroisoferulic acid (Booth, Robbins & Gagne, 1965).

2.3.2 Toxicological studies

Studies of genotoxicity in vitro have been reported for two flavouring agents currently being evaluated (Nos 2170 and 2172), three flavouring agents previously evaluated in this group (Nos 706–708) and one structurally related substance (neohesperidin dihydrochalcone). For neohesperidin dihydrochalcone, studies of acute toxicity, short-term studies of toxicity, long-term studies of toxicity and carcinogenicity and studies of reproductive and developmental toxicity were also available.

(a) Acute toxicity

An oral median lethal dose (LD_{50} value) of greater than 5000 mg/kg bw has been reported for neohesperidin dihydrochalcone (Nutrilite Products Inc., 1980).

(b) Short-term studies of toxicity

Seven short-term repeated-dose studies on neohesperidin dihydrochalcone were available. The results of these studies are described below, and studies for which NOAELs were derived are summarized in Table 3.

In a published study, groups of 20 male and 20 female Wistar rats were given neohesperidin dihydrochalcone at a dietary concentration of 0, 2000, 10 000 or 50 000 mg/kg feed, equal to mean doses of 0, 150, 760[2] and 4010 mg/kg bw per day

---

[2] Previously rounded to 750 mg/kg bw per day (Annex 1, reference *202*).

Table 3. Results of short-term studies of toxicity with phenol and phenol derivatives used as flavouring agents and structurally related substances

No.	Flavouring agent	Species; sex	No. of test groups[a] / no. per group[b]	Route	Duration (days)	NOAEL (mg/kg bw per day)	Reference
NA	Neohesperidin dihydrochalcone	Rat; M, F	3/40	Oral, diet	91	760	Lina, Dreef-van der Meulen & Leegwater (1990)
NA	Neohesperidin dihydrochalcone	Rat; M, F	1/15	Oral, diet	M: 92 F: 113	500[c]	Booth, Robbins & Gagne (1965)
NA	Neohesperidin dihydrochalcone	Rat; M, F	3/12	Oral, diet	365	2500[d]	Gumbmann et al. (1978)

F, female; M, male; NA, not applicable

[a] Total number of test groups does not include control animals.
[b] Total number per test group includes both male and female animals.
[c] This is the overall NOAEL from three short-term studies of toxicity by Booth, Robbins & Gagne (1965).
[d] Highest dose tested.

for males and 0, 170, 850 and 4330 mg/kg bw per day for females, respectively, for 91 days. Feed and water were provided ad libitum. The rats were weighed weekly and observed daily for external signs of toxicity. Feed intake was measured over 1-week periods, and water intake was recorded in weeks 7 and 12. Ophthalmoscopy was performed in the animals of the control and high-dose groups before the start and at the end of the treatment. Haematological and biochemical analyses and urine analysis were performed at the end of the exposure period. The rats were killed by exsanguination and subjected to macroscopic examination. Any abnormalities were recorded, and the adrenals, brain, caecum, gonads, heart, kidneys, liver, spleen, thymus and thyroid (with parathyroids) were weighed. Organs and tissues from all animals in the control and high-dose groups were examined microscopically. One female of the intermediate-dose group that died intercurrently was subjected to histopathological examination as well.

No treatment-related mortalities or clinical symptoms were observed in the low- or mid-dose groups. One mid-dose female was found dead on day 30. Macroscopic and microscopic examinations revealed renal failure, presumably by obstruction of the urinary tract. As such a lesion was not observed in any of the other animals, including high-dose animals, or in other studies, the Committee did not consider this death to be treatment related. In the high-dose group, both males and females showed soft stools in weeks 2 and 3.

In the high-dose group, body weight gains were decreased in males throughout the study and in females in the first 2 weeks. Feed intake was only slightly decreased in high-dose males in the first 2 weeks. Ophthalmoscopy and haematological examinations did not reveal treatment-related effects. Plasma alkaline phosphatase activity (+20%) and bilirubin concentration (+225%) were increased in high-dose females. Total plasma protein concentration was decreased (–6%) in high-dose males. Urine analysis revealed a decrease (up to 10%) in urinary pH in both sexes in the high-dose group. The macroscopic examinations at autopsy revealed no treatment-related changes. Relative caecum weight was markedly increased in the high-dose group in both sexes (up to 76% for caecum with content and 52% for caecum without content). In the males of the low- and mid-dose groups, increases in relative caecum weight were also observed (up to 28%), but there was no clear dose-dependent effect. The relative weights of the brain and testicles were increased in males of the high-dose group, although the absolute weights of these organs (data not shown) were comparable to those of the control group. The changes were therefore considered to be due to the decreased body weights in this group. Microscopic examination did not reveal any treatment-related changes in the caecum or in other organs or tissues.

The NOAEL was 760 mg/kg bw per day (Lina, Dreef-van der Meulen & Leegwater, 1990).

Three consecutively performed short-term studies of toxicity and reproductive toxicity with neohesperidin dihydrochalcone were (limitedly) reported by Booth, Robbins & Gagne (1965). In the first study, groups of five male and five female rats were provided neohesperidin dihydrochalcone at dietary concentrations of 0, 6.4, 64, 640 or 1280 mg/kg feed (equivalent to 0, 0.64, 6.4, 64 and 128 mg/kg bw per day) for a total of 148 days. After 90 days of treatment, a reproductive toxicity study was initiated, during which all rats of both sexes continued to be fed the same neohesperidin dihydrochalcone–enriched diet. Reproductive performance was evaluated on the basis of the number of litters cast, the number of pups, and the number and body weight of the pups when weaned at 3 weeks of age. Haematology data, including red and white blood cell counts and haemoglobin concentrations, were determined for the parental rats that were fed at the two highest neohesperidin dihydrochalcone concentrations and for controls. All rats were autopsied after a total of 148 days (including the time for the reproductive toxicity study). Organ weights of liver, heart, kidneys, spleen, brain, testes, adrenals and thyroids were recorded at autopsy for the control rats and the rats from the highest-dose group. These organs and ovaries, lung, pancreas, stomach, intestine, bladder and pituitary were examined microscopically.

Effects observed included a lower body weight (–12%), compared with the control animals, in male rats fed 128 mg/kg bw per day after 148 days. In addition, histopathological examination revealed liver lipidosis in the female rats fed the neohesperidin dihydrochalcone–enriched diet (no data provided on the incidences at the different dose levels). Also, a decrease in the number of pups weaned was observed (21 in the control group compared with 11, 6, 15 and 8 in the four treatment groups, respectively).

The second study was performed to study in more detail the liver lipidosis observed in females. Groups of six female rats were fed a diet with 0 or 1280 mg neohesperidin dihydrochalcone per kilogram (equivalent to 0 and 128 mg/kg bw per day) for 90 days (Booth, Robbins & Gagne, 1965). As liver lipidosis was also detected in other experiments with other compounds using the basal diet used in the first study, a different basal diet, which was also reported to be adequate for breeding, was used in this second study. After the experimental period, rats were autopsied, and tissues were evaluated histopathologically. A decreased body weight (−11%), compared with controls, was observed in the treated females. Histopathological examinations did not reveal any liver lipidosis.

In the third study, groups of 5 male and 10 female rats were provided a diet containing 0 or 5000 mg neohesperidin dihydrochalcone per kilogram (equivalent to 0 and 500 mg/kg bw per day) for over 90 days. The basal diet was the same as in the second study. After 70 days of treatment, the rats had reached sexual maturity, and a reproductive toxicity study was initiated, comparable to that described for the first study. Males were autopsied after 92 days, and females were autopsied after 113 days, which included 2–3 weeks of the lactation period. Measurement of organ weights, haematological analyses and histopathological examinations were performed. No significant changes were observed, and no effects on reproductive parameters were observed.

The depressed body weights observed in male rats of the highest-dose group after 148 days in the first study and in female rats in the second study were not observed in males or females of the third study that were exposed to a higher dose of neohesperidin dihydrochalcone. The liver lipidosis observed in female rats in the first study was not observed in the subsequent studies. The observed apparent mortality, in terms of the number of pups weaned, in the first study could not be confirmed in the third study, in which a higher dose level was tested. The Committee is therefore of the opinion that the effects observed in the first two studies are not related to the exposure to neohesperidin dihydrochalcone.

The overall NOAEL from these three studies was 500 mg/kg bw per day (Booth, Robbins & Gagne, 1965).

Neohesperidin dihydrochalcone (purity >99%) was fed to groups of five male and six female weanling Fischer rats at dietary concentrations of 0 or 50 000 mg/kg feed (equivalent to 0 and 5000 mg/kg bw per day). After 90 days of treatment, a reproductive toxicity test was initiated. The results of this study are described in section (e) below (reproductive/developmental toxicity). The F_0 generation was examined after 122 days and 170 days for the males and females, respectively. No adverse effects were observed on body weight gain or feed intake. No changes in liver, heart, kidneys, spleen, pancreas, testes or adrenal weight were observed. An increase in thyroid weight was observed in the neohesperidin dihydrochalcone–treated group, which was significant in the male animals. Gross and histological examinations revealed no lesions that could be specifically related to the ingestion of neohesperidin dihydrochalcone. No impaired reproductive performance was observed.

A NOAEL cannot be derived based on the results of this study (Gumbmann et al., 1978).

Groups of six male and six female Fischer rats (from the F_2 generation obtained as part of the three-generation reproductive toxicity study, see section (e) below on reproductive/developmental toxicity) were given neohesperidin dihydrochalcone (purity >99%) at dietary concentrations of 0, 5000, 25 000 or

50 000 mg/kg feed (equivalent to doses of 0, 250, 1250 and 2500 mg/kg bw per day) for 1 year. In mid-dose females, a slight, but statistically significant, decrease in body weight gain was observed. As body weight gain was not affected at the high dose, the Committee considered this effect not treatment related. In high-dose males, a statistically significant decrease in absolute liver weight was observed. However, the reduction was small (6%), and the Committee considered it not toxicologically relevant. None of the histological lesions that were observed could be related to neohesperidin dihydrochalcone exposure.

The NOAEL was 2500 mg/kg bw per day, the highest dose tested (Gumbmann et al., 1978).

Groups of weanling male Sprague-Dawley rats were fed a diet containing 0 or 100 000 mg neohesperidin dihydrochalcone per kilogram (equivalent to 0 and 5000 mg/kg bw per day) for a period of 11 months. In the treatment group, the final body weights were significantly lower and the relative testes weights were significantly higher compared with the control group (absolute testes weights were not reported). None of the histological lesions that were observed in the experimental group could be related to the neohesperidin dihydrochalcone exposure.

A NOAEL cannot be derived based on the results of this study (Gumbmann et al., 1978).

(c) Long-term studies of toxicity and carcinogenicity

Two long-term repeated-dose studies on neohesperidin dihydrochalcone were available. The results of these studies are described below, and the study for which a NOAEL was derived is summarized in Table 4.

Groups of 24 male and 24 female Fischer rats were administered neohesperidin dihydrochalcone in the diet at concentrations of 0, 5000, 25 000 or 50 000 mg/kg (equivalent to 0, 250, 1250 and 2500 mg/kg bw per day) for their lifespan or 2 years. Decreased body weight gain in high-dose animals became apparent within the first 10 weeks of the study and was more pronounced in females by the 60th week. At this time, the diet of the high-dose group for one half of the rats of both sexes was enriched with two supplements (containing minerals and brewers' dried yeast). These two supplements produced a response in growth, such that there was no significant difference in body weights at 100 weeks between the animals fed the high-dose diet with the two supplements and the animals of the control group.

Mortality was highest in the control groups in both sexes at 100 weeks (66% survival), excluding the low-dose males. The higher number of deaths, 50%, in this latter group is considered to be unrelated to the exposure to neohesperidin dihydrochalcone, as the mortality in the mid- and high-dose males was quite low (i.e. 4/24 and 5/24, respectively). Results of haematology and urine analysis obtained at 6-month intervals did not reveal any treatment-related abnormalities. Plasma cholesterol, determined at autopsy, was reported to be lower in both sexes of the high-dose group compared with controls, whether or not the high-dose diet was supplemented with yeast and minerals. No quantitative data were provided,

Table 4. Results of long-term studies of toxicity with phenol and phenol derivatives used as flavouring agents and structurally related substances

No.	Flavouring agent	Species; sex	No. of test groups[a] / no. per group[b]	Route	Duration (days)	NOAEL (mg/kg bw per day)	Reference
NA	Neohesperidin dihydrochalcone	Dog; M, F	3/6	Oral, diet	730	1000	Gumbmann et al. (1978)

F, female; M, male; NA, not applicable

[a] Total number of test groups does not include control animals.
[b] Total number per test group includes both male and female animals.

so the extent of the decrease in cholesterol levels is not clear. It is reported that histological examination indicated that the incidence of tumours was fairly uniformly distributed among the groups and not treatment related. The incidence of non-neoplastic lesions also showed no relation to the neohesperidin dihydrochalcone exposure. An exception to this was the occurrence of focal renal cortical atrophy, which tended to be higher in the low-dose, mid-dose and non-supplemented high-dose groups of females, compared with the female controls and the female animals in the supplemented high-dose group. Except for animals in the supplemented high-dose group, all animals exhibited diffuse thyroid follicular hyperplasia and hypertrophy consistent with a dietary iodine deficiency. The Committee noted that the description of the study was brief and often lacked quantitative data. The Committee did not derive a NOAEL from this study, because there were confounding effects due to dietary deficiencies (Gumbmann et al., 1978).

In a 2-year chronic study, groups of three male and three female Beagle dogs received neohesperidin dihydrochalcone in the diet. The doses were 0, 200, 1000 and 2000 mg/kg bw per day. At 6-month intervals, the dogs were given general physical examinations, which included the collection and analysis of blood and urine samples. No treatment-related effects on growth or haematological parameters were observed. Plasma alkaline phosphatase activity was consistently elevated in the males of the high-dose group at 12, 18 and 24 months, but this increase could not be related to any other clinical or histological change. In females, plasma thyroxine concentration tended to be decreased (extent not reported) in the high-dose group from 6 months until the end of the treatment. A slight dose-related increase in relative liver weight was observed in male and female animals. Testes weights for one dog in the mid-dose group and one in the high-dose group were unusually low (0.08% and 0.04% of body weight, compared with 0.13% in controls). This was accompanied by marked testicular atrophy and degeneration (confirmed histologically) in these two dogs. Thyroid weight was significantly increased in males of the high-dose group, and mild thyroid hypertrophy and hyperplasia were detected in two of the three dogs of each sex in this high-dose group. Other observed histological lesions could not be related to the dietary treatment.

The NOAEL was 1000 mg/kg bw per day, based on the effects on the thyroid observed in the high-dose group (Gumbmann et al., 1978).

(d) Genotoxicity

(i) In vitro

The results of studies of genotoxicity in vitro for two flavouring agents in this group (Nos 2170 and 2172), three flavouring agents previously evaluated in this group (Nos 706–708) and one structurally related substance (neohesperidin dihydrochalcone) are summarized in Table 5 and described below. Only the study of Leuschner (2009) was certified for compliance with good laboratory practice (GLP) and quality assurance and reported to be performed according to Organisation for Economic Co-operation and Development (OECD) Test Guideline 471.

No evidence of mutagenicity was observed in Ames assays when 3′,7-dihydroxy-4′-methoxyflavan (No.2170), (±)-eriodictyol (No.2172), 2,6-dimethylphenol (No. 707), 3,4-dimethylphenol (No. 708) and neohesperidin dihydrochalcone were incubated with *Salmonella typhimurium* strains TA97, TA98, TA100, TA102, TA1535, TA1537 and/or TA1538, with or without metabolic activation, at concentrations up to 10 000 µg/plate (Batzinger, Ou & Bueding, 1977; Brown, Dietrich & Brown, 1977; MacGregor & Jurd, 1978; Brown & Dietrich, 1979; Nagao et al., 1981; Zeiger et al., 1987; National Toxicology Program, 1988b,c; Leuschner, 2009). In the National Toxicology Program study with 2,5-dimethylphenol (No. 706), negative results were obtained in strains TA97, TA98 and TA1535 at concentrations up to 1000 µg/ml. Toxicity was observed at 2000 µg/ml. In the presence of metabolic activation, negative results were obtained in strain TA100, but in the presence of metabolic activation, equivocal results were obtained in strain TA100 at concentrations up to 1000 µg/ml. Toxicity was observed at 2000 µg/ml (National Toxicology Program, 1988a). In the second study, negative results were obtained for all strains, in the presence and absence of metabolic activation (National Toxicology Program, 1989).

(ii) Conclusion

The available data for the current evaluation are in line with the conclusion in the previous evaluations of the flavouring agents belonging to the group of phenol and phenol derivatives that these compounds are unlikely to be genotoxic in vivo (Annex 1, references *150* and *203*).

(e) Reproductive/developmental toxicity

In a three-generation reproductive toxicity study, groups of six male and six female Fischer rats were given neohesperidin dihydrochalcone at dietary concentrations of 0, 5000, 25 000 or 50 000 mg/kg feed (equivalent to doses of 0, 250, 1250 and 2500 mg/kg bw per day). The study was started at the time of weaning of the F_0 generation. This F_0 generation was mated at sexual maturity to produce an F_1 generation, which, in turn, produced an F_2 generation. After producing an F_3 generation, the F_2 rats were continued on the same diets for 1 year and subjected to histological examination (see section (b) on short-term studies of toxicity). When the F_1 rats were weaned, the F_0 rats were used to produce second litters (F_{1b}), the embryos being removed by caesarean section on the 20th day of gestation for teratological evaluation. No detrimental effects on reproduction related

Table 5. Results of studies of genotoxicity in vitro with phenol and phenol derivatives used as flavouring agents and structurally related substances

No.	Flavouring agent	End-point	Test system	Concentration	Results	Reference
2170	3',7-Dihydroxy-4'-methoxyflavan	Reverse mutation	Salmonella typhimurium TA98, TA100, TA102, TA1535 and TA1537	10–1000 µg/plate, ±S9[a]	Negative[b]	Leuschner (2009)
2172	(±)-Eriodictyol	Reverse mutation	S. typhimurium TA98, TA100, TA1535, TA1537 and TA1538	100 µg/plate, ±S9[c]	Negative	Brown & Dietrich (1979)
2172	(±)-Eriodictyol	Reverse mutation	S. typhimurium TA98 and TA100	Up to 200 µg/plate, ±S9[d,e]	Negative	Nagao et al. (1981)
706	2,5-Dimethylphenol	Reverse mutation	S. typhimurium TA97, TA98, TA100 and TA1535	10–2000 µg/plate[d], ±S9	Equivocal[f]	National Toxicology Program (1988a)
706	2,5-Dimethylphenol	Reverse mutation	S. typhimurium TA97, TA98, TA100 and TA1535	10–500 µg/plate[d]	Negative[g]	National Toxicology Program (1989)
707	2,6-Dimethylphenol	Reverse mutation	S. typhimurium TA97, TA98, TA100 and TA1535	10–2000 µg/plate, ±S9[d,h]	Negative[i]	National Toxicology Program (1988b)
708	3,4-Dimethylphenol	Reverse mutation	S. typhimurium TA97, TA98, TA100 and TA1535	10–667, –S9[d] 10–3333, +S9[d,j]	Negative[k]	National Toxicology Program (1988c)
NA	Neohesperidin dihydrochalcone	Reverse mutation	S. typhimurium TA98 and TA100	40–120 mg/plate, ±S9	Negative	Batzinger, Ou & Bueding (1977)
NA	Neohesperidin dihydrochalcone	Reverse mutation	S. typhimurium TA98 and TA100	0.1–0.25 ml urine excreted by mice over 24 hours after administration of 2500 mg/kg bw[l]	Negative	Batzinger, Ou & Bueding (1977)
NA	Neohesperidin dihydrochalcone	Reverse mutation	S. typhimurium TA98 and TA100	2500 mg/kg bw[m]	Negative	Batzinger, Ou & Bueding (1977)
NA	Neohesperidin dihydrochalcone	Reverse mutation	S. typhimurium (no data on strains used)	No data	Negative	Brown, Dietrich & Brown (1977)
NA	Neohesperidin dihydrochalcone	Reverse mutation	S. typhimurium TA98, TA100, TA1535, TA1537 and TA1538	200 µg/plate, ±S9[c]	Negative	Brown & Dietrich (1979)

NA	Neohesperidin dihydrochalcone	Reverse mutation	S. typhimurium TA98, TA100, TA1535, TA1537 and TA1538	500 µg/plate, ±S9[c,n]	Negative	Brown & Dietrich (1979)
NA	Neohesperidin dihydrochalcone	Reverse mutation	S. typhimurium TA98 and TA100	820–8200 nmol/plate, ±S9[c]	Negative	MacGregor & Jurd (1978)
NA	Neohesperidin dihydrochalcone	Reverse mutation	S. typhimurium TA98 and TA100	50–1670 nmol/plate, ±S9[o]	Negative	MacGregor & Jurd (1978)
NA	Neohesperidin dihydrochalcone	Reverse mutation	S. typhimurium TA97, TA98, TA100 and TA1535	100–10 000 µg/plate, ±S9[d]	Negative	Zeiger et al. (1987)

S9, 9000 × g supernatant fraction of rat liver homogenate

[a] Two independent experiments using the preincubation method and the plate incorporation method, respectively.
[b] Precipitation was observed in all strains, with and without metabolic activation, at the highest concentration tested.
[c] Plate incorporation method.
[d] Preincubation method.
[e] The test compound was isolated from *Miscanthus sinensis*.
[f] (Slight) toxicity was observed in all strains, with and without metabolic activation, at concentrations from 667 µg/plate upwards. In strain TA97, slight toxicity was observed from 333 µg/plate upwards. Results were negative in all strains, except for strain TA100, for which equivocal results were obtained with metabolic activation.
[g] Slight toxicity and precipitation were observed at a concentration of 100 µg/plate in strain TA97.
[h] The highest concentration of 2000 µg/plate was tested only in strains TA98 and TA100, with metabolic activation. For strains TA97 and TA1535, 1000 µg/plate was the highest concentration tested.
[i] Toxicity was observed at the highest concentration tested (2000 µg/plate) and in strain TA97 without metabolic activation at 1000 µg/plate. Slight toxicity was observed in strain TA97 with metabolic activation and in all other strains, with and without metabolic activation, at 1000 µg/plate.
[j] For strains TA97 and TA1535, 1000 µg/plate was the highest concentration tested (3333 µg/plate). Slight toxicity was observed at the highest dose tested without metabolic activation in all strains except TA98. Slight toxicity was observed with metabolic activation in all strains at 1000 µg/plate.
[k] Toxicity was observed at the highest concentration tested (3333 µg/plate). Slight toxicity was observed at the highest dose tested without metabolic activation in all strains except TA98. Slight toxicity was observed with metabolic activation in all strains at 1000 µg/plate.
[l] Urine was tested with and without β-glucuronidase.
[m] Host-mediated assay. The bacterial strains were incubated for 6 hours in the peritoneal cavity of mice that were administered the test compound by gavage.
[n] Test compound was the aglycone hesperetin dihydrochalcone. Concentration expressed on hesperetin dihydrochalcone basis.
[o] Test compound was the aglycone hesperetin dihydrochalcone.

to the exposure to neohesperidin dihydrochalcone via the diet were observed, except for slightly decreased survival of pups in the two highest dose groups of the F_3 generation. The F_{1b} fetuses were examined grossly, after which one third were prepared for visualization of the skeletal system and two thirds were fixed for serial sectioning. Few abnormalities were detected (Gumbmann et al., 1978).

The Committee noted that quantitative data were not reported and did not derive a NOAEL from this study.

The embryotoxicity and teratogenicity of neohesperidin dihydrochalcone were evaluated in a GLP-compliant study from public literature performed in accordance with OECD Test Guideline 414 (Prenatal Developmental Toxicity Study) (Waalkens-Berendsen, Kuilman-Wahls & Bär, 2004). Female and male Wistar Crl:(WI)WU BR rats were placed together for mating (female:male ratio 2:1). Females showing evidence of mating were randomly assigned to each experimental group. The day on which there was evidence of mating was recorded as day 0 of gestation. At that time, the females were about 12–13 weeks old. During gestation, groups of 28 individually housed dams were administered neohesperidin dihydrochalcone at dietary concentrations of 0, 12 500, 25 000 or 50 000 mg/kg feed (equal to 0, 800–900, 1600–1700 and 3100–3400 mg/kg bw per day) from day 0 up to and including day 21 of gestation. The general condition and behaviour of the animals were observed twice daily. Body weight was determined on days 0, 7, 14 and 21 of gestation. Feed consumption was determined during three consecutive periods (days 0–7, 7–14 and 14–21 of gestation). On day 21 of gestation, the females were killed by decapitation. Macroscopic abnormalities of major organs of the abdominal and thoracic cavity were recorded. The ovaries, uterus and caecum were removed and weighed. The number of corpora lutea in each ovary was recorded. The fetuses were removed from the uterus, dried of amniotic fluid, weighed, sexed by observing the anogenital distance and examined for gross abnormalities. The placentas of the live fetuses were weighed and examined for macroscopic abnormalities. Early and late resorptions and dead fetuses were counted. Resorption was classified as "early" when only placental tissue was visible and "late" when placental tissue as well as embryonic tissue were visible. The number of implantation sites in both uterine horns was recorded. Finally, the empty uterus was weighed. The fetuses were randomly divided into two groups and processed for either visceral or skeletal microscopic examination. Skeletal and visceral examinations were conducted on fetuses of the control and high-dose groups only.

The consumption of neohesperidin dihydrochalcone was well tolerated. Maternal feed consumption, when expressed in grams per kilogram of body weight per day, was slightly but significantly increased in the mid- and high-dose groups during the last week of pregnancy. No clinical signs, abnormal behaviour or intolerance to the diet were noted in any treatment group. None of the females aborted or died during the study. Two rats (one of the low-dose group and one of the high-dose group) delivered just before the caesarean section. At necropsy, the absolute and relative weights of the full caecum were significantly increased in all neohesperidin dihydrochalcone treatment groups. The weight of the empty caecum was significantly increased in the high-dose group. No other gross changes were observed that could be attributed to the treatment.

At caesarean section, all pregnant females had litters with viable fetuses. No statistically significant differences were observed in gravid or empty uterus weight, ovary weight or placenta weight. The mean numbers of corpora lutea and implantation sites, the preimplantation loss, the number of live fetuses, the mean number of early and late resorptions, and the post-implantation loss did not differ between any of the neohesperidin dihydrochalcone–treated groups and the controls.

The sex ratio was, within normal limits, close to 1 in all groups. One fetus of the mid-dose group was dead. Mean fetal body weights of the viable fetuses were similar in all groups. Large fetuses were found in the control group (two fetuses from two litters), in the low-dose group (9/2) and in the high-dose group (2/2). Small fetuses were found in the low-dose group (2/2). One fetus of the mid-dose group showed a subcutaneous haemorrhagic area. Visceral malformations were not found in any group. None of the observed visceral anomalies and variations were considered to be treatment related. Skeletal malformations and anomalies were not observed. The few skeletal variations included mainly irregularly shaped sternebrae, but the incidence was low and not different between the high-dose group and the controls. Variations in ossification (incomplete or absent ossification) were observed. The incidences of 1–4 incompletely ossified digits and 3–6 unossified digits of the hind proximal phalanges were significantly increased in the high-dose group. However, when the categories 5–8 incompletely ossified digits and 7–10 unossified digits were also taken into account, no differences in ossification were observed. Therefore, the Committee considered the differences in ossification incidental and not related to treatment.

The NOAEL was 3100–3400 mg/kg bw per day, the highest dose tested (Waalkens-Berendsen, Kuilman-Wahls & Bär, 2004).

Additional information on reproductive toxicity is described in the combined short-term studies of toxicity and reproductive toxicity by Booth, Robbins & Gagne (1965) and Gumbmann et al. (1978), described in section (b) above (short-term studies of toxicity).

3. REFERENCES

Batzinger RP, Ou SYL, Bueding E (1977). Saccharin and other sweeteners: mutagenic properties. *Science*, 198:944–946.

Booth AN, Robbins DJ, Gagne WE (1965). Toxicity study of two flavanone dihydrochalcones (potential artificial sweetening agents). Unpublished report from Western Regional Research Laboratory. Submitted to WHO by the International Organization of the Flavor Industry, Brussels, Belgium.

Brown JP, Dietrich PS (1979). Mutagenicity of plant flavonols in the *Salmonella* mammalian microsome test. Activation of flavonol glycosides by mixed glycosidases from rat cecal bacteria and other sources. *Mutation Research*, 66:223–240.

Brown JP, Dietrich PS, Brown RJ (1977). Frameshift mutagenicity of certain naturally occurring phenolic compounds in the "*Salmonella*/microsome" test: activation of anthraquinone and flavonol glycosides by gut bacterial enzymes. *Biochemical Society Transactions*, 5:1489–1452.

Cramer GM, Ford RA, Hall RL (1978). Estimation of toxic hazard—a decision tree approach. *Food and Cosmetics Toxicology*, 16:255–276.

Day AJ et al. (2000). Dietary flavonoid and isoflavone glycosides are hydrolysed by the lactase site of lactase phlorizin hydrolase. *FEBS Letters*, 468:166–170.

Donovan JL et al. (2006). Flavonoids. In: Crozier A, Clifford MN, Ashihara H, eds. *Plant secondary metabolites: occurrence, structure and role in the human diet.* Oxford, England, Blackwell Publishing, pp. 303–351.

European Flavour and Fragrance Association (2004). European inquiry on volume use. Private communication to the Flavor and Extract Manufacturers Association of the United States, Washington, DC, USA. Submitted to WHO by the International Organization of the Flavor Industry, Brussels, Belgium.

Gavin CL, Williams MC, Hallagan JB (2008). *Flavor and Extract Manufacturers Association of the United States 2005 poundage and technical effects update survey.* Washington, DC, USA, Flavor and Extract Manufacturers Association of the United States.

Gee JM et al. (2000). Intestinal transport of quercetin glycosides in rats involves both deglycosylation and interaction with the hexose transport pathway. *Journal of Nutrition*, 130:2765–2771.

Gumbmann MR et al. (1978). Toxicity studies of neohesperidin dihydrochalcone. In: Shaw JH, Roussos GG, eds. *Sweeteners and dental caries.* Washington, DC, USA, Information Retrieval Inc., pp. 301–310.

International Organization of the Flavor Industry (2011). Interim inquiry on volume use and added use levels for flavoring agents to be presented at the 76th JECFA meeting. Private communication to the Flavor and Extract Manufacturers Association of the United States, Washington, DC, USA. Submitted to WHO by the International Organization of the Flavor Industry, Brussels, Belgium.

Japan Flavor and Fragrance Materials Association (2005). Japanese inquiry on volume use. Private communication to the Flavor and Extract Manufacturers Association of the United States, Washington, DC, USA. Submitted to WHO by the International Organization of the Flavor Industry, Brussels, Belgium.

Leuschner J (2009). Mutagenicity study of 3',7-dihydroxy-4'-methoxyflavan in the *Salmonella typhimurium* reverse mutation assay (in vitro). Unpublished report no. 18432/25/04 from LPT Laboratory of Pharmacology and Toxicology GmbH & Co., Hamburg, Germany. Submitted to WHO by the International Organization of the Flavor Industry, Brussels, Belgium.

Lina BAR, Dreef-van der Meulen HC, Leegwater DC (1990). Subchronic (13-week) oral toxicity of neohesperidin dihydrochalcone in rats. *Food and Chemical Toxicology*, 28:507–513.

MacGregor JT, Jurd L (1978). Mutagenicity of plant flavonoids: structural requirements for mutagenic activity in *Salmonella typhimurium. Mutation Research*, 54:297–309.

Manach C et al. (2005). Bioavailability and bioefficacy of polyphenols in humans. I. Review of 97 bioavailability studies. *American Journal of Clinical Nutrition*, 81(Suppl. 1):230S–242S.

Nagao M et al. (1981). Mutagenicities of 61 flavonoids and 11 related compounds. *Environmental Mutagenesis*, 3:401–419.

National Toxicology Program (1988a). *2,5-Dimethyl phenol* Salmonella *study detail.* Study No. 068561. Bethesda, MD, USA, United States Department of Health and Human Services, National Institutes of Health, National Institute of Environmental Health Sciences, National Toxicology Program (http://ntp-apps.niehs.nih.gov/ntp_tox/index.cfm?fuseaction=salmonella.salmonellaData&endpointlist=SA&study%5Fno=068561&cas%5Fno=95%2D87%2D4&activetab=detail).

National Toxicology Program (1988b). *2,6-Dimethyl phenol* Salmonella *study detail.* Study No. 391551. Bethesda, MD, USA, United States Department of Health and Human Services, National Institutes of Health, National Institute of Environmental Health Sciences, National Toxicology Program (http://ntp-apps.niehs.nih.gov/ntp_tox/index.cfm?fuseaction=salmonella.salmonellaData&endpointlist=SA&study%5Fno=391551&cas%5Fno=576%2D26%2D1&activetab=detail).

National Toxicology Program (1988c). *3,4-Dimethyl phenol* Salmonella *study detail.* Study No. 244345. Bethesda, MD, USA, United States Department of Health and Human

Services, National Institutes of Health, National Institute of Environmental Health Sciences, National Toxicology Program (http://ntp-apps.niehs.nih.gov/ntp_tox/index.cfm?fuseaction=salmonella.salmonellaData&endpointlist=SA&study%5Fno=244345&cas%5Fno=95%2D65%2D8&activetab=detail).

National Toxicology Program (1989). *2,5-Dimethyl phenol* Salmonella *study detail*. Study No. 755471. Bethesda, MD, USA, United States Department of Health and Human Services, National Institutes of Health, National Institute of Environmental Health Sciences, National Toxicology Program (http://ntp-apps.niehs.nih.gov/ntp_tox/index.cfm?fuseaction=salmonella.salmonellaData&endpointlist=SA&study%5Fno=755471&cas%5Fno=95%2D87%2D4&activetab=detail).

Nutrilite Products Inc. (1980). The safety of neohesperidin dihydrochalcone. Study No. 78559. March. Unpublished report. Submitted to WHO by the International Organization of the Flavor Industry, Brussels, Belgium.

United States Department of Agriculture (2006). USDA database for the flavonoid content of selected foods. Release 2. Beltsville, MD, USA, United States Department of Agriculture, Agricultural Research Service, Beltsville Human Nutrition Research Center, Food Composition Laboratory and Nutrient Data Laboratory.

Waalkens-Berendsen DH, Kuilman-Wahls MEM, Bär A (2004). Embryotoxicity and teratogenicity study with neohesperidin dihydrochalcone in rats. *Regulatory Toxicology and Pharmacology*, 40:74–79.

Yáñez JA et al. (2007). Pharmacokinetics of selected chiral flavonoids: hesperetin, naringenin and eriodictyol in rats and their content in fruit juices. *Biopharmaceutics and Drug Disposition*, 29:63–82.

Zeiger E et al. (1987). *Salmonella* mutagenicity tests: III. Results from the testing of 255 chemicals. *Environmental Mutagenesis*, 9(Suppl. 9):1–110.

PYRAZINE DERIVATIVES (addendum)

First draft prepared by

I.G. Sipes[1], M. DiNovi[2] and P. Sinhaseni[3]

[1] Department of Pharmacology, College of Medicine, University of Arizona, Tucson, Arizona, United States of America (USA)
[2] Center for Food Safety and Applied Nutrition, Food and Drug Administration, College Park, Maryland, USA
[3] Community Risk Analysis Research and Development Center, Bangkok, Thailand

1. Evaluation 223
   1.1 Introduction 223
   1.2 Assessment of dietary exposure 224
   1.3 Absorption, distribution, metabolism and elimination 224
   1.4 Application of the Procedure for the Safety Evaluation of Flavouring Agents 232
   1.5 Consideration of combined intakes from use as flavouring agents 233
   1.6 Conclusions 233
2. Relevant background information 233
   2.1 Explanation 233
   2.2 Additional considerations on dietary exposure 234
   2.3 Biological data 234
      2.3.1 Biochemical data: absorption, distribution, metabolism and elimination 234
      2.3.2 Toxicological studies 234
         (a) Acute toxicity 234
         (b) Short-term studies of toxicity 234
         (c) Genotoxicity 237
3. References 243

1. EVALUATION

1.1 Introduction

The Committee evaluated a group of eight pyrazine derivatives used as flavouring agents. This group includes five alkyl-substituted pyrazine derivatives (Nos 2125–2128 and 2130), two alkoxy-substituted pyrazine derivatives (Nos 2129 and 2131) and one sulfide-substituted pyrazine derivative (No. 2132). The evaluations were conducted according to the Procedure for the Safety Evaluation of Flavouring Agents (see Figure 1, Introduction) (Annex 1, reference *131*). None of these agents have been evaluated previously.

The Committee previously evaluated 41 other members of this group of flavouring agents at its fifty-seventh meeting (Annex 1, reference *154*). All 41 substances in that group were concluded to be of no safety concern based on estimated dietary exposures.

Seven of the eight pyrazine derivatives (Nos 2125–2130 and 2132) in this group have been reported to occur naturally and can be found in chicken, cocoa, oats, malt, peanut, sesame seed, barley, beef, beer, coconut, coffee, pork, potato, shrimp, wild rice, popcorn, hazelnut, *Capsicum* species and beans (Nijssen, van Ingen-Visscher & Donders, 2011).

1.2 Assessment of dietary exposure

The total annual volumes of production of the eight pyrazine derivatives are approximately 1 kg in each of Europe, the USA and Japan (European Flavour and Fragrance Association, 2004; Japan Flavor and Fragrance Materials Association, 2005; Gavin, Williams & Hallagan, 2008; International Organization of the Flavor Industry, 2011). More than 90% of the total annual volumes of production are accounted for by the mixture of 2,5-dimethyl-6,7-dihydro-5H-cyclopentapyrazine and 2,7-dimethyl-6,7-dihydro-5H-cyclopentapyrazine (No. 2128) in Europe, 2-methyl-5-vinylpyrazine (No. 2127) in the USA and 2-ethyl-3-methylthiopyrazine (No. 2132) in Japan.

Dietary exposures were estimated using the maximized survey-derived intake (MSDI) method and the single portion exposure technique (SPET), with the highest values reported in Table 1. The estimated daily dietary exposure is highest for the mixture of 3,5-dimethyl-2-isobutylpyrazine and 3,6-dimethyl-2-isobutylpyrazine (No. 2130) (5000 µg, the SPET value obtained from seasonings and flavours). For the other flavouring agents, the estimated daily dietary exposures range from 0.01 to 3000 µg, with the SPET yielding the highest estimates.

Annual volumes of production of this group of flavouring agents as well as the daily dietary exposures calculated using both the MSDI method and the SPET are summarized in Table 2.

1.3 Absorption, distribution, metabolism and elimination

Information on the absorption, distribution, metabolism and elimination of flavouring agents belonging to the group of pyrazine derivatives has been described in the report of the fifty-seventh meeting (Annex 1, reference *154*).

The pyrazine derivatives in this group are predicted to be absorbed rapidly from the gastrointestinal tract and excreted. The biotransformation of alkyl-substituted pyrazine derivatives (Nos 2125–2128 and 2130) is predicted to occur primarily by oxidation of the side-chains to yield the corresponding secondary alcohols. The resulting alcohols are predicted to undergo conjugation with glucuronic acid. Additional products of oxidative metabolism can be excreted unchanged or conjugated with glycine, glucuronic acid or sulfate before excretion. Hydroxylation of the pyrazine ring may also occur.

Additionally, the alkoxy side-chains are predicted to undergo *O*-dealkylation reactions, followed by excretion as glucuronic acid conjugates in the urine. The presence of sulfur in the side-chain (No. 2132) permits rapid oxidation to sulfoxides and then to sulfones, which are metabolically stable and predicted to be excreted in the urine.

PYRAZINE DERIVATIVES (addendum)

Table 1. Summary of the results of the safety evaluations of pyrazine derivatives used as flavouring agents[a,b,c]

Flavouring agent	No.	CAS No. and structure	Step A3/ B3[d] Does estimated dietary exposure exceed the threshold of concern?	Step A4 Is the flavouring agent or are its metabolites endogenous?	Step A5[e] Adequate margin of exposure for the flavouring agent or a related substance? Follow-on from step B3[e] Are additional data available for flavouring agent with an estimated dietary exposure exceeding the threshold of concern?	Comments on predicted metabolism	Related structure name (No.) and structure (if applicable)	Conclusion based on current estimated dietary exposure
Structural class II								
Isopropenylpyrazine	2125	38713-41-6	B3: Yes, SPET: 3000	NR	Yes. The NOAEL of 14 mg/kg bw per day in a 92-day study in rats for the structurally related 2-vinylpyridine (Vlaovic, 1984) is 280 times the estimated dietary exposure to No. 2125 relative to the SPET value and 84 million times relative to the MSDI (0.01 µg/day) when used as a flavouring agent.	Note 1	2-Vinylpyridine	No safety concern
5-Ethyl-2,3-dimethylpyrazine	2126	15707-34-3	A3: No, SPET: 400	A4: NR	A5: NR	Note 1		No safety concern

continued

Table 1 *(continued)*

Flavouring agent	No.	CAS No. and structure	Step A3/B3[1] Does estimated dietary exposure exceed the threshold of concern?	Step A4 Is the flavouring agent or are its metabolites endogenous?	Step A5[a] Adequate margin of exposure for the flavouring agent or a related substance? *Follow-on from step B3[a]* Are additional data available for flavouring agent with an estimated dietary exposure exceeding the threshold of concern?	Comments on predicted metabolism	Related structure name (No.) and structure (if applicable)	Conclusion based on current estimated dietary exposure
2-Methyl-5-vinylpyrazine	2127	13925-08-1	B3: Yes, SPET: 2000	NR	Yes. The NOAEL of 14 mg/kg bw per day in a 92-day study in rats for the structurally related 2-vinylpyridine (Vlaovic, 1984) is 420 times the estimated dietary exposure to No. 2127 relative to the SPET value and 8.4 million times relative to the MSDI (0.1 µg/day) when used as a flavouring agent.	Note 1	2-Vinylpyridine	No safety concern
Mixture of 2,5-dimethyl-6,7-dihydro-5H-cyclopentapyrazine and 2,7-dimethyl-6,7-dihydro-5H-cyclopentapyrazine	2128	38917-61-2; 38917-62-3	A3: Yes, SPET: 3000	A4: No	A5: Yes. The NOAEL of 50 mg/kg bw per day in a 90-day study in rats for the structurally related 5-methyl-6,7-dihydro-5H-cyclopentapyrazine (No. 781) (Wheldon et al., 1967) is 1000 times the estimated dietary exposure to No. 2128 when used as a flavouring agent.	Note 1	5-Methyl-6,7-dihydro-5H-cyclopentapyrazine (No. 781)	No safety concern

PYRAZINE DERIVATIVES (addendum)

2-Ethoxy-3-isopropylpyrazine	2129	72797-16-1	A3: No, SPET: 1	A4: NR	A5: NR	Note 2		No safety concern

Structural class III

Mixture of 3,5-dimethyl-2-isobutylpyrazine and 3,6-dimethyl-2-isobutylpyrazine	2130	38888-81-2; 70303-42-3	A3: Yes, SPET: 5000	A4: No	A5: Yes. The NOEL of 44 mg/kg bw per day in a 90-day study in rats for the structurally related 2,3,5,6-tetramethylpyrazine (No. 780) (Oser, 1969) is 530 times the estimated dietary exposure to No. 2130 relative to the SPET value and 260 million times relative to the MSDI (0.01 µg/day) when used as a flavouring agent.	Note 1	2,3,5,6-Tetramethylpyrazine (No. 780)	No safety concern
2-Ethoxy-3-ethylpyrazine	2131	35243-43-7	A3: No, SPET: 2	A4: NR	A5: NR	Note 2		No safety concern

Table 1 (continued)

Flavouring agent	No.	CAS No. and structure	Step A3/B3[d] Does estimated dietary exposure exceed the threshold of concern?	Step A4 Is the flavouring agent or are its metabolites endogenous?	Step A5[e] Adequate margin of exposure for the flavouring agent or a related substance? Follow-on from step B3[e] Are additional data available for flavouring agent with an estimated dietary exposure exceeding the threshold of concern?	Comments on predicted metabolism	Related structure name (No.) and structure (if applicable)	Conclusion based on current estimated dietary exposure
2-Ethyl-3-methylthiopyrazine	2132	72987-62-3	A3: No, SPET: 5	A4: NR	A5: NR	Note 3		No safety concern

bw, body weight; CAS, Chemical Abstracts Service; NOAEL, no-observed-adverse-effect level; NR, not required for evaluation because dietary exposure to the flavouring agent was determined to be of no safety concern at step A3 of the Procedure

[a] Forty-one flavouring agents in this group were previously evaluated by the Committee (Annex 1, reference *154*).
[b] *Step 1*: Five flavouring agents in this group (Nos 2125–2129) are in structural class II. Three flavouring agents in this group (Nos 2130–2132) are in structural class III (Cramer, Ford & Hall, 1978).
[c] *Step 2*: Six of the flavouring agents in this group can be predicted to be metabolized to innocuous products.
[d] The thresholds for human dietary exposures for structural classes II and III are 540 and 90 µg/person per day, respectively. All dietary exposures are expressed in µg/day. The dietary exposure values listed represent the highest estimated dietary exposures calculated using either the SPET or the MSDI method. The SPET gave the highest estimated dietary exposures in all cases.
[e] The margins of exposure were calculated based on the estimated dietary exposures calculated using the SPET.

Notes:
1. The biotransformation of substituted pyrazines is expected to occur primarily via oxidation of the side-chain. Alkyl-ring substituents (>C1) are expected to undergo oxidation to the corresponding secondary alcohol, which may be further oxidized to the corresponding ketone for excretion unchanged or conjugated in the urine. An alternative pathway for substituted pyrazines and primary pathway for pyrazine involves hydroxylation of the pyrazine ring. Methyl-substituted pyrazines are oxidized to yield the corresponding pyrazine-2-carboxylic acid derivatives. Products of oxidative metabolism may be excreted unchanged or conjugated with glycine, glucuronic acid or sulfate prior to excretion.
2. Pyrazine or pyrazine derivatives with a ring-activating alkoxy side-chain primarily undergo ring hydroxylation. Additionally, the methoxy groups will undergo demethylation, and the resulting hydroxyl groups will undergo conjugation with glucuronic acid.
3. The presence of sulfur in the side-chain permits rapid oxidation. Alkyl and aromatic sulfides are oxidized to sulfoxides and then to sulfones, which are excreted in the urine.

Table 2. Annual volumes of production and daily dietary exposures for pyrazine derivatives used as flavouring agents in Europe, the USA and Japan

Flavouring agent (No.)	Most recent annual volume of production (kg)[a]	Dietary exposure				Annual volume of consumption via natural occurrence in foods (kg)[d]	Consumption ratio[e]
		MSDI[b]		SPET[c]			
		µg/day	µg/kg bw per day	µg/day	µg/kg bw per day		
Isopropenylpyrazine (2125)							
Europe	0.1	0.01	0.0002	3000	50	+	NA
USA	ND	ND	ND				
Japan	ND	ND	ND				
5-Ethyl-2,3-dimethylpyrazine (2126)							
Europe	ND	ND	ND	400	7	13 359	133 590
USA	0.1	0.01	0.0002				
Japan	ND	ND	ND				
2-Methyl-5-vinylpyrazine (2127)							
Europe	ND	ND	ND	2000	33	3860	3860
USA	1	0.1	0.002				
Japan	ND	ND	ND				
Mixture of 2,5-dimethyl-6,7-dihydro-5H-cyclopentapyrazine and 2,7-dimethyl-6,7-dihydro-5H-cyclopentapyrazine (2128)							
Europe	1	0.1	0.002	3000	50	6605	66 050
USA	0.1	0.01	0.0002				
Japan	ND	ND	ND				

continued

Table 2 (continued)

Flavouring agent (No.)	Most recent annual volume of production (kg)[a]	Dietary exposure				Annual volume of consumption via natural occurrence in foods (kg)[d]	Consumption ratio[e]
		MSDI[b]		SPET[c]			
		µg/day	µg/kg bw per day	µg/day	µg/kg bw per day		
2-Ethoxy-3-isopropylpyrazine (2129)							
Europe	ND	ND	ND	1	0.01	+	NA
USA	ND	ND	ND				
Japan	0.1	0.03	0.0005				
Mixture of 3,5-dimethyl-2-isobutylpyrazine and 3,6-dimethyl-2-isobutylpyrazine (2130)							
Europe	0.1	0.01	0.0002	5000	83	+	NA
USA	0.1	0.01	0.0002				
Japan	ND	ND	ND				
2-Ethoxy-3-ethylpyrazine (2131)							
Europe	ND	ND	ND	2	0.03	–	NA
USA	ND	ND	ND				
Japan	0.1	0.03	0.0005				
2-Ethyl-3-methylthiopyrazine (2132)							
Europe	ND	ND	ND	5	0.1	+	NA
USA	ND	ND	ND				
Japan	1	0.2	0.003				

Total	
Europe	1.2
USA	1.3
Japan	1.2

NA, not applicable; ND, no data reported; +, reported to occur naturally in foods (Nijssen, van Ingen-Visscher & Donders, 2011), but no quantitative data; –, not reported to occur naturally in foods

[a] From European Flavour and Fragrance Association (2004), Japan Flavor and Fragrance Materials Association (2005), Gavin, Williams & Hallagan (2008) and International Organization of the Flavor Industry (2011). Values greater than 0 kg but less than 0.1 kg were reported as 0.1 kg.

[b] MSDI (µg/person per day) calculated as follows:
(annual volume, kg) × (1 × 10^9 µg/kg)/(population × survey correction factor × 365 days), where population (10%, "eaters only") = 32 × 10^6 for Europe, 31 × 10^6 for the USA and 13 × 10^6 for Japan; and where survey correction factor = 0.8 for the surveys in Europe, the USA and Japan, representing the assumption that only 80% of the annual flavour volume was reported in the poundage surveys (European Flavour and Fragrance Association, 2004; Japan Flavor and Fragrance Materials Association, 2005; Gavin, Williams & Hallagan, 2008; International Organization of the Flavor Industry, 2011).

MSDI (µg/kg bw per day) calculated as follows:
(µg/person per day)/body weight, where body weight = 60 kg. Slight variations may occur from rounding.

[c] SPET (µg/person per day) calculated as follows:
(standard food portion, g/day) × (highest usual use level) (International Organization of the Flavor Industry, 2011). The dietary exposure from the single food category leading to the highest dietary exposure from one portion is taken as the SPET estimate.
SPET (µg/kg bw per day) calculated as follows:
(µg/person per day)/body weight, where body weight = 60 kg. Slight variations may occur from rounding.

[d] Quantitative data for the USA reported by Stofberg & Grundschober (1987).

[e] The consumption ratio is calculated as follows:
(annual consumption via natural occurrence in foods in the USA, kg)/(most recent reported annual volume of production as a flavouring agent in the USA, kg).

1.4 Application of the Procedure for the Safety Evaluation of Flavouring Agents

Step 1. In applying the Procedure for the Safety Evaluation of Flavouring Agents to these eight flavouring agents, the Committee assigned five flavouring agents (Nos 2125–2129) to structural class II and three flavouring agents (Nos 2130–2132) to structural class III (Cramer, Ford & Hall, 1978).

Step 2. Six flavouring agents (Nos 2126 and 2128–2132) are predicted to be metabolized to innocuous products. Therefore, the safety evaluation for these flavouring agents proceeded via the A-side of the Procedure. The remaining two flavouring agents (Nos 2125 and 2127) are not predicted to be metabolized to innocuous products. The safety evaluation for these two flavouring agents proceeded down the B-side of the Procedure.

Step A3. Estimated dietary exposures to two of the three flavouring agents in this group in structural class II (Nos 2126 and 2129) are below the threshold of concern (i.e. 540 µg/person per day for class II). These flavouring agents would not be expected to be of concern at current estimated dietary exposures. The estimated dietary exposure to the other flavouring agent in this group (No. 2128) in structural class II is above the threshold of concern, and therefore its safety evaluation proceeded to step A4. The estimated daily dietary exposures to two of the three flavouring agents in structural class III (Nos 2131 and 2132) are below the threshold of concern (i.e. 90 µg/person per day for class III). These flavouring agents would not be expected to be of concern at current estimated dietary exposures. The estimated dietary exposure to the remaining flavouring agent (No. 2130) in structural class III is above the threshold of concern, and the safety evaluation of this flavouring agent proceeded to step A4.

Step A4. The two flavouring agents (Nos 2128 and 2130) and their respective metabolites considered at this step are not endogenous, and the safety evaluations for these flavouring agents therefore proceeded to step A5.

Step A5. For the mixture of 2,5-dimethyl-6,7-dihydro-5H-cyclopentapyrazine and 2,7-dimethyl-6,7-dihydro-5H-cyclopentapyrazine (No. 2128), the no-observed-adverse-effect level (NOAEL) of 50 mg/kg of body weight (bw) per day for the structurally related 5-methyl-6,7-dihydro-5H-cyclopentapyrazine (No. 781) from a 90-day dietary study in rats (Wheldon et al., 1967) provides a margin of exposure of 1000 in relation to the current estimated dietary exposure to No. 2128 (SPET = 3000 µg/day) when used as a flavouring agent.

For the mixture of 3,5-dimethyl-2-isobutylpyrazine and 3,6-dimethyl-2-isobutylpyrazine (No. 2130), the no-observed-effect level (NOEL) of 44 mg/kg bw per day for the structurally related 2,3,5,6-tetramethylpyrazine (No. 780) from a 90-day dietary study in rats (Oser, 1969) provides a margin of exposure of 530 in relation to the current estimated dietary exposure to No. 2130 calculated using the SPET (5000 µg/day) or 260 million in relation to the MSDI (0.01 µg/day) when used as a flavouring agent.

Step B3. The estimated dietary exposures to two flavouring agents (Nos 2125 and 2127) in structural class II are above the threshold of concern (540 µg/person per day for class II). Accordingly, for these flavouring agents, data are required on the flavouring agent or a closely related substance in order to perform a safety evaluation.

Consideration of flavouring agents with high exposure evaluated via the B-side of the decision-tree:

For isopropenylpyrazine (No. 2125), the NOAEL of 14 mg/kg bw per day (based on a dose of 20 mg/kg bw per day administered 5 days/week) for the structurally related 2-vinylpyridine from a 92-day oral toxicity study in rats (Vlaovic, 1984) provides a margin of exposure of 280 in relation to the current estimated dietary exposure to No. 2125 calculated using the SPET (3000 µg/day) or 84 million compared with the MSDI (0.01 µg/day) when used as a flavouring agent.

For 2-methyl-5-vinylpyrazine (No. 2127), the NOAEL of 14 mg/kg bw per day (based on a dose of 20 mg/kg bw per day administered 5 days/week) for the structurally related 2-vinylpyridine from a 92-day oral toxicity study in rats (Vlaovic, 1984) provides a margin of exposure of 420 in relation to the current estimated dietary exposure to No. 2127 calculated using the SPET (2000 µg/day) or 8.4 million compared with the MSDI (0.1 µg/day) when used as a flavouring agent.

The Committee therefore concluded that none of the eight additional flavouring agents (Nos 2125–2132) in this group of pyrazine derivatives would pose a safety concern at current estimated dietary exposures.

Table 1 summarizes the evaluations of all eight flavouring agents in this group.

1.5 Consideration of combined intakes from use as flavouring agents

The highest MSDI for any of these eight pyrazine derivatives is 0.2 µg/day. Consideration of combined intakes is not deemed necessary, because the additional flavouring agents would not contribute significantly to the combined intake of this flavouring group.

1.6 Conclusions

In the previous evaluation of the flavouring agents in this group of pyrazine derivatives, biochemical data, metabolism and acute toxicity studies, short-term studies of toxicity, long-term studies of toxicity and carcinogenicity, and studies of genotoxicity and reproductive toxicity were available (Annex 1, reference *154*). The additional toxicity data on subchronic toxicity (No. 784 and the structurally related 2-vinylpyridine) and on genotoxicity (several previously evaluated pyrazine derivatives and 2-vinylpyridine) considered at this meeting support the previous evaluation.

The Committee concluded that these eight flavouring agents, which are additions to the group of pyrazine derivatives evaluated previously, would not give rise to safety concerns at current estimated dietary exposures.

2. RELEVANT BACKGROUND INFORMATION

2.1 Explanation

This monograph summarizes the key data relevant to the safety evaluation of five alkyl-substituted pyrazine derivatives (Nos 2125–2128 and 2130), two alkoxy-substituted pyrazine derivatives (Nos 2129 and 2131) and one sulfide-substituted pyrazine derivative (No. 2132) (see Table 1), which are additions to the group of pyrazine derivatives evaluated previously.

2.2 Additional considerations on dietary exposure

Seven of the eight pyrazine derivatives (Nos 2125–2130 and 2132) in this group have been reported to occur naturally and can be found in chicken, cocoa, oats, malt, peanut, sesame seed, barley, beef, beer, coconut, coffee, pork, potato, shrimp, wild rice, popcorn, hazelnut, *Capsicum* species and beans (Nijssen, van Ingen-Visscher & Donders, 2011).

2.3 Biological data

2.3.1 Biochemical data: absorption, distribution, metabolism and elimination

No relevant information additional to that available and described in the monograph of the fifty-seventh meeting (Annex 1, reference *154*) was available on the absorption, distribution, metabolism or elimination of flavouring agents belonging to the group of pyrazine derivatives.

2.3.2 Toxicological studies

Short-term and long-term studies of the toxicity of any of the eight flavouring agents evaluated in this group have not been reported. The results of studies on the toxicity of 2-acetylpyrazine (No. 784) and of a structurally related substance, 2-vinylpyridine, as well as a number of genotoxicity studies, were made available and are summarized below.

(a) Acute toxicity

Although acute oral median lethal doses (LD_{50} values) in rats have not been reported for the eight flavouring agents evaluated in this group, the LD_{50} value for 2-vinylpyridine, a structural relative of isopropenylpyrazine (No. 2125) and 2-methyl-5-vinylpyrazine (No. 2127), was determined to be 336 mg/kg bw in rats (Vlaovic, 1984). As reported previously, in the mouse, the acute oral LD_{50} for 2-methylpyrazine (No. 761) was greater than 2000 mg/kg bw (Hope, 1983) and for 2-methoxy-(3, 5 or 6)-isopropylpyrazine (No. 790) was greater than 5000 mg/kg bw (Johnson, 1978). All the available acute oral LD_{50} values indicate a low level of toxicity for the substituted pyrazine derivatives, as summarized in Table 3.

(b) Short-term studies of toxicity

The results of short-term studies of the toxicity of pyrazine derivatives used as flavouring agents are summarized in Table 4 and described below.

(i) 2-Acetylpyrazine (No. 784)

In a 90-day dietary toxicity study, 30 rats of each sex were given 2-acetylpyrazine (No. 784) in the diet, sufficient to provide a dose of approximately 18 mg/kg bw per day. There were no mortalities, no changes in feed intake or body

Table 3. Results of acute oral toxicity studies with pyrazine derivatives used as flavouring agents

No.	Flavouring agent	Species; sex	LD_{50} (mg/kg bw)	Reference
761	2-Methylpyrazine	Mouse; M, F	>2000	Hope (1983)
790	2-Methoxy-(3, 5 or 6)-isopropylpyrazine	Mouse; M, F	>5000	Johnson (1978)

F, female; M, male

Table 4. Results of short-term studies of the toxicity of pyrazine derivatives used as flavouring agents

No.	Flavouring agent	Species; sex	No. of test groups[a] / no. per group[b]	Route	Duration (days)	NOEL (mg/kg bw per day)	Reference
784	2-Acetylpyrazine	Rat; M, F	1/60	Diet	90	18[c]	Oser (1970)

F, female; M, male
[a] Total number of test groups does not include control animals.
[b] Total number per test group includes both male and female animals.
[c] The only dose tested.

weight, no adverse physical signs and no events indicative of unusual behaviour during the study period. Similarly, there were no significant differences in the haematological values, blood chemistry or urine analysis. There were no significant differences found in the relative weights of the liver, kidney, lung, uterus, bladder, thyroid, pancreas, prostate or bladder.

As no responses indicative of toxicological or pathological effects were associated with administration of 2-acetylpyrazine to rats at a dose of 18 mg/kg bw per day, the author concluded that this dose, the only dose tested, represented a NOEL (Oser, 1970).

(ii) Structurally related substance: 2-vinylpyridine

Three short-term studies of toxicity have been reported for 2-vinylpyridine, which is structurally related to isopropenylpyrazine (No. 2125) and 2-methyl-5-vinylpyrazine (No. 2127).

In a 28-day repeated-dose toxicity study, five rats of each sex per dose were administered 2-vinylpyridine at doses of 0, 12.5, 50 or 200 mg/kg bw per day by corn oil gavage. Another group of five rats of each sex was administered 2-vinylpyridine at 200 mg/kg bw per day for 28 days followed by a 14-day recovery period prior to termination of the study. The rats were observed daily for clinical signs of toxicity. Body weight and feed consumption were measured weekly. Haematology and blood chemistry parameters were measured at the end of the study. A complete necropsy and histopathological examination were performed.

Sialorrhoea (hypersalivation) was observed in the 50 and 200 mg/kg bw per day treatment groups. Body weight gain and feed consumption were both decreased in the 200 mg/kg bw per day group relative to controls. Urine analysis revealed decreases in specific gravity for the females receiving 50 and 200 mg/kg bw per day, accompanied by an increase in urine volume for the 200 mg/kg bw per day females. Testes weights, relative to controls, were increased in males in the 200 mg/kg bw per day group. Decreases in relative and absolute spleen weights and increases in relative liver weights were reported in high-dose females. Statistically significant increases in squamous cell hyperplasia were observed in both sexes at 200 mg/kg bw per day. Males administered 200 mg/kg bw per day showed erosion and cellular infiltration of the forestomach. Females receiving 50 and 200 mg/kg bw per day showed submucosal oedema and/or erosion in the glandular stomach.

Based on these observations, the authors set the NOAEL at 12.5 mg/kg bw per day (Oba et al., 2009).

In an oral toxicity study, five Sprague-Dawley rats of each sex per dose were administered 1–13 doses of 2-vinylpyridine at 0, 80, 200 or 500 mg/kg bw per day via corn oil gavage over a period of 1–17 days. The 500 mg/kg bw per day groups were killed in extremis after 1–2 doses; however, tissue samples were not collected or evaluated. The 200 mg/kg bw per day male group showed a slight reduction in feed consumption and body weight gain during days 1–4. Both sexes of the 200 mg/kg bw per day group showed weakness, hypersalivation, depressed activity, tremors, a slight increase in absolute and relative liver weights, and hyperkeratosis and acanthosis of the non-glandular stomach (usually associated with the administration of a gastric irritant). Females at this dose level also showed a slight increase in alanine aminotransferase activity, minimal levels of atypical lymphocytes in the blood, and haemorrhage, oedema, acute inflammation and focal necrosis of the non-glandular gastric mucosa. The 80 mg/kg bw per day males showed sialorrhoea, a slight increase in alkaline phosphatase level, a slight increase in relative liver weight, and hyperkeratosis, acanthosis and oedema of the non-glandular stomach. Sialorrhoea and hyperkeratosis and acanthosis of the non-glandular gastric mucosa were also observed in females administered 80 mg/kg bw per day.

A NOAEL for 2-vinylpyridine could not be determined (Vlaovic, 1984).

In a subchronic oral toxicity study, groups of 30 rats of each sex per dose were administered 2-vinylpyridine via corn oil gavage at doses of 0, 20, 60 or 180 mg/kg bw per day, 5 days/week, for 42 or 92 days. Any signs of toxicity and changes in body weights and feed consumption were recorded. Haematology and clinical chemistry were examined at the time of necropsy. Interim necropsies were performed for 10 rats of each sex per dose at day 42. Liver, spleen, heart, ovary, testes, kidney, brain and adrenal gland weights were measured at necropsy and examined for gross lesions.

Four rats died during the study, which was attributed to gavage administration error. Males in the group receiving 180 mg/kg bw per day showed a reduction in body weight and feed consumption compared with controls. None of the doses caused significant differences in haematology parameters between test and control groups. Clinical chemistry evaluations revealed a significant, dose-dependent decrease in mean aspartate aminotransferase activities in the males, the meaning of which remains uncertain. The absolute liver weights were significantly increased for both sexes in the 180 mg/kg bw per day group, and the relative liver weights were significantly increased for both sexes in the 60 and 180 mg/kg bw per day groups. The relative kidney weights of all treated males and the 180 mg/kg bw per day group females were significantly increased. The absolute brain weights were significantly decreased, the relative brain weights were significantly increased and the absolute heart weights were significantly decreased in the 180 mg/kg bw per day group males. In males of the 20 and 180 mg/kg bw per day groups, absolute and relative adrenal weights were increased, but as no significant increases were reported in the 60 mg/kg bw per day group, the effect was likely not related to administration of the test substance. Increases were observed in relative testes and ovary weights in the 180 mg/kg bw per day group males and females, respectively. Pathological evaluation of tissues from the 180 mg/kg bw per day group animals obtained at 42 days (interim group) revealed lesions of the non-glandular stomach, including acanthosis, hyperkeratosis, congestion, acute inflammatory cell infiltrates, oedema, haemorrhage, coagulation, necrosis of the superficial mucosa and mucosal erosion. At day 42, the following were observed in the 60 mg/kg bw per day group: acanthosis, hyperkeratosis, congestion and acute inflammatory cell infiltrates of the non-glandular stomach. Lesions were not observed at 42 days in the tissues obtained from the rats treated with 20 mg/kg bw per day. Evaluation after 92 days revealed lesions of the non-glandular stomach that included acanthosis, hyperkeratosis, oedema, chronic inflammation and congestion with the 180 mg/kg bw per day treatments and acanthosis and hyperkeratosis of the non-glandular stomach at 60 mg/kg bw per day.

Based on increased absolute and relative liver weights and gross evidence of gastric irritation at 60 mg/kg bw per day, and their absence at 20 mg/kg bw per day, the author of the study determined the NOAEL for 2-vinylpyridine to be 20 mg/kg bw per day (Vlaovic, 1984). However, it was noted that the doses were administered 5 days/week instead of 7 days/week. Based on dosing on 5 days/week, the NOAEL was adjusted to 14 mg/kg bw per day.

(c)   Genotoxicity

Studies of the genotoxicity of pyrazine derivatives used as flavouring agents are summarized in Table 5 and described below.

(i)   In vitro

Reverse mutation assays have been reported for the structurally related substance, 2-vinylpyridine. No evidence of genotoxicity was observed when *Salmonella typhimurium* strains TA98, TA100 and TA1535 were incubated with 0, 0.1 or 0.5 ml of the highly volatile 2-vinylpyridine in a desiccator for 7 hours with or without S9 metabolic activation. However, the 0.5 ml atmosphere was cytotoxic to the bacteria (Simmon & Baden, 1980). In a standard plate incorporation and preincubation assay protocol, *S. typhimurium* strains TA98, TA100, TA1535 and TA1537 and *Escherichia coli* WP2*uvrA* were incubated with 0, 39.1, 78.1, 156, 313, 625, 1250, 2500 or 5000 μg of 2-vinylpyridine per plate in the presence and absence of rat S9 metabolic activation. 2-Vinylpyridine was toxic at the highest doses. Results

Table 5. Studies of the genotoxicity of pyrazine derivatives used as flavouring agents

No.	Flavouring agent	End-point	Test object	Concentration	Results	Reference
In vitro						
771	2,3-Diethylpyrazine	Reverse mutation	Salmonella typhimurium TA98, TA100, TA1535, TA1537 and TA1538; Escherichia coli WP2uvrA	1.5, 5, 15, 50, 150, 500, 1500 and 5000 µg/plate	Negative[a]	Kawamura (2004)
771	2,3-Diethylpyrazine	Chromosomal aberration	Chinese hamster lung cells	180, 350, 700 and 1400 µg/ml	Positive[a,b,c]	Kouzi (2004)
771	2,3-Diethylpyrazine	Chromosomal aberration	Chinese hamster lung cells	180, 350, 700 and 1400 µg/ml	Negative[c,d,e]	Kouzi (2004)
775	2-Ethyl-3,(5 or 6)-dimethylpyrazine	Reverse mutation	S. typhimurium TA98, TA100, TA1535, TA1537 and TA1538	0.02, 0.04, 0.07, 0.15, 0.29, 0.59, 1.17, 2.34, 4.69, 9.38, 18.75, 37, 50, 75 and 150 µg/plate	Negative[a,f]	Jagannath (1983)
775	2-Ethyl-3,(5 or 6)-dimethylpyrazine	Reverse mutation	S. typhimurium TA98, TA100, TA1535, TA1537 and TA1538	50 000 µg/plate	Negative[a]	Heck, Vollmuth & Cifone (1989)
775	2-Ethyl-3,(5 or 6)-dimethylpyrazine	Unscheduled DNA synthesis	Rat hepatocytes	100 µg/ml	Negative[a]	Heck, Vollmuth & Cifone (1989)
790	2-Methoxy-(3, 5 or 6)-isopropylpyrazine	Reverse mutation	S. typhimurium TA98, TA100, TA1535, TA1537 and TA1538	15, 50, 150, 500 and 1500 µg/plate	Negative[a]	Richold, Jones & Fenner (1983a)
791	2-Methoxy-3-(1-methylpropyl)pyrazine	Reverse mutation	S. typhimurium TA98, TA100, TA102, TA1535 and TA1537	0.5, 1.5, 5, 15 and 50 µg/plate	Negative[a]	Richold, Jones & Fenner (1983b)
792	2-Isobutyl-3-methoxypyrazine	Reverse mutation	S. typhimurium TA98, TA100, TA1535 and TA1537	15, 50, 150, 500 and 1500 µg/plate	Negative[a]	Richold, Jones & Hales(1983)
951	Pyrazine	Reverse mutation	S. typhimurium TA98 and TA100	Not reported	Negative[a]	Takahashi & Ono (1993)

951	Pyrazine	Reverse mutation	S. typhimurium TA98, TA100, TA1535 and TA1537	333, 1000, 3333, 6666 and 10 000 µg/plate	Negative[a]	Fung et al. (1988)
951	Pyrazine	Mouse lymphoma assay	L517PYTK(+/−)	4857, 6143, 7429, 8714 and 10 000 µg/ml	Negative	Fung et al. (1988)
In vivo						
771	2,3-Diethylpyrazine	Micronucleus induction	ICR mice; M	62.5, 125 and 250 mg/kg bw per day	Positive[g]	Hatano Research Institute (2004)

M, male

[a] With and without metabolic activation.
[b] Six-hour treatment followed by an 18-hour recovery period.
[c] Toxicity could not be assessed because mitotic indices were not reported for the control treatment.
[d] Without metabolic activation.
[e] Continuous treatment for 24 hours.
[f] Preincubation.
[g] Bone marrow harvested 24 hours after administration of the test article.

were negative for the *S. typhimurium* strains and positive at the highest non-toxic dose of 2500 µg/plate in *E. coli* WP2*uvrA* without metabolic activation and at 1250, 2500 and 5000 µg/plate with metabolic activation. A confirmatory assay in *E. coli* WP2*uvrA* incubated with 0, 2500, 3000, 3500, 4000, 4500 or 5000 µg of 2-vinylpyridine per plate in the presence of S9 showed an increase in reverse mutants that was not dose dependent, with toxicity at the highest dose (Nakajima et al., 2004a). Thus, the positive result was reproduced in a second independent experiment, but was not concentration dependent; therefore, the biological significance of these results remains unclear. No mutagenic activity was observed in the reverse mutation assay in *E. coli* WP2*uvrA* at 2-vinylpyridine concentrations of 625 µg/plate or below.

A chromosomal aberration study has also been reported for 2-vinylpyridine using concentrations up to 300 µg/ml in Chinese hamster lung cells. Aberrations were observed in the presence and absence of metabolic activation with 24- and 48-hour exposures and with continuous exposure (Nakajima et al., 2004b). An English translation was not available to evaluate whether the chromosomal aberrations observed were the result of induction of cytotoxicity.

An Ames assay has been reported for 2,3-diethylpyrazine (No. 771) in *S. typhimurium* strains TA98, TA100, TA1535, TA1537 and TA1538 and *E. coli* WP2*uvrA* at 0, 1.5, 5, 15, 50, 150, 500, 1500 and 5000 µg/plate with and without S9 metabolic activation. 2,3-Diethylpyrazine was not mutagenic under the conditions of this assay (Kawamura, 2004).

To evaluate the in vitro clastogenic potential of 2,3-diethylpyrazine (No. 771), a cytogenetic analysis was performed in Chinese hamster lung cells treated for 6 hours with 2,3-diethylpyrazine at 0, 180, 350, 700 and 1400 µg/ml with and without S9 metabolic activation, followed by an 18-hour recovery period. An additional treatment in which cells were exposed to 2,3-diethylpyrazine continuously for 24 hours without metabolic activation was included in the study. Increased mitotic indices were reported for the highest concentration; however, these observations are of limited value in determining the clastogenicity of the test material, as no mitotic indices were provided for the vehicle control assays. After a 6-hour treatment in the presence and absence of metabolic activation, a statistically significant increase in the percentage of aberrant cells (excluding gaps) was obtained at the highest test concentration of 1400 µg/ml (19.5% aberrant cells in the absence of metabolic activation and 16.5% in the presence of S9 with a vehicle control level of 1.5–3.0% aberrant cells). In addition, an increase in polyploidal cells was observed after treatment with 700 µg/ml in the absence of metabolic activation and treatment with 700 and 1400 µg/ml in the presence of metabolic activation. No significant increases in aberration frequency or polyploidy were obtained after 24-hour treatment in the absence of S9. The clastogenic effects at 1400 µg/ml may be related to the relatively high toxicity at this dose (approximately 40–50% reduction in cell growth relative to the vehicle control), and the measure of cytotoxicity used (cell monolayer confluence) may have underestimated any cytotoxic effect. However, it was concluded that treatment of Chinese hamster lung cells with 2,3-diethylpyrazine is associated with statistically significant increases in chromosomal aberration frequency and numerical aberrations (exclusively polyploidy), in both the presence and absence of metabolic activation (Kouzi, 2004). The biological significance of this finding is unclear, although it is recognized that a number of compounds have been reported to induce polyploidy in the Chinese hamster lung cell line (Ishidate, 1988). In this particular case, the in vitro clastogenic effect in Chinese hamster lung cells and the increase in polyploidy may be related to cytotoxicity rather than true genotoxicity, as other alkylpyrazines have been reported to induce clastogenic effects in vitro in Chinese hamster cells at high, cytotoxic concentrations of pyrazine (Stich et al., 1980). At a previous JECFA meeting (Annex 1, reference *155*), the Committee stated that the relevance of genotoxic events associated with pyrazine and other alkylpyrazine derivatives in *Saccharomyces cerevisiae* and Chinese hamster ovary cells is difficult to interpret because of the use of high, nearly toxic concentrations.

A reverse mutation assay was conducted for 2-ethyl-3,(5 or 6)-dimethylpyrazine (No. 775) in *S. typhimurium* strains TA98, TA100, TA1535, TA1537 and TA1538 treated with 0.02–150 µg/plate with and without metabolic activation. No significant increases in reverse mutants resulted (Jagannath, 1983). 2-Ethyl-3,(5 or 6)-dimethylpyrazine was also reported to be not mutagenic at 50 000 µg/plate with and without metabolic liver activation in *S. typhimurium* strains TA98, TA100, TA1535, TA1537 and TA1538. It was also negative for unscheduled DNA synthesis in rat hepatocytes at a concentration of 100 µg/ml (Heck, Vollmuth & Cifone, 1989).

Negative Ames assay results were reported in *S. typhimurium* strains TA98, TA100, TA1535, TA1537 and TA1538 for 2-methoxy-(3, 5 or 6)-isopropylpyrazine (No. 790) at 15–1500 µg/plate, 2-methoxy-3-(1-methylpropyl)pyrazine (No. 791) at 0.5–50 µg/plate and 2-isobutyl-3-methoxypyrazine (No. 792) at 15–1500 µg/plate, with or without metabolic activation (Richold, Jones & Fenner, 1983a,b; Richold, Jones & Hales, 1983). Genotoxic potential was not observed in *S. typhimurium* strains TA98 and TA100 incubated in the presence of pyrazine (No. 951) with and without metabolic activation (no concentration reported; Takahashi & Ono, 1993) or in strains TA98, TA100, TA1535 and TA1537 up to concentrations of 10 000 µg/plate (Fung et al., 1988). Pyrazine was also negative in a mouse lymphoma L517PYTK(+/–) mutation assay at concentrations up to 10 000 µg/ml (Fung et al., 1988).

(ii) In vivo

To further elucidate the genotoxic profile of 2,3-diethylpyrazine (No. 771), an in vivo mouse bone marrow micronucleus study was conducted with oral administration of 62.5, 125 and 250 mg/kg bw per day. These doses were chosen on the basis of a preliminary toxicity study in which mortality occurred at doses of 500 mg/kg bw and higher. Groups of five male CD-1 ICR mice were dosed by oral gavage for 2 days at a 24-hour interval. Mice were euthanized approximately 24 hours after the second and final dose, and bone marrow was collected for the estimation of micronucleus frequency by scoring polychromatic erythrocytes (PCEs; 2000 per mouse). Cytotoxicity was assessed by determining the ratio of polychromatic to total erythrocytes (% PCE). The study resulted in a statistically significant increase in micronucleus frequency after treatment with all three doses, with a 6- to 7-fold increase over concurrent control in the two highest doses. A dose-related response trend was observed where the effect seemed to plateau at the highest dose, which can be explained by a significant reduction in the per cent PCE, indicative of bone marrow toxicity. It was concluded that oral administration of 2,3-diethylpyrazine to mice is associated with clastogenic/aneugenic effects in the bone marrow (Hatano Research Institute, 2004). This result contrasts with negative results obtained in this same assay for a range of other structurally related alkylpyrazines administered at very high oral doses (1–3 g/kg bw). The positive results of the mouse micronucleus test may occur by substance-induced hypothermia (Asanami & Shimono, 1997) or hyperthermia (King & Wild, 1983). Based on the absence of an increase in per cent PCE with 2,3-diethylpyrazine treatment in the in vivo micronucleus assay, it is assumed that the positive results could not be attributed to an effect on erythropoiesis. In fact, 2,3-diethylpyrazine treatment led to a decrease in per cent PCE at the highest dose. With an effect on erythropoiesis reasonably discounted, the possibility of the positive result in the mouse micronucleus assay may be due to disturbances in thermoregulation. In a study by Boulet (2011), this phenomenon was evaluated in a core body temperature measurement. In this study, 2,3-diethylpyrazine (0, 125 or 250 mg/kg bw per day on days 1 and 2 for two total doses) was administered by oral gavage to male ICR mice ($n = 5$ per dose group). Doses chosen for the study were the highest doses evaluated in the positive in vivo micronucleus assay. The low dose was considered to be in the range of the NOAEL. Core body temperatures were measured for all treatment groups at 0, 2, 4, 8 and 16 hours after each daily administration of 2,3-diethylpyrazine. A dose-dependent toxicity was observed with 2,3-diethylpyrazine, with one death resulting at the 250 mg/kg bw per day dose on the second day prior to body temperature measurement. The body temperatures were significantly reduced in a dose-dependent manner by treatment with 2,3-diethylpyrazine (Boulet, 2011). The results of this study could confirm that 2,3-diethylpyrazine treatment in mice at high doses causes disturbances in homeostasis (i.e. thermoregulation), resulting in induction of hypothermia. As referenced above (King & Wild, 1983), as well as by others (Shelby, Tice & Witt, 1997; Asanami, Shimon & Kaneda, 1998; Spencer, Gollapudi & Waechter, 2007), alterations in thermoregulation resulting from substance administration can be associated with the positive results in the in vivo micronucleus assay. In studies reported by Hatano Research Institute (2004), 2,3-diethylpyrazine (125 and 250 mg/kg bw per day) causes substantial reductions in body temperature in mice and shows positive results in the micronucleus test in male and female CD-1 (ICR) mice. No definite conclusion about mechanisms of toxicity can be drawn, because no confirmatory tests have been conducted to determine whether the maintenance of normal body temperature can prevent the induction of micronuclei by these specific compounds. Maintaining body temperature has been effective in preventing the induction of micronuclei by other compounds that can reduce body temperature, but results of such studies on these pyrazine derivatives were not available.

The weight of evidence obtained in genotoxicity studies performed on pyrazine derivatives and structurally related substances supports the conclusion that these substances are not mutagenic, which suggests that these pyrazine derivatives are not directly genotoxic. Results of in vitro bacterial mutagenicity tests are negative except at concentrations approaching toxic levels, as were the results of clastogenicity studies in Chinese hamster lung cells. The atypical positive results in one in vivo micronucleus test may be related to chemical-induced physiological alterations (hypothermia) and bone marrow toxicity associated with the high doses of 2,3-diethylpyrazine administered in these studies. Studies with other compounds of this group as well as control of hypothermia during dosing are needed to better understand the relevance of these positive findings in the mouse micronucleus test.

3. REFERENCES

Asanami S, Shimono K (1997). Hypothermia induces micronuclei in mouse bone marrow cells. *Mutation Research*, 393(1–2):91–98.
Asanami S, Shimon K, Kaneda S (1998). Transient hypothermia induces micronuclei in mice. *Mutation Research*, 413(1):7–14.
Boulet J (2011). 2,3-Diethylpyrazine: a 2-day oral toxicity study in mice. Unpublished report no. 32679 from Product Safety Labs, Dayton, NJ, USA. Submitted to WHO by the International Organization of the Flavor Industry, Brussels, Belgium.
Cramer GM, Ford RA, Hall RL (1978). Estimation of toxic hazard—a decision tree approach. *Food and Cosmetics Toxicology*, 16:255–276.
European Flavour and Fragrance Association (2004). European inquiry on volume use. Private communication to the Flavor and Extract Manufacturers Association of the United States, Washington, DC, USA. Submitted to WHO by the International Organization of the Flavor Industry, Brussels, Belgium.
Fung VA et al. (1988). Mutagenic activity of some coffee ingredients. *Mutation Research*, 204(2):219–228.
Gavin CL, Williams MC, Hallagan JB (2008). *Flavor and Extract Manufacturers Association of the United States 2005 poundage and technical effects update survey*. Washington, DC, USA, Flavor and Extract Manufacturers Association of the United States.
Hatano Research Institute (2004). The micronucleus test of 2,3-diethylpyrazine in male and female CD-1 (1CR) mice. Unpublished report from Hatano Research Institute, Hadano, Japan. Submitted to WHO by the International Organization of the Flavor Industry, Brussels, Belgium.
Heck JD, Vollmuth TA, Cifone MA (1989). An evaluation of food flavoring ingredients in a genetic toxicity screening battery. *Toxicologist*, 9(1):257.
Hope J (1983). Acute oral toxicity test of methyl pyrazine. Unpublished report no. 43000 from Unilever Research Laboratory, Bedfordshire, England. Submitted to WHO by the International Organization of the Flavor Industry, Brussels, Belgium.
International Organization of the Flavor Industry (2011). Interim inquiry on volume use and added use levels for flavoring agents to be presented at the 76th JECFA meeting. Private communication to the Flavor and Extract Manufacturers Association of the United States, Washington, DC, USA. Submitted to WHO by the International Organization of the Flavor Industry, Brussels, Belgium.
Ishidate M Jr (1988). *Data book of chromosomal aberration test in vitro*, revised ed. Tokyo, Japan, Life-Science Information Center, pp. 118–119.
Jagannath DR (1983). Mutagenic potential of B166 (2-ethyl-3,5(6) dimethyl pyrazine). Unpublished report no. 20988 from Litton Bionetics, Inc., Kensington, MD, USA. Submitted to WHO by the International Organization of the Flavor Industry, Brussels, Belgium.
Japan Flavor and Fragrance Materials Association (2005). Japanese inquiry on volume use. Private communication to the Flavor and Extract Manufacturers Association of the United States, Washington, DC, USA. Submitted to WHO by the International Organization of the Flavor Industry, Brussels, Belgium.
Johnson AW (1978). Acute oral toxicity test 2-isopropyl-3-methoxy pyrazine. Unpublished report no. 11571 from Unilever Research Laboratory, Bedfordshire, England. Submitted to WHO by the International Organization of the Flavor Industry, Brussels, Belgium.
Kawamura K (2004). 2,3-Diethylpyrazine: bacterial mutation assay (Ames-type test). Unpublished report no. SR03198 from Hatano Research Institute, Hadano, Japan. Submitted to WHO by the International Organization of the Flavor Industry, Brussels, Belgium.
King MT, Wild D (1983). The mutagenic potential of hyperthermia and fever in mice. *Mutation Research*, 111(2):219–226.
Kouzi MS (2004). The chromosomal aberration test 2,3-diethylpyrazine using Chinese hamster cultured cells. Unpublished report no. G-03-080 from Hatano Research Institute, Hadano,

Japan. Submitted to WHO by the International Organization of the Flavor Industry, Brussels, Belgium.

Nakajima M et al. (2004a) Reverse mutation test of 2-vinylpyridine on bacteria. Unpublished report from Biosafety Research Center, Shizuoka, Japan. Submitted to WHO by the International Organization of the Flavor Industry, Brussels, Belgium.

Nakajima M et al. (2004b) In vitro chromosomal aberration test of 2-vinylpyridine on cultured Chinese hamster cells. Unpublished report from Biosafety Research Center, Shizuoka, Japan. Submitted to WHO by the International Organization of the Flavor Industry, Brussels, Belgium.

Nijssen LM, van Ingen-Visscher CA, Donders JJH (2011). Volatile Compounds in Food (VCF) database, version 13.1. Zeist, the Netherlands, TNO Triskelion (http://www.vcf-online.nl/VcfHome.cfm).

Oba K et al. (2009). Twenty-eight-day repeat dose oral toxicity test of 2-vinylpyridine in rats. Unpublished report from Biosafety Research Center, Shizuoka, Japan. Submitted to WHO by the International Organization of the Flavor Industry, Brussels, Belgium.

Oser BL (1969). 90-day feeding study with 2,3,5-tetramethyl pyrazine in rats. Unpublished report no. 90614 from Food and Drug Research Laboratories, Maspeth, NY, USA. Submitted to WHO by the International Organization of the Flavor Industry, Brussels, Belgium.

Oser BL (1970). 90-day feeding study with 2-acetyl pyrazine. Unpublished report no. 90617 from Food and Drug Research Laboratories, Maspeth, NY, USA. Submitted to WHO by the International Organization of the Flavor Industry, Brussels, Belgium.

Richold M, Jones E, Fenner LA (1983a). Ames metabolic activation test to assess the potential mutagenic effect of PM 535. Unpublished report no. ABR149 from Huntingdon Life Sciences, Cambridgeshire, England. Submitted to WHO by the International Organization of the Flavor Industry, Brussels, Belgium.

Richold M, Jones E, Fenner LA (1983b). Ames metabolic activation test to assess the potential mutagenic effect of PM 517. Unpublished report no. ULR 96E/8368 from Huntingdon Life Sciences, Cambridgeshire, England. Submitted to WHO by the International Organization of the Flavor Industry, Brussels, Belgium.

Richold M, Jones E, Hales JF (1983). Ames metabolic activation test to assess the potential mutagenic effect of isobutyl methoxy pyrazine. Unpublished report no. ULR 93A/82908 from Huntingdon Life Sciences, Cambridgeshire, England. Submitted to WHO by the International Organization of the Flavor Industry, Brussels, Belgium.

Shelby MD, Tice RR, Witt KL (1997). Hypothermia induces micronuclei in mouse bone marrow cells. *Mutation Research*, 393(1):91–98.

Simmon VF, Baden JM (1980). Mutagenic activity of vinyl compounds and derived epoxides. *Mutation Research*, 78(3):227–231.

Spencer PJ, Gollapudi BB, Waechter JM Jr (2007). Induction of micronuclei by phenol in the bone marrow: I. Association with chemically induced hypothermia. *Toxicological Sciences*, 97(1):120–127.

Stich HF et al. (1980). Mutagenic activity of pyrazine derivatives: a comparative study with *Salmonella typhimurium*, *Saccharomyces cerevisiae* and Chinese hamster ovary cells. *Food and Cosmetics Toxicology*, 18(6):581–584.

Stofberg J, Grundschober F (1987). Consumption ratio and food predominance of flavoring materials. *Perfumer & Flavorist*, 12:27–68.

Takahashi A, Ono H (1993). Mutagenicity assessment in 44 epoxy resins hardeners in *Salmonella typhimurium* tester strains. *Chemistry Express*, 6(9):785–788.

Vlaovic MS (1984). Subchronic oral toxicology of 2-vinylpyridine. Unpublished report no. 180295/TX-84-19 from Eastman Kodak Company, Rochester, NY, USA. Submitted to WHO by the International Organization of the Flavor Industry, Brussels, Belgium.

Wheldon GH et al. (1967). The effects of ten food-flavouring additives administered to rats over a period of thirteen weeks. Unpublished report no. 1903/67/103 from Huntingdon Life Sciences, Cambridgeshire, England. Submitted to WHO by the International Organization of the Flavor Industry, Brussels, Belgium.

PYRIDINE, PYRROLE AND QUINOLINE DERIVATIVES (addendum)

First draft prepared by

S.M.F. Jeurissen[1], M. DiNovi[2], A. Mattia[2] and A.G. Renwick[3]

[1] Centre for Substances and Integrated Risk Assessment, National Institute for Public Health and the Environment, Bilthoven, the Netherlands
[2] Center for Food Safety and Applied Nutrition, Food and Drug Administration, College Park, Maryland, United States of America (USA)
[3] School of Medicine, University of Southampton, Southampton, England

1. Evaluation 245
   1.1 Introduction 245
   1.2 Assessment of dietary exposure 246
   1.3 Absorption, distribution, metabolism and elimination 246
   1.4 Application of the Procedure for the Safety Evaluation of
       Flavouring Agents 258
   1.5 Consideration of combined intakes from use as flavouring
       agents 259
   1.6 Conclusions 260
2. Relevant background information 260
   2.1 Explanation 260
   2.2 Additional considerations on dietary exposure 260
   2.3 Biological data 260
       2.3.1 Biochemical data: absorption, distribution, metabolism
             and elimination 260
             (a) Pyrroles 261
       2.3.2 Toxicological studies 261
             (a) Acute toxicity 261
             (b) Short-term and long-term studies of toxicity and
                 carcinogenicity 261
             (c) Genotoxicity 269
             (d) Reproductive/developmental toxicity 273
3. References 274

1. EVALUATION

1.1 Introduction

The Committee evaluated 11 additional flavouring agents belonging to the group of pyridine, pyrrole and quinoline derivatives. The additional flavouring agents included two pyrroles (Nos 2150 and 2152), eight alkylated pyridines (Nos 2151, 2153–2156 and 2158–2160) and one quinoline (No. 2157). The Committee decided to evaluate two imidazolidines that were originally submitted in this group (Nos 2161 and 2162) as additional flavouring agents belonging to the group of miscellaneous nitrogen-containing substances (see relevant monograph in this volume). The evaluations were conducted according to the Procedure for the Safety Evaluation of Flavouring Agents (see Figure 1, Introduction) (Annex 1, reference 131). None of these flavouring agents have previously been evaluated by the Committee. Three of the flavouring agents in this group (Nos 2158–2160) that were evaluated at this meeting are reported to be flavour modifiers.

The Committee previously evaluated 22 other members of this group of flavouring agents at its sixty-third meeting (Annex 1, reference 173). The Committee concluded that all 22 flavouring agents in that group were of no safety concern at estimated dietary exposures.

Three of the 11 flavouring agents evaluated at the current meeting are natural components of food (Nos 2150–2152) and have been detected in black and green teas, coffee, cocoa, mate, skim milk powder, beer, wine, cognac, onion, popcorn, raisin, clam, cheese, egg, oats, pork, shoyu, wheaten bread, liquorice, honey, peanut, potato, okra, soya bean and tamarind (Nijssen, van Ingen-Visscher & Donders, 2011).

1.2 Assessment of dietary exposure

The total annual volumes of production of the 11 pyridine, pyrrole and quinoline derivatives are approximately 0.4 kg in the USA, 0.2 kg in Europe and 64 kg in Japan (European Flavour and Fragrance Association, 2004; Japan Flavor and Fragrance Materials Association, 2005; Gavin, Williams & Hallagan, 2008; International Organization of the Flavor Industry, 2011). Approximately 98% of the total annual volume of production in Japan is accounted for by one flavouring agent in this group—namely, 2-methoxypyridine (No. 2156).

Dietary exposures were estimated using the maximized survey-derived intake (MSDI) method and the single portion exposure technique (SPET), with the highest values reported in Table 1. The estimated dietary exposure is highest for 2,4-dimethylpyridine (No. 2151) (4000 μg/day, the SPET value obtained from soups). For the other flavouring agents, the estimated daily dietary exposures, calculated using either the MSDI method or the SPET, range from 0.01 to 1500 μg, with the SPET yielding the highest estimates.

Annual volumes of production of this group of flavouring agents as well as the daily dietary exposures calculated using both the MSDI method and the SPET are summarized in Table 2.

1.3 Absorption, distribution, metabolism and elimination

Information on the absorption, distribution, metabolism and elimination of the flavouring agents belonging to the group of pyridine, pyrrole and quinoline derivatives has previously been described in the monograph of the sixty-third meeting (Annex 1, reference 174).

New structural elements not evaluated previously are aldehyde-substituted pyrroles (Nos 2150 and 2152). The aldehyde group of aldehyde-substituted pyrroles can be expected to be oxidized to the corresponding carboxylic acid, as was shown for pyrrole-2-carboxaldehyde.

Table 1. Summary of the results of the safety evaluations of pyridine, pyrrole and quinoline derivatives used as flavouring agents[a,b,c]

Flavouring agent	No.	CAS No. and structure	Step A3/B3[d] Does estimated dietary exposure exceed the threshold of concern?	Step A4 Is the flavouring agent or are its metabolites endogenous?	Step A5/B4[e] Adequate margin of exposure for the flavouring agent or a related substance? Follow-on from step B3[e] Are additional data available for flavouring agent with an estimated dietary exposure exceeding the threshold of concern?	Step B5 Do the conditions of use result in a dietary exposure greater than 1.5 μg/day?	Comments on predicted metabolism	Related structure name (No.) and structure (if applicable)	Conclusion based on current estimated dietary exposure
Structural class II									
1-Ethyl-2-pyrrole-carboxaldehyde	2150	2167-14-8	B3: No, SPET: 510	NR	B4: No	Yes	Notes 1–3	NR	Additional data required to complete evaluation

continued

Table 1 (continued)

Flavouring agent	No.	CAS No. and structure	Step A3/B3[1] Does estimated dietary exposure exceed the threshold of concern?	Step A4 Is the flavouring agent or are its metabolites endogenous?	Step A5/B4[e] Adequate margin of exposure for the flavouring agent or a related substance? Follow-on from step B3[e] Are additional data available for flavouring agent with an estimated dietary exposure exceeding the threshold of concern?	Step B5 Do the conditions of use result in a dietary exposure greater than 1.5 μg/day?	Comments on predicted metabolism	Related structure name (No.) and structure (if applicable)	Conclusion based on current estimated dietary exposure
2,4-Dimethylpyridine	2151	108-47-4	A3: Yes, SPET: 4000	No	A5: Yes. The NOAEL of 30 mg/kg bw per day for the related substance 5-ethyl-2-methylpyridine (No. 1318), based on an abstract describing a 28-day study in rats (Biomedizinische Forschungsanstalt m.b.H., 1988), is 450 (based on the SPET) and 180 million (based on the MSDI) times the estimated daily dietary exposure to No. 2151 when used as a flavouring agent.	NR	Notes 3 and 4	5-Ethyl-2-methylpyridine (No. 1318)	No safety concern (temporary)

PYRIDINE, PYRROLE AND QUINOLINE DERIVATIVES (addendum)

							Additional data required to complete evaluation
1-Methyl-1H-pyrrole-2-carboxaldehyde	2152	1192-58-1	B3: No, SPET: 300	NR	B4: No	Yes	Notes 1–3
Structural class III							
2-Acetyl-4-isopropenylpyridine	2153	142896-11-5	A3: No, SPET: 3	NR	NR	NR	Notes 4–6 No safety concern
4-Acetyl-2-isopropenylpyridine	2154	142896-12-6	A3: No, SPET: 3	NR	NR	NR	Notes 4–6 No safety concern

continued

Table 1 (continued)

Flavouring agent	No.	CAS No. and structure	Step A3/B3[1] Does estimated dietary exposure exceed the threshold of concern?	Step A4 Is the flavouring agent or are its metabolites endogenous?	Step A5/B4[e] Adequate margin of exposure for the flavouring agent or a related substance? Follow-on from step B3[e] Are additional data available for flavouring agent with an estimated dietary exposure exceeding the threshold of concern?	Step B5 Do the conditions of use result in a dietary exposure greater than 1.5 µg/day?	Comments on predicted metabolism	Related structure name (No.) and structure (if applicable)	Conclusion based on current estimated dietary exposure
2-Acetyl-4-isopropylpyridine	2155	142896-09-1	A3: No, SPET: 3	NR	NR	NR	Notes 3–5		No safety concern
2-Methoxypyridine	2156	1628-89-3	B3: No, SPET: 40	NR	B4: No	Yes	Notes 3, 4 and 7		Additional data required to complete evaluation

PYRIDINE, PYRROLE AND QUINOLINE DERIVATIVES (addendum)

Name	No.	CAS No.	Specified use level / SPET	Additional data			Conclusion based on current intake
6-Methoxyquinoline	2157	5263-87-6	B3: No, SPET: 25	NR	B4. Yes. The NOAEL of 140 mg/kg bw per day for the structurally related substance 8-hydroxyquinoline in a 2-year study in rats and mice (National Toxicology Program, 1985) is 340 000 times the estimated daily dietary exposure to No. 2157 when used as a flavouring agent.	Notes 4, 7 and 8	8-Hydroxyquinoline — No safety concern
1-(2-Hydroxyphenyl)-3-(pyridin-4-yl)propan-1-one	2158	1186004-10-3	B3: Yes, SPET: 1500	NR	Additional data: Genotoxic potential in vitro was demonstrated for No. 2158.	Note 7	Additional data required to complete evaluation

continued

Table 1 (continued)

Flavouring agent	No.	CAS No. and structure	Step A3/B3[d] Does estimated dietary exposure exceed the threshold of concern?	Step A4 Is the flavouring agent or are its metabolites endogenous?	Step A5/B4[e] Adequate margin of exposure for the flavouring agent or a related substance? Follow-on from step B3[e] Are additional data available for flavouring agent with an estimated dietary exposure exceeding the threshold of concern?	Step B5 Do the conditions of use result in a dietary exposure greater than 1.5 µg/day?	Comments on predicted metabolism	Related structure name (No.) and structure (if applicable)	Conclusion based on current estimated dietary exposure
1-(2-Hydroxy-4-isobutoxyphenyl)-3-(pyridin-2-yl)propan-1-one	2159	1190230-47-7	B3: Yes, SPET: 1000	NR	Additional data: Genotoxic potential in vitro was demonstrated for the structurally related flavouring agent No. 2158.	NR	Note 7		Additional data required to complete evaluation
1-(2-Hydroxy-4-methoxyphenyl)-3-(pyridin-2-yl)propan-1-one	2160	1190229-37-8	B3: Yes, SPET: 1000	NR	Additional data: Genotoxic potential in vitro was demonstrated for the structurally related flavouring agent No. 2158.	NR	Note 7		Additional data required to complete evaluation

bw, body weight; CAS, Chemical Abstracts Service; NOAEL, no-observed-adverse-effect level; NR, not required for evaluation

a Twenty-two flavouring agents belonging to the group of pyridine, pyrrole and quinoline derivatives were previously evaluated by the Committee at its sixty-third meeting (Annex 1, reference *173*).
b *Step 1*: Three flavouring agents in this group (Nos 2150–2152) are in structural class II. The other eight flavouring agents in this group (Nos 2153–2160) are in structural class III.
c *Step 2*: Four of the flavouring agents in this group can be predicted to be metabolized to innocuous products.
d The thresholds for human dietary exposure for structural classes II and III are 540 and 90 µg/person per day, respectively. All dietary exposure values are expressed in µg/day. The dietary exposure values listed represent the highest estimated dietary exposures calculated using either the SPET or the MSDI method. The SPET gave the highest estimated dietary exposures in all cases.
e The margins of exposure were calculated based on the estimated dietary exposures calculated using the SPET.

Notes:
1. The pyrrole ring undergoes hydroxylation and is excreted in the urine as the corresponding glucuronic acid conjugate.
2. The aldehyde group can be expected to be oxidized to the corresponding carboxylic acid. Alkyl side-chain oxidation may also occur.
3. Alkyl side-chain oxidation followed by glucuronic acid conjugation and excretion or oxidation to the corresponding carboxylic acid.
4. The pyridine ring system undergoes hydroxylation and is excreted in the urine as the corresponding glucuronic acid conjugate.
5. The acetyl group is reduced and conjugated with glucuronic acid.
6. Alkenyl side-chain oxidation followed by glucuronic acid conjugation and excretion or oxidation to the corresponding carboxylic acid.
7. *N*-oxidation and *O*-dealkylation followed by glucuronic acid or sulfate conjugation.
8. Forms a reactive epoxide metabolite that is detoxified through glutathione conjugation.

Table 2. Annual volumes of production and daily dietary exposures for pyridine, pyrrole and quinoline derivatives used as flavouring agents in Europe, the USA and Japan

Flavouring agent (No.)	Most recent annual volume of production (kg)[a]	Dietary exposure				Annual volume of consumption via natural occurrence in foods (kg)[d]	Consumption ratio[e]
		MSD[b]		SPET[c]			
		µg/day	µg/kg bw per day	µg/day	µg/kg bw per day		
1-Ethyl-2-pyrrolecarboxaldehyde (2150)							
Europe	ND	ND	ND			2692	NA
USA	ND	ND	ND	510	8.5		
Japan	0.1	0.03	0.0005				
2,4-Dimethylpyridine (2151)							
Europe	0.1	0.01	0.0002			0.4	4
USA	0.1	0.01	0.0002	4000	67		
Japan	ND	ND	ND				
1-Methyl-1H-pyrrole-2-carboxaldehyde (2152)							
Europe	0.1	0.01	0.0002			22 052	NA
USA	ND	ND	ND	300	5		
Japan	ND	ND	ND				

PYRIDINE, PYRROLE AND QUINOLINE DERIVATIVES (addendum)

2-Acetyl-4-isopropenylpyridine (2153)					
Europe	ND	ND		3	0.05	–	NA
USA	ND	ND					
Japan	0.1	0.03	0.0005				
4-Acetyl-2-isopropenylpyridine (2154)					
Europe	ND	ND		3	0.05	–	NA
USA	ND	ND					
Japan	0.1	0.03	0.0005				
2-Acetyl-4-isopropylpyridine (2155)					
Europe	ND	ND		3	0.05	–	NA
USA	ND	ND					
Japan	0.1	0.03	0.0005				
2-Methoxypyridine (2156)					
Europe	ND	ND		40	0.7	–	NA
USA	ND	ND					
Japan	63	18	0.3				
6-Methoxyquinoline (2157)					
Europe	ND	ND		25	0.4	–	NA
USA	ND	ND					
Japan	0.1	0.03	0.0005				

continued

Table 2 (continued)

Flavouring agent (No.)	Most recent annual volume of production (kg)[a]	Dietary exposure				Annual volume of consumption via natural occurrence in foods (kg)[d]	Consumption ratio[e]
		MSDI[b]		SPET[c]			
	µg/day	µg/day	µg/kg bw per day	µg/day	µg/kg bw per day		
1-(2-Hydroxyphenyl)-3-(pyridin-4-yl)propan-1-one (2158)				1500	25	–	NA
Europe	ND	ND					
USA	0.1	0.01	0.0002				
Japan	ND	ND					
1-(2-Hydroxy-4-isobutoxyphenyl)-3-(pyridin-2-yl)propan-1-one (2159)				1000	17	–	NA
Europe	ND	ND					
USA	0.1	0.01	0.0002				
Japan	ND	ND					
1-(2-Hydroxy-4-methoxyphenyl)-3-(pyridin-2-yl)propan-1-one (2160)				1000	17	–	NA
Europe	ND	ND					
USA	0.1	0.01	0.0002				
Japan	ND	ND					

	Total
Europe	0.2
USA	0.4
Japan	64

bw, body weight; NA, not applicable; ND, no data reported; –, not reported to occur naturally in foods

[a] From European Flavour and Fragrance Association (2004), Japan Flavor and Fragrance Materials Association (2005), Gavin, Williams & Hallagan (2008) and International Organization of the Flavor Industry (2011). Values greater than 0 kg but less than 0.1 kg were reported as 0.1 kg.

[b] MSDI (µg/person per day) calculated as follows:
(annual volume, kg) × (1 × 10^9 µg/kg)/(population × survey correction factor × 365 days), where population (10%, "eaters only") = 32 × 10^6 for Europe, 31 × 10^6 for the USA and 13 × 10^6 for Japan; and where survey correction factor = 0.8 for the surveys in Europe, the USA and Japan, representing the assumption that only 80% of the annual flavour volume was reported in the poundage surveys (European Flavour and Fragrance Association, 2004; Japan Flavor and Fragrance Materials Association, 2005; Gavin, Williams & Hallagan, 2008; International Organization of the Flavor Industry, 2011).

MSDI (µg/kg bw per day) calculated as follows:
(µg/person per day)/body weight, where body weight = 60 kg. Slight variations may occur from rounding.

[c] SPET (µg/person per day) calculated as follows:
(standard food portion, g/day) × (highest usual use level) (International Organization of the Flavor Industry, 2011). The dietary exposure from the single food category leading to the highest dietary exposure from one portion is taken as the SPET estimate.

SPET (µg/kg bw per day) calculated as follows:
(µg/person per day)/body weight, where body weight = 60 kg. Slight variations may occur from rounding.

[d] Quantitative data for the USA reported by Stofberg & Grundschober (1987).

[e] The consumption ratio is calculated as follows:
(annual consumption via natural occurrence in foods in the USA, kg)/(most recent reported annual volume of production as a flavouring agent in the USA, kg).

1.4 Application of the Procedure for the Safety Evaluation of Flavouring Agents

Step 1. In applying the Procedure for the Safety Evaluation of Flavouring Agents to the 11 flavouring agents in this group of pyridine, pyrrole and quinoline derivatives, the Committee assigned 3 flavouring agents to structural class II (Nos 2150–2152) and 8 flavouring agents to structural class III (Nos 2153–2160) (Cramer, Ford & Hall, 1978).

Step 2. Four of the flavouring agents (Nos 2151 and 2053–2155) in this group can be predicted to be metabolized to innocuous products. The evaluation of these flavouring agents therefore proceeded via the A-side of the Procedure. The remaining flavouring agents (Nos 2050, 2052 and 2156–2160) in this group cannot be predicted to be metabolized to innocuous products. Therefore, the evaluation of these flavouring agents proceeded via the B-side of the Procedure.

Step A3. The highest estimated dietary exposure to one flavouring agent in structural class II (No. 2151) is above the relevant threshold of concern (i.e. 540 μg/person per day for class II). Accordingly, the evaluation of this flavouring agent proceeded to step A4.

The highest estimated dietary exposures to all three flavouring agents in structural class III (Nos 2153–2155) are below the threshold of concern (i.e. 90 μg/person per day for class III). The Committee therefore concluded that these flavouring agents are not of safety concern at current estimated dietary exposures.

Step A4. Neither the flavouring agent 2,4-dimethylpyridine (No. 2151) nor its metabolites are endogenous substances. Accordingly, the evaluation of this flavouring agent proceeded to step A5.

Step A5. The no-observed-adverse-effect level (NOAEL) for the structurally related substance 5-ethyl-2-methylpyridine (No. 1318) of 30 mg/kg of body weight (bw) per day, based on an abstract describing a 28-day study in rats (Biomedizinische Forschungsanstalt m.b.H.,1988), provides a margin of exposure of 450 in relation to the highest estimated dietary exposure to 2,4-dimethylpyridine (No. 2151; SPET = 4000 μg/day) when used as a flavouring agent. The Committee noted that the margin of exposure for No. 2151 based on the MSDI of 0.01 μg/day is 180 million and concluded that the margins of exposure of 450 (based on the SPET) and 180 million (based on the MSDI) are adequate.

The Committee therefore concluded that this flavouring agent is not of safety concern at current estimated dietary exposures. However, because the full study was not available, the safety evaluation for No. 2151 was considered temporary.

Step B3. The highest estimated dietary exposures to the two flavouring agents in structural class II (Nos 2150 and 2152) and two flavouring agents in structural class III (Nos 2156 and 2157) are below the relevant threshold of concern (i.e. 540 μg/person per day for class II and 90 μg/person per day for class III). Accordingly, the evaluation of these four flavouring agents proceeded to step B4.

The highest estimated dietary exposures to the three remaining flavouring agents in structural class III (Nos 2158–2160) are above the relevant threshold of concern. Therefore, additional data are necessary for the evaluation of these flavouring agents.

Step B4. The NOAEL of 140 mg/kg bw per day for the structurally related substance 8-hydroxyquinoline in a 2-year study in rats and mice (National Toxicology Program, 1985) provides a margin of exposure of 340 000 for 6-methoxyquinoline (No. 2157; SPET = 25 μg/day) when used as a flavouring agent.

For 1-ethyl-2-pyrrolecarboxaldehyde (No. 2150), 1-methyl-1H-pyrrole-2-carboxaldehyde (No. 2152) and 2-methoxypyridine (No. 2156), no toxicological data are available on the flavouring agents or adequate related substances with which to calculate a margin of exposure. Therefore, the evaluation of these flavouring agents proceeded to step B5.

Step B5. The conditions of use for 1-ethyl-2-pyrrolecarboxaldehyde (No. 2150), 1-methyl-1H-pyrrole-2-carboxaldehyde (No. 2152) and 2-methoxypyridine (No. 2156) result in dietary exposures greater than 1.5 μg/day. Therefore, the Committee determined that additional data would be necessary to complete the evaluation of these flavouring agents.

Consideration of flavouring agents with high exposure evaluated via the B-side of the decision-tree:

In accordance with the Procedure, additional data were evaluated for 1-(2-hydroxyphenyl)-3-(pyridin-4-yl)propan-1-one (No. 2158), 1-(2-hydroxy-4-isobutoxyphenyl)-3-(pyridin-2-yl)propan-1-one (No. 2159) and 1-(2-hydroxy-4-methoxyphenyl)-3-(pyridin-2-yl)propan-1-one (No. 2160), as the estimated dietary exposures exceeded the threshold of concern for structural class III (90 μg/person per day).

For the three structurally related compounds Nos 2158–2160, the data available indicate potential genotoxicity. No. 2158 was negative in two bacterial reverse mutation tests with and without metabolic activation (Sokolowski, 2009). It was positive for clastogenicity in an in vitro chromosomal aberration assay using human lymphocytes at concentrations of 245 μg/ml or more without metabolic activation in two independent tests (Bohnenberger, 2009). No. 2159 was negative in a bacterial reverse mutation test with and without metabolic activation and a forward mutation test in mouse lymphoma L5178Y cells (Wollny, 2009; Sokolowski, 2010). For No. 2160, no genotoxicity data are available. The data available for No. 2158 demonstrate that this compound has genotoxic potential in vitro. As in vivo genotoxicity data are not available for No. 2158 and as no tests on clastogenicity are available for the structurally related compounds Nos 2159 and 2160, the Committee determined that additional data on genotoxicity would be necessary to complete the safety evaluations of Nos 2158–2160.

Table 1 summarizes the evaluations of the 11 pyridine, pyrrole and quinoline derivatives (Nos 2150–2160) in this group.

1.5 Consideration of combined intakes from use as flavouring agents

The 11 additional flavouring agents in this group of pyridine, pyrrole and quinoline derivatives all have MSDI values less than or equal to 20% of the threshold of concern for structural class III. Consideration of combined intakes is therefore not deemed necessary.

1.6 Conclusions

In the previous evaluation of flavouring agents in the group of pyridine, pyrrole and quinoline derivatives, studies of acute toxicity, short-term and long-term studies of toxicity and carcinogenicity (21–460 days) and studies of genotoxicity were available (Annex 1, reference 173). The toxicity data available for this evaluation generally supported those from previous evaluations. However, in the previous evaluation of this group, it was concluded, on the basis of the available evidence, that the 22 pyridine, pyrrole and quinoline derivatives evaluated would not demonstrate genotoxic potential. On the basis of the data available for the current evaluation, this conclusion cannot be extended to Nos 2158–2160. The Committee concluded that, due to concerns with genotoxicity, additional data on genotoxicity would be necessary to complete the evaluation of Nos 2158–2160 at current estimated dietary exposures.

The Committee concluded that additional toxicity data on the flavouring agents or adequate related substances would be necessary to complete the evaluation of Nos 2150, 2152 and 2156 at current estimated dietary exposures.

The Committee concluded that five flavouring agents (Nos 2151, 2153–2155 and 2157), which are additions to the group of pyridine, pyrrole and quinoline derivatives evaluated previously, would not give rise to safety concerns at current estimated dietary exposures. However, for No. 2151, the safety evaluation was temporary, pending the submission of the full report of the critical study for the next JECFA meeting at which flavouring agents are evaluated.

2. RELEVANT BACKGROUND INFORMATION

2.1 Explanation

This monograph summarizes key aspects relevant to the safety evaluation of 11 pyridine, pyrrole and quinoline derivatives, which are additions to a group of 22 flavouring agents evaluated previously by the Committee at its sixty-third meeting (Annex 1, reference 173).

2.2 Additional considerations on dietary exposure

Annual volumes of production and dietary exposures estimated using both the MSDI method and the SPET for each flavouring agent are reported in Table 2.

Three of the 11 flavouring agents in the group are natural components of food (Stofberg & Grundschober, 1987; Nijssen, van Ingen-Visscher & Donders, 2011) (Table 2).

2.3 Biological data

2.3.1 Biochemical data: absorption, distribution, metabolism and elimination

Information on the hydrolysis, absorption, distribution, metabolism and elimination of flavouring agents belonging to the group of pyridine, pyrrole and quinoline derivatives has been described in the report of the sixty-third meeting

(Annex 1, reference 173). An additional study on pyrrole-2-carboxaldehyde is summarized below.

(a) Pyrroles

Male Wistar rats received a single dose of 50 mg of pyrrole-2-carboxaldehyde per kilogram of body weight via oral gavage (in water). In urine samples collected 24 hours after administration, pyrrole-2-carboxylic acid was found, and its formation from pyrrole-2-carboxaldehyde was reported to be stoichiometric (Kagami et al., 2008).

2.3.2 Toxicological studies

(a) Acute toxicity

No oral median lethal doses (LD_{50} values) were available for the 11 flavouring agents in this group. For the structurally related material 5-ethyl-2-methylpyridine (No. 1318), oral LD_{50} values ranging from 459 to 2295 mg/kg bw were reported in rats, mice and rabbits (Table 3). These results support the findings in the previous evaluation (Annex 1, reference 173) that the oral acute toxicity of flavouring agents belonging to the group of pyridine, pyrrole and quinoline derivatives is low.

(b) Short-term and long-term studies of toxicity and carcinogenicity

Short-term and long-term studies of the toxicity and carcinogenicity of pyridine, pyrrole and quinoline derivatives used as flavouring agents and structurally related compounds are described below. Those studies for which NOAELs were derived are summarized in Table 4.

Table 3. Results of acute oral toxicity studies with pyridine, pyrrole and quinoline derivatives used as flavouring agents

No.	Flavouring agent	Species; sex	LD_{50} (mg/kg bw)	Reference
1318	5-Ethyl-2-methylpyridine	Rat; M, F	1737	Hazleton Laboratories Europe Ltd (1986)
1318	5-Ethyl-2-methylpyridine	Rat; NR	710	Bio-Toxicology Laboratories, Inc. (1976)
1318	5-Ethyl-2-methylpyridine	Rat; F	918–2295	Consultox Laboratories Ltd (1973)
1318	5-Ethyl-2-methylpyridine	Mouse; NR	569	Consultox Laboratories Ltd (1973)
1318	5-Ethyl-2-methylpyridine	Rabbit; NR	459–918	Consultox Laboratories Ltd (1973)

F, female; M, male; NR, not reported

Table 4. Results of oral short-term and long-term studies of toxicity and carcinogenicity with pyridine, pyrrole and quinoline derivatives used as flavouring agents and structurally related compounds

No.	Flavouring agent	Species; sex	No. of test groups[a] / no. per group[b]	Route	Duration (days)	NOAEL (mg/kg bw per day)	Reference
2158	1-(2-Hydroxyphenyl)-3-(pyridin-4-yl)-propan-1-one	Rat; M, F	4/10–20	Diet	28	340	Korgaonkar (2009a)
2159	1-(2-Hydroxy-4-isobutoxyphenyl)-3-(pyridin-2-yl)-propan-1-one	Rat; M, F	4/10–20	Diet	28	340	Korgaonkar (2009b)
NA	N-Methylpyrrole	Rat; M, F	3/10	Diet	28	30	Dhinsa, Watson & Brooks (2008)
NA	8-Hydroxyquinoline	Rat; M, F Mouse; M, F	5/20	Diet	91	Rat: 170 Mouse: 410	National Toxicology Program (1985)
NA	8-Hydroxyquinoline	Rat; M, F Mouse; M, F	2/100	Diet	730	Rat: 140[c] Mouse: 400[c]	National Toxicology Program (1985)

F, female; M, male; NA, not applicable

[a] Total number of test groups does not include control animals.
[b] Total number per test group includes both male and female animals.
[c] Highest dose tested.

(i)   1-(2-Hydroxyphenyl)-3-(pyridin-4-yl)propan-1-one (No. 2158)

Groups of male and female Crl:CD(SD) rats were given 1-(2-hydroxyphenyl)-3-(pyridin-4-yl)propan-1-one at dietary concentrations of 0, 400, 1000, 4000 or 10 000 mg/kg feed (equal to 0, 40, 90, 340 and 850 mg/kg bw per day for males and 0, 40, 90, 380 and 1000 mg/kg bw per day for females) for 28 days. Group sizes were 5 rats of each sex per group for the 400 and 1000 mg/kg feed groups and 10 rats of each sex per group for the 0, 4000 and 10 000 mg/kg feed groups. After 28 days of exposure, five males and five females from each group were killed. The remaining rats were killed following a 14-day recovery period. The study was performed according to Organisation for Economic Co-operation and Development (OECD) Test Guideline 407 (Repeated Dose 28-Day Oral Toxicity Study in Rodents; 2008). All animals were observed twice daily for mortality and moribundity. Clinical examinations were performed daily, and detailed physical examinations were performed weekly. Individual body weights and feed consumption were recorded weekly. Functional observational battery and locomotor activity tests were performed during weeks 3 and 5 (recovery period). Haematological, biochemical, urinary and thyroid hormone parameters were determined in week 4 and week 6. Complete necropsies were conducted on all animals, and selected organs were weighed at the scheduled necropsies. Selected tissues were examined microscopically from all animals in the control and the highest-dose groups at primary necropsy. Gross lesions and livers were examined from all animals at primary and recovery necropsies. Adrenal glands were examined from all animals at primary necropsy and from females of the 4000 and 10 000 mg/kg feed groups at recovery necropsy.

No effects on survival of the rats were observed. Body weight gains were significantly decreased in rats of the highest-dose group during study weeks 0–1, but the cumulative body weight gain at the end of the dosing period was not significantly different from controls. Slightly lower feed consumption was noted in the highest-dose group in study weeks 0–1 and during the remainder of the dosing period for males only.

Statistically significantly higher rearing counts were noted in males of the highest-dose group. Because rearing counts were unusually low in the controls compared with other studies and because of the lack of any other related effects, this was not considered to be of toxicological importance.

Urine, haematological and biochemical analyses did not reveal treatment-related effects. Reticulocyte counts were higher in females of the 4000 and 10 000 mg/kg feed groups in the recovery period. However, these changes fell within the historical control range and were not noted at the end of the dosing period. In males, lower albumin (10 000 mg/kg feed group only, 4.1 g/dl compared with 4.4 g/dl in controls) and total bilirubin values (0.10 and 0.08 mg/dl in 4000 and 1000 mg/kg feed group, respectively, compared with 0.14 mg/dl in controls) were noted in week 4. These changes were small, fell within the historical control range and, for total bilirubin, involved a change in a direction of no relevant biological importance. Lower mean total thyroxine (−37%) and reverse triiodothyronine levels (−42%) were noted in the 10 000 mg/kg feed group males at study week 4. As no correlating changes in organ weights, clinical pathology or histopathology were observed and as the values were within the historical control range, these changes were not considered toxicologically relevant.

Macroscopic examinations did not reveal treatment-related effects. Some changes in organ weights were observed. Absolute and relative liver weights were increased (up to 16%) in males and females of the highest-dose group. These changes were not observed in the recovery group. At the recovery necropsy, higher absolute and relative adrenal gland weights (up to 34%) were noted in females of the 4000 and 10 000 mg/kg feed groups compared with controls.

Minimal to mild hepatocellular vacuolation was observed in males of the 4000 and 10 000 mg/kg feed groups (in 1/5 and 5/5 animals, respectively) and in females of the 400, 4000 and 10 000 mg/kg feed groups (in 2/5, 1/5 and 4/5 animals, respectively) at primary necropsy and in livers of males and females of the highest-dose group at recovery necropsy (2/5 and 1/5 animals, respectively). Additionally, minimal to mild hypertrophy of the adrenal cortex was noted in females in the 4000 and 10 000 mg/kg feed groups (1/5 and 2/5 animals, respectively) at recovery necropsy.

Based on the increased incidences of minimal to mild hepatocellular vacuolation, the NOAEL for 1-(2-hydroxyphenyl)-3-(pyridin-4-yl)propan-1-one is 4000 mg/kg feed, equal to 340 mg/kg bw per day (Korgaonkar, 2009a).

(ii)   1-(2-Hydroxy-4-isobutoxyphenyl)-3-(pyridin-2-yl)propan-1-one (No. 2159)

Crl:CD(SD) rats were administered 1-(2-hydroxy-4-isobutoxyphenyl)-3-(pyridin-2-yl)propan-1-one at dietary concentrations of 0, 400, 1000, 4000 or 10 000 mg/kg feed (equal to 0, 30, 80, 340 and 900 mg/kg bw per day for males and 0, 40, 100, 370 and 940 mg/kg bw per day for females) for 28 days. Group sizes were 5 rats of each sex per group for the 400 and 1000 mg/kg feed groups and 10 rats of each sex per group for the 0, 4000 and 10 000 mg/kg feed groups. The study was performed according to OECD Test Guideline 407. After 28 days of exposure, five males and five females of each group were killed. The remaining rats were killed following a 14-day recovery period. Animals were observed twice daily for mortality and moribundity. Clinical examinations were performed daily, and detailed physical examinations were performed weekly. Individual body weights and feed consumption were recorded weekly. Functional observational battery and locomotor activity tests were performed during weeks 3 and 5. Haematological, biochemical, urinary and thyroid hormone parameters were determined in week 4 and week 6. Complete macroscopic examinations and microscopic examinations of selected tissues were performed for all animals. Selected organs were weighed at the scheduled necropsies.

No treatment-related deaths or clinical signs were observed. At the end of the treatment period, mean body weights were statistically significantly lower in highest-dose females compared with controls (up to −11%). In females of the highest-dose group, statistically significantly lower feed consumption was noted from study week 3 to study week 4 (18 g/animal per day compared with 21 g/animal per day in controls), and no faecal pellets were observed during home cage observations in weeks 3 and 5.

A statistically significant increase in mean ambulatory motor activity counts was observed in females of the highest-dose group in study week 3. The change was reported to be outside the relevant historical control range. Consistently, mean total motor activity counts were also (not significantly) increased in this group. Both values were also higher (although not statistically significantly) in week 5 (recovery period).

Urine analysis did not reveal treatment-related changes. Some statistically significant changes were observed in haematological parameters (haemoglobin distribution width, mean corpuscular haemoglobin levels, mean corpuscular volume, platelet counts, basophil counts, red cell counts and red cell width) that were observed in one sex only, were within the historical control range, were present at the recovery necropsy but not at the primary necropsy and/or were not associated with gross or microscopic findings.

Biochemical analyses indicated that phosphorus levels were increased in males of the two highest dose groups. Glucose levels were decreased in females of the highest-dose group. These changes showed a dose–response profile, but as they were within historical control values, and in the absence of relevant organ weight changes or gross or microscopic pathological findings, these changes were not considered toxicologically relevant. Statistically significantly higher calcium levels were noted in males of the two highest dose groups, but this was due to extremely low calcium levels in controls. Thyroid stimulating hormone levels were increased in males of the highest-dose group, but this effect was observed only in week 6. Reverse triiodothyronine levels were elevated in females of the 4000 mg/kg feed group only. These changes were not considered to be of toxicological importance.

At the primary necropsy, spleen weight was increased in females of the 1000 mg/kg feed group and decreased in females of the 4000 and 10 000 mg/kg feed groups. Absolute liver weights were increased in males of the 4000 mg/kg feed group, and relative liver weights were increased in males of the 4000 and 10 000 mg/kg feed groups and females of the 10 000 mg/kg feed group. However, there were no correlating serum chemistry differences or macroscopic or microscopic findings; therefore, these organ weight changes were considered non-adverse.

Additional changes in (relative) organ weights (epididymides, spleen, brain, ovary/oviduct weights) were not considered toxicologically relevant, as they were observed at intermediate doses or at recovery autopsy only, were a result of a decrease in body weight and/or were not associated with gross or microscopic findings.

Based on the decrease in feed consumption and the resulting decrease in body weight, the NOAEL for 1-(2-hydroxy-4-isobutoxyphenyl)-3-(pyridin-2-yl)-propan-1-one via the diet in rats is 4000 mg/kg feed, equal to 340 mg/kg bw per day (Korgaonkar, 2009b).

(iii)   5-Ethyl-2-methylpyridine (No. 1318)

A 28-day oral toxicity study was conducted in male and female Sprague-Dawley rats (numbers per group unspecified) receiving 5-ethyl-2-methylpyridine via oral gavage at doses of 0, 30, 95 or 300 mg/kg bw per day. The study was reported to be certified for compliance with good laboratory practice (GLP) and performed according to OECD Test Guideline 407. No mortality was observed at any dose level. The high-dose group showed reduced body weight gain and feed intake. In addition, elevated blood urea nitrogen levels, creatinine levels, aspartate aminotransferase activities and relative liver and kidney weights were observed. The mid-dose group showed minor deviations of clinical chemistry parameters and increased liver weight. Males of both the mid- and high-dose groups showed hyaline droplet nephropathy.

The authors derived a NOAEL of 30 mg/kg bw per day (Biomedizinische Forschungsanstalt m.b.H., 1988). Only a summary of this study was available, lacking detailed information on the results. Because the full study report was not available, the Committee could not confirm this NOAEL.

(iv) N-Methylpyrrole, structurally related compound

In a 14-day dietary palatability/range-finding study, groups of three male and three female Sprague-Dawley rats were administered microencapsulated test material (composed of a 7.65% concentration of N-methylpyrrole in a maltodextrin base) at dietary concentrations of 0, 10 000 or 20 000 mg/kg feed. Animals were observed daily for ill-health, toxicity and behavioural changes. Body weights and feed consumption were recorded on days 1 (body weight only), 4, 8, 11 and 15. External and internal macroscopic and microscopic examinations were performed on all animals at termination. A reduction in body weight gain was observed in both treatment groups (up to −16%). Actual body weight losses were observed in males and females of the high-dose group (up to −7%) and in females of the low-dose group (−4%) on day 4. This was accompanied by a decreased feed intake during the first week of the study (except for high-dose females). Sloughing of the glandular gastric epithelium for all females and two males of the high-dose group and two females and one male of the low-dose group was observed (Dhinsa, Watson & Brooks, 2008).

In the subsequent 28-day study of toxicity, groups of five male and five female Sprague-Dawley rats were given microencapsulated test material (composed of a 7.65% concentration of N-methylpyrrole in a maltodextrin base) at a dietary concentration of 0, 1500, 4000 or 10 000 mg/kg feed (equal to mean daily N-methylpyrrole doses of 0, 10, 30 and 60 mg/kg bw per day for males and 0, 10, 30 and 70 mg/kg bw per day for females). The highest dose was set based on the results of the 14-day range-finding study. The study was performed according to OECD Test Guideline 407 and was certified for compliance with GLP and quality assurance. Observations included mortality, clinical signs, behaviour, body weight, feed intake, water consumption, a functional observational battery, motor activity, haematology, clinical chemistry, organ weights, and macroscopic and microscopic (selected organs only) pathology.

No treatment-related deaths or clinical signs occurred. Males of the mid- and high-dose groups displayed significantly higher overall motor activity in the functional performance test, but no dose–response relationship was observed. Males in the high-dose group displayed a statistically significant increase in hindlimb grip strength, whereas females in the same dose group showed a significant decrease. These incidental findings were not considered to be of toxicological significance.

Mean body weights and body weight gain were decreased in males and females of the two highest dose groups throughout the treatment period. Body weight gain was statistically significantly decreased in males and females of the high-dose group compared with controls in weeks 1, 2 (males only) and 4 (females only) and in males of the mid-dose group in week 1. These observations were accompanied by reductions (not statistically significant) in feed consumption for females of the mid- and high-dose groups throughout the study and for males of the mid- and high-dose groups in week 1 only. No treatment-related effects were detected in the haematological parameters. Significant reductions in total protein, calcium and albumin levels, with a number of individual values outside the normally expected ranges, were observed in males in all treatment groups. No such effects were evident for females. Males in the high-dose group displayed a statistically significant reduction in cholesterol level, but the observed difference was considered by the authors to have arisen incidentally as a consequence of one substantially high control value. Females displayed statistically significant increases in alanine aminotransferase activity at all dose levels (30–50%) and in creatinine levels (11%) at the highest dose level. As these clinical chemistry findings for the most part occurred in one sex only, lacked a dose–response relationship and were not accompanied by histopathological findings, they were considered to be unrelated to the treatment. Absolute and relative liver weights were significantly increased in males in the two highest dose groups, but not in females. Slight, but statistically significant, reductions in heart weights (both absolute and relative) were observed in males of the low-dose group, but not at higher dose levels. Macroscopic and microscopic examinations revealed no treatment-related changes.

The Committee noted that, in contrast to the 14-day study, no histopathological effects on the stomach were reported in this study. However, based on the effects on feed consumption and body weight gain, which were also observed in the 10 000 mg/kg feed group in the range-finding study, the NOAEL was 4000 mg/kg feed per day, equal to 30 mg/kg bw per day (Dhinsa, Watson & Brooks, 2008).

(v) 8-Hydroxyquinoline, structurally related compound

In a 15-day tolerance study, groups of five male and five female F344/N rats and groups of five male and five female B6C3F1 mice were administered 8-hydroxyquinoline at dietary concentrations of 0, 3000, 6000, 12 000, 25 000 and 50 000 mg/kg feed (equivalent to 0, 300, 600, 1200, 2500 and 5000 mg/kg bw per day for rats and 0, 450, 1200, 1800, 3750 and 7500 mg/kg bw per day for mice). Feed and water were provided ad libitum. Animals were observed twice daily for signs of moribundity or mortality and were weighed on days 0, 14 and 16. Necropsies were performed on all animals.

Two male rats in the 50 000 mg/kg feed group died, on days 12 and 13, respectively, and one male in the 25 000 mg/kg feed group died during the necropsy period. Males and females of the highest-dose group as well as males of the 25 000 mg/kg feed group lost weight, and rats from the highest-dose group appeared emaciated. Although feed intake was not measured, rats administered feed containing 8-hydroxyquinoline at concentrations of 12 000 mg/kg feed and higher ate noticeably less than controls.

All mice that received 25 000 or 50 000 mg/kg feed died, and the females that died appeared to be emaciated. Four of five male mice that received 12 000 mg/kg feed lost weight during the study; although feed intake was not measured, they ate noticeably less than controls (National Toxicology Program, 1985).

In the subsequent 13-week study, groups of 10 male and 10 female F344/N rats and groups of 10 male and 10 female B6C3F1 mice were administered 8-hydroxyquinoline at dietary concentrations of 0, 400 (mice only), 800, 1500, 3000, 6000 or 12 000 (rats only) mg/kg feed (equal to 0, 50, 90, 170, 340 and 660 mg/kg bw per day for male rats and 0, 70, 130, 180, 320 and 660 mg/kg bw per day for female rats; 0, 60, 110, 200, 410 and 770 mg/kg bw per day for male mice and 0, 80, 170, 280, 1180 and 890 mg/kg bw per day for female mice).

Animals were examined twice daily for moribundity and mortality. Weight and feed consumption were measured weekly. Necropsies were performed on all animals. Histopathological examinations were performed on the following organs and tissues of control animals and of rats and mice of the highest-dose groups: gross lesions and tissue masses, mandibular lymph nodes, mammary glands, salivary glands, sternebrae, thyroid gland, skin, parathyroids, small intestine, colon, liver, prostate/testes or ovaries/uterus, gall bladder (mice only), lungs and bronchi, heart, brain, oesophagus, stomach, thymus, trachea, pancreas, spleen, kidneys, adrenal gland, urinary bladder and pituitary gland.

No unscheduled deaths occurred in rats. Final mean body weights were decreased in male rats of the 12 000 mg/kg feed group (−18%) and female rats of the 6000 and 12 000 mg/kg feed groups (−11% and −10%, respectively). In females, this was accompanied by a reduction in feed consumption of approximately 25% in the three highest dose groups. No compound-related histopathological lesions were found in male rats. Lymphoid hyperplasia was found in two females of the highest-dose group (and none of the control animals). This was not considered a treatment-related effect by the authors.

No unscheduled deaths occurred in mice. Final mean body weights were decreased in males (−11%) and females (−10%) of the highest-dose group. This was accompanied by reductions in feed intake of 18% and 26%, respectively. No compound-related histopathological effects were observed.

The NOAEL was 3000 mg/kg feed for rats (equal to 170 mg/kg bw per day) and 3000 mg/kg feed for mice (equal to 410 mg/kg bw per day), based on effects on body weight (National Toxicology Program, 1985).

In the subsequent 2-year study, groups of 50 male and 50 female F344/N rats and B6C3F1 mice were given 8-hydroxyquinoline at dietary concentrations of 0, 1500 or 3000 mg/kg feed for 103 weeks. These dietary concentrations were equal to 0, 70 and 140 mg/kg bw per day for male rats, 0, 90 and 170 mg/kg bw per day for female rats, 0, 220 and 400 mg/kg bw per day for male mice and 0, 350 and 620 mg/kg bw per day for female mice. All animals were observed twice daily for mortality and signs of moribundity. Clinical signs were recorded once per week. Body weights were recorded weekly for the first 12 weeks and every 4 weeks thereafter. Feed consumption was monitored every 4 weeks from the beginning of the study. Besides the tissues mentioned in the 13-week study, regional lymph nodes, skin, sciatic nerve, bone marrow, costochondral junction, larynx, trachea, nasal cavity, brain, eyes, external and middle ear and spinal cord were also subject to histopathological examinations.

Mean body weights of high-dose rats of both sexes were slightly lower than those of controls. The average daily feed consumption in the high-dose groups was 88% and 78% of that of the control group, and in the low-dose groups, 93% and 89%, for males and females, respectively. Survival of rats was not affected. Alveolar/bronchiolar adenomas or carcinomas (combined) in male rats occurred with a statistically significant positive trend, and the incidence in the high-dose group was marginally, but significantly, greater than that in controls. It was concluded that these effects in the lungs were not associated with the administration of 8-hydroxyquinoline, as neither the individual nor the combined incidences of adenomas and adenocarcinomas were greater than those previously observed, focal epithelial hyperplasia was not increased in males and there were no lung tumours with a statistically significant incidence in female rats or in mice of both sexes.

Thyroid gland C-cell adenomas and C-cell adenomas and carcinomas (combined) in male rats and C-cell adenomas in female rats were significantly increased by trend tests. The incidences in the treatment groups were not significantly different from those in the controls by survival-adjusted tests. Marginal decreases were observed in the incidences of neoplastic nodules and mononuclear cell leukaemia in males. These differences were not considered treatment related. Neurological or neuropathological lesions were not observed in this study.

Mean body weights of male mice exposed to 3000 mg/kg feed and female mice exposed to 1500 and 3000 mg/kg feed were (slightly) lower than those of the controls. The average daily feed consumption by low- and high-dose mice was 81% and 72%, respectively, of that of the controls in male mice and 86% and 71%, respectively, of that of the controls in female mice. Survival of mice was not affected by the presence of 8-hydroxyquinoline in the feed when compared with controls. Haemangiomas and haemangiomas or haemangiosarcomas (combined) occurred in mice with significant negative trends, and the incidences in the dosed groups were significantly lower than those in the controls. However, the difference was not significant by methods that adjusted for survival. Marginal decreases in malignant lymphoma (males) and hepatocellular carcinomas (females) were not considered to be chemically related. Further, male and female mice showed increased incidences of lung tumours, although they were not statistically significant and were within the range of historical values. In female mice, necrotizing inflammation of the ovary, uterus and thoracic or abdominal cavities was found in 20 of 26 control animals, 11 of 24 low-dose animals and 10 of 21 high-dose animals. Microscopic analysis confirmed that these lesions were related to a *Klebsiella* infection.

The authors concluded that under the conditions of these studies, there was no evidence of carcinogenicity for male and female F344/N rats or male and female B6C3F1 mice given 8-hydroxyquinoline in feed at concentrations of 1500 or 3000 mg/kg for 103 weeks. The NOAEL was 3000 mg/kg feed, equal to 140 mg/kg bw per day in rats and 400 mg/kg bw per day in mice (National Toxicology Program, 1985).

(c)   Genotoxicity

Studies of genotoxicity in vitro and in vivo have been reported for 3 of the 11 pyridine, pyrrole and quinoline derivatives in this group (Nos 2151, 2158 and 2159) and for one compound evaluated previously (5-ethyl-2-methylpyridine, No. 1318) For the studies with 5-ethyl-2-methylpyridine, only short summaries are available. The results of these studies are summarized in Table 5 and described below.

(i)   In vitro

No evidence of mutagenicity was observed in bacterial reverse mutation assays when 2,4-dimethylpyridine (No. 2151), 1-(2-hydroxyphenyl)-3-(pyridin-4-yl)-propan-1-one (No. 2158), 1-(2-hydroxy-4-isobutoxyphenyl)-3-(pyridin-2-yl)propan-

Table 5. Studies of genotoxicity in vitro and in vivo with pyridine, pyrrole and quinoline derivatives used as flavouring agents

No.	Flavouring agent	End-point	Test system	Concentration/dose	Results	Reference
In vitro						
2151	2,4-Dimethylpyridine	Reverse mutation	*Salmonella typhimurium* TA98, TA100, TA1535, TA1537, TA1538	10–1000 μg/plate, ±S9	Negative[a]	Ho et al. (1981)
2151	2,4-Dimethylpyridine	Mitotic aneuploidy	*Saccharomyces cerevisiae* D61.M	4.3–6.5 mg/ml (0.4–0.6%)[b]	Positive[c]	Zimmermann et al. (1986)
2158	1-(2-Hydroxyphenyl)-3-(pyridin-4-yl)propan-1-one	Reverse mutation	*S. typhimurium* TA98, TA100, TA1535, TA1537; *Escherichia coli* WP2uvrA	3–5000 μg/plate, ±S9	Negative[d]	Sokolowski (2009)
2158	1-(2-Hydroxyphenyl)-3-(pyridin-4-yl)propan-1-one	Chromosomal aberration	Human lymphocytes	1st experiment: 140, 245, 429 μg/ml, –S9; 80, 140, 245 μg/ml, +S9 2nd experiment: 350, 400, 450 μg/ml, –S9	Positive/ negative[e]	Bohnenberger (2009)
2159	1-(2-Hydroxy-4-isobutoxyphenyl)-3-(pyridin-2-yl)propan-1-one	Reverse mutation	*S. typhimurium* TA98, TA100, TA1535, TA1537; *E. coli* WP2uvrA	3–5000 μg/plate, ±S9[d]	Negative[f]	Sokolowski (2010)
2159	1-(2-Hydroxy-4-isobutoxyphenyl)-3-(pyridin-2-yl)propan-1-one	Forward mutation	Mouse lymphoma L5178Y TK+/– cells	1st experiment: 2.5–30 μg/ml, –S9 10–120 μg/ml, +S9 2nd experiment: 2.5–20 μg/ml, –S9 10–100 μg/ml, +S9	Negative[g]	Wollny (2009)
1318	5-Ethyl-2-methylpyridine	Reverse mutation	*S. typhimurium* TA98, TA100, TA1535, TA1537, TA1538	100–5000 μg/plate, ±S9	Negative[h]	Notox (1986)

1318	5-Ethyl-2-methylpyridine	Chromosomal aberration	Human lymphocytes	100, 200, 300, 400 µg/ml, −S9	Negative[h,i]	Life Science Research Ltd (1989)
1318	5-Ethyl-2-methylpyridine	Chromosomal aberration	Human lymphocytes	78–5000 µg/ml, ±S9	Positive/negative[h,j]	Microtest Research Ltd (1987)

In vivo

1318	5-Ethyl-2-methylpyridine	Micronucleus induction	Mice; M, F	156, 313, 625 mg/kg bw[k]	Negative[h]	Life Science Research Ltd (1990)

F, female; M, male; S9, 9000 × g supernatant fraction of rat liver homogenate

[a] Two independent experiments using the plate incorporation method. No quantitative data were reported.
[b] Calculated using the density of 2,4-dimethylpyridine = 1.08 g/ml.
[c] The previously evaluated compounds 2-acetylpyridine (No. 1309), 3-acetylpyridine (No. 1316), 2,6-dimethylpyridine (No. 1317) and 5-ethyl-2-methylpyridine (No. 1318) also gave positive results when tested at concentrations ranging from 0.15% to 1.11%.
[d] Two independent experiments using the preincubation method and the plate incorporation method, respectively. In the plate incorporation experiment, toxicity was reported in all strains with and without S9 from 2500 µg/plate upwards. Also, toxicity was observed at 1000 µg/plate for *S. typhimurium* strains TA98 and TA1537 and *E. coli* WP2uvrA without metabolic activation. In the preincubation experiment, toxicity was observed in all strains from 1000 µg/plate upwards (with and without metabolic activation).
[e] Cells were analysed 18 hours after 4 hours of treatment. With metabolic activation, statistically significant increases in aberration frequency were observed from 245 µg/ml upwards. Without metabolic activation, negative results were obtained.
[f] Two independent experiments using the preincubation method and the plate incorporation method, respectively. In the plate incorporation experiment, toxicity was reported in *S. typhimurium* strain TA1537 from 1000 and 2500 µg/plate upwards (with and without metabolic activation, respectively), in *S. typhimurium* strain TA1535 from 1000 µg/plate upwards with metabolic activation and in *E. coli* WP2uvrA at 5000 µg/plate with metabolic activation. In the preincubation experiment, toxicity was observed in *S. typhimurium* strain TA1537 from 2500 and 1000 µg/plate upwards (with and without metabolic activation, respectively) and at 5000 µg/plate in *S. typhimurium* strain TA1535 without metabolic activation.
[g] Two independent tests in each experiment. In experiment 1, cells were treated for 4 hours with concentrations manifesting up to 50% cytotoxicity. In experiment 2, cells were exposed for 24 hours (without metabolic activation) or 4 hours (with metabolic activation). Minimal bacterial toxicity was reported at 5000 µg/plate.
[h] No quantitative data reported.
[i] Two independent experiments using 5-ethyl-2-methylpyridine with a purity of ≥96% and ≥99%, respectively.
[j] Clastogenic at near-toxic doses without metabolic activation.
[k] Single dose administered by gavage. Examinations after 24, 48 and 72 hours.

1-one (No. 2159) and 5-ethyl-2-methylpyridine (No. 1318) were incubated with *Salmonella typhimurium* strains TA98, TA100, TA1535, TA1537 and/or TA1538 and/ or *Escherichia coli* WP2*uvrA*, with or without metabolic activation, at concentrations up to 5000 µg/plate (Ho et al., 1981; Notox, 1986; Sokolowski, 2009, 2010). Three of these studies (Notox, 1986; Sokolowski, 2009, 2010) were reported to be certified for compliance with GLP and/or quality assurance and were performed according to OECD Test Guideline 471 (Bacterial Reverse Mutation Test).

In non-standardized assays, 2,4-dimethylpyridine (No. 2151) at 0.4–0.6% (4.3–6.5 mg/ml) caused a dose-dependent increase in mitotic aneuploidy in strain D61.M of *Saccharomyces cerevisiae* (Zimmermann et al., 1986). At the higher test concentrations, growth of D61.M was strongly or completely inhibited. In the same study, 2-acetylpyridine (No. 1309), 3-acetylpyridine (No. 1316), 2,6-dimethylpyridine (No. 1317) and 5-ethyl-2-methylpyridine (No. 1318) also gave positive results when tested at comparable high concentrations. The Committee concluded that this is unlikely to occur at low doses because yeast is generally believed to have a threshold for the induction of aneuploidy (Zimmermann et al., 1985a,b,c).

Assays in mammalian cell lines were available for Nos 2158, 2159 and 1318. In a chromosomal aberration assay using human lymphocytes, 1-(2-hydroxyphenyl)-3-(pyridin-4-yl)propan-1-one (No. 2158) was positive for clastogenicity from 245 µg/ml upwards without metabolic activation in two independent tests after an exposure time of 4 hours. No evidence for an increase in polyploid metaphases was observed (Bohnenberger, 2009). 1-(2-Hydroxy-4-isobutoxyphenyl)-3-(pyridin-2-yl)propan-1-one (No. 2159) was negative in a forward mutation assay in mouse lymphoma L5178Y cells (Wollny, 2009). These two studies were certified for compliance with GLP and quality assurance and performed according to OECD Test Guidelines (No. 476, In Vitro Mammalian Cell Gene Mutation Test; or No. 473, In Vitro Mammalian Chromosome Aberration Test).

5-Ethyl-2-methylpyridine (No. 1318) was clastogenic at near-toxic doses without metabolic activation in one chromosomal aberration assay (78–5000 µg/ml) and gave negative results in another chromosomal aberration assay with much lower concentrations (100–400 µg/ml). With metabolic activation, 5-ethyl-2-methylpyridine gave negative results. The studies were reported to be conducted under GLP and according to OECD Test Guideline 473. No numerical data were provided (Microtest Research Ltd, 1987; Life Science Research Ltd, 1989).

(ii) In vivo

No genotoxic potential was demonstrated in a mouse micronucleus assay with 5-ethyl-2-methylpyridine (No. 1318) (Life Science Research Ltd, 1990). The study was reported to be certified for compliance with GLP and/or quality assurance and was performed according to OECD Test Guideline 474 (Mammalian Erythrocyte Micronucleus Test).

(iii) Conclusion

In the previous evaluation of this group of flavouring agents, it was concluded that, on the basis of the available evidence, the 22 pyridine, pyrrole and quinoline derivatives in that group do not demonstrate genotoxic potential. On the basis of the data available for the current evaluation, this conclusion can be extended to the compounds currently under evaluation, except for Nos 2158–2160.

1-(2-Hydroxyphenyl)-3-(pyridin-4-yl)propan-1-one (No. 2158) was negative in two bacterial reverse mutation tests with and without metabolic activation (Sokolowski, 2009). In an in vitro chromosomal aberration assay using human lymphocytes, 1-(2-hydroxyphenyl)-3-(pyridin-4-yl)propan-1-one (No. 2158) was positive for clastogenicity in concentrations of 245 µg/ml or more without metabolic activation in two independent tests (Bohnenberger, 2009). 1-(2-Hydroxy-4-isobutoxyphenyl)-3-(pyridin-2-yl)propan-1-one (No. 2159) was negative in a bacterial reverse mutation test with and without metabolic activation and a forward mutation test in mouse lymphoma L5178Y cells (Wollny, 2009; Sokolowski, 2010). For No. 2160, no genotoxicity data are available. The data available for No. 2158 demonstrate that this compound has genotoxic potential in vitro. As in vivo genotoxicity data are not available for No. 2158 and as no tests on clastogenicity are available for the structurally related flavouring agents Nos 2159 and 2160, additional data on genotoxicity are necessary to conclude on the genotoxicity of these flavouring agents.

(d) Reproductive/developmental toxicity

In a one-generation reproductive toxicity study, male and female Sprague-Dawley rats (number unspecified) were administered 0, 30, 95 or 300 mg of 5-ethyl-2-methylpyridine (No. 1318) per kilogram of body weight per day via oral gavage for 15 days prior to pairing, during mating, gestation and lactation and until postpartum day 4. Only a summary from this study was available, which lacked detailed information on the results. This study was reported to be certified for compliance with GLP and was performed according to draft OECD Test Guideline 421 (Reproduction/Developmental Toxicity Screening Test, version January 2003).

Animals in all treatment groups showed increased salivation after dosing, which was most marked at the mid- and high-dose levels. Animals receiving 300 mg/kg bw per day showed reduction in body temperature and abnormal respiration during weeks 2–4 after dosing. In addition to these effects, a small number of other signs were seen infrequently in the high-dose group. Two high-dose males were killed in extremis, with signs including ataxia, partially closed eyes, prostrate posture and underactivity. Terminal investigations revealed reduced/dehydrated gastrointestinal contents, accentuated lobular liver patterns, apparently reduced testes, epididymides, prostate glands and seminal vesicles and a small mass on one epididymis with the presence of spermatozoal granuloma. The deaths were considered to be treatment related.

Males of the high-dose group showed reduced body weight gain. Female body weight was reduced during gestation in the mid- and high-dose groups and during lactation in the high-dose group. This was accompanied by slightly lower feed consumption by high-dose females during lactation. Estrous cycles were essentially unaffected by the test material; however, one high-dose pair failed to mate. Gestation length was unaffected by the treatment. One female receiving 30 mg/kg bw per day and three receiving 300 mg/kg bw per day were terminated due to total litter loss. All females displayed inactive mammary tissues upon examination. Two high-dose females showed liver changes, small spleen and pale areas in the kidneys. The numbers of implantations, survival and growth in utero, litter size, pup viability indices, sex ratios and body weights at day 1 of age and body weight gain at day 4 were unaffected in the 30 and 95 mg/kg bw per day groups. In the 300 mg/kg bw per day group, the offspring showed reduced body weights at day 1 and poor subsequent body weight gains associated with a decrease in viability of the offspring. Necropsy revealed no changes that could be attributed to maternal treatment with the test substance. Further, no macroscopic or microscopic changes considered to be related to treatment were observed at necropsy of the parental males and females.

On the basis of this summary, it can be concluded that reproductive performance was not affected at levels up to 300 mg/kg bw per day. The overall NOAEL for general toxicity was 30 mg/kg bw per day (based on the weight reduction observed in mid-dose females during gestation). For offspring parameters, the NOAEL was considered to be 95 mg/kg bw per day, based on the reduced body weights at birth and the poor subsequent body weight gain of pups of the 300 mg/kg bw per day group (Pharmaco-LSR Ltd, 1994).

3. REFERENCES

Biomedizinische Forschungsanstalt m.b.H. (1988). P0072: 4 week oral toxicity study in rats. Unpublished report no. 87/077 from Biomedizinische Forschungsanstalt m.b.H [cited in Organisation for Economic Co-operation and Development, 1995].

Bio-Toxicology Laboratories, Inc. (1976). Lonza report 0191 [cited in Organisation for Economic Co-operation and Development, 1995].

Bohnenberger S (2009). Chromosome aberration test in human lymphocytes in vitro with York. Unpublished report no. 1251702 from Harlan Laboratories Ltd, Itingen, Switzerland. Submitted to WHO by the International Organization of the Flavor Industry, Brussels, Belgium.

Consultox Laboratories Ltd (1973). Lonza report 0192 [cited in Organisation for Economic Co-operation and Development, 1995].

Cramer GM, Ford RA, Hall RL (1978). Estimation of toxic hazard—a decision tree approach. *Food and Cosmetics Toxicology*, 16:255–276.

Dhinsa NK, Watson P, Brooks P (2008). *N*-Methylpyrrole (NMP): twenty-eight day repeated dose oral (dietary) toxicity study in the rat. Unpublished report no. 1834-0009 from Safepharm Laboratories, Derbyshire, England. Submitted to WHO by the International Organization of the Flavor Industry, Brussels, Belgium.

European Flavour and Fragrance Association (2004). European inquiry on volume use. Private communication to the Flavor and Extract Manufacturers Association of the United States, Washington, DC, USA. Submitted to WHO by the International Organization of the Flavor Industry, Brussels, Belgium.

Gavin CL, Williams MC, Hallagan JB (2008). *Flavor and Extract Manufacturers Association of the United States 2005 poundage and technical effects update survey.* Washington, DC, USA, Flavor and Extract Manufacturers Association of the United States.

Hazleton Laboratories Europe Ltd (1986). P0072: acute oral toxicity study in the rat. Unpublished report no. 5165-733/266 [cited in Organisation for Economic Co-operation and Development, 1995].

Ho C et al. (1981). Analytical and biological analyses of test materials from the synthetic fuel technologies. IV. Studies of chemical structure–mutagenic activity relationships of aromatic nitrogen compounds relevant to synfuels. *Mutation Research*, 85:335–345.

International Organization of the Flavor Industry (2011). Interim inquiry on volume use and added use levels for flavoring agents to be presented at the 76th JECFA meeting. Private communication to the Flavor and Extract Manufacturers Association of the United States, Washington, DC, USA. Submitted to WHO by the International Organization of the Flavor Industry, Brussels, Belgium.

Japan Flavor and Fragrance Materials Association (2005). Japanese inquiry on volume use. Private communication to the Flavor and Extract Manufacturers Association of the United States, Washington, DC, USA. Submitted to WHO by the International Organization of the Flavor Industry, Brussels, Belgium.

Kagami K et al. (2008). Suppression of blood lipid concentrations by volatile Maillard reaction products. *Nutrition*, 24:1159–1166.

Korgaonkar CK (2009a). A 28-day oral (dietary) toxicity study of GR-72-4062 in Sprague Dawley rats with a 14-day recovery period. Unpublished report no. WIL-529017 from WIL Research Laboratories, LLC, Ashland, OH, USA. Submitted to WHO by the International Organization of the Flavor Industry, Brussels, Belgium.

Korgaonkar CK (2009b). A 28-day oral (dietary) toxicity study of GR-72-5251 in Sprague Dawley rats with a 14-day recovery period. Unpublished report no. WIL-529019 from WIL Research Laboratories, LLC, Ashland, OH, USA. Submitted to WHO by the International Organization of the Flavor Industry, Brussels, Belgium.

Life Science Research Ltd (1989). In vitro assessment of the clastogenic activity of P0072/F2 and P0072 reference material in cultured human lymphocytes. Unpublished report no. 89/LZA032/0876 [cited in Organisation for Economic Co-operation and Development, 1995].

Life Science Research Ltd (1990). P0072: assessment of clastogenic action on bone marrow erythrocytes in the micronucleus test. Unpublished report no. 90/LZA063/0792 [cited in Organisation for Economic Co-operation and Development, 1995].

Microtest Research Ltd (1987). Study to evaluate the chromosome damaging potential of P0072 by its effects on cultured human lymphocytes using an in vitro cytogenetics assay. Unpublished report no. LOB 7/HLC, September [cited in Organisation for Economic Co-operation and Development, 1995].

National Toxicology Program (1985). *Toxicology and carcinogenesis studies of 8-hydroxyquinoline (CAS No. 148-24-30) in F344/N rats and B6C3F1 mice (feed studies)*. NTP TR 276; NIH Publication No. 85-2532. Research Triangle Park, NC, USA, United States Department of Health and Human Services, National Institutes of Health, National Institute of Environmental Health Sciences, National Toxicology Program (http://ntp.niehs.nih.gov/).

Nijssen LM, van Ingen-Visscher CA, Donders JJH (2011). Volatile Compounds in Food (VCF) database, version 13.1. Zeist, the Netherlands, TNO Triskelion (http://www.vcf-online.nl/VcfHome.cfm).

Notox (1986). Evaluation of the mutagenic activity of P0072 in the Ames *Salmonella*/microsome test. Unpublished report no. 0321/ES 184 [cited in Organisation for Economic Co-operation and Development, 1995].

Organisation for Economic Co-operation and Development (1995). *5-Ethyl-2-picoline: SIDS initial assessment report for SIAM 3*. Geneva, Switzerland, United Nations Environment Programme, Chemicals Branch. Submitted to WHO by the International Organization of the Flavor Industry, Brussels, Belgium.

Pharmaco-LSR Ltd (1994). P0072: reproduction/developmental toxicity screening test. Unpublished report no. 94/LZA124/0292 [cited in Organisation for Economic Co-operation and Development, 1995].

Sokolowski A (2009). *Salmonella typhimurium* and *Escherichia coli* reverse mutation assay with York. Unpublished report no. 1251701 from Harlan Laboratories Ltd, Itingen, Switzerland. Submitted to WHO by the International Organization of the Flavor Industry, Brussels, Belgium.

Sokolowski A (2010). *Salmonella typhimurium* and *Escherichia coli* reverse mutation assay with GR-72-5251. Unpublished report no. 1291601 from Harlan Laboratories Ltd, Itingen,

Switzerland. Submitted to WHO by the International Organization of the Flavor Industry, Brussels, Belgium.

Stofberg J, Grundschober F (1987). Consumption ratio and food predominance of flavoring materials. *Perfumer & Flavorist*, 12:27–68.

Wollny HE (2009). Cell mutation assay at the thymidine kinase locus (TK+/–) in mouse lymphoma L5178Y cells with GR-72-5251. Unpublished report no. 1291602 from Harlan Laboratories Ltd, Itingen, Switzerland. Submitted to WHO by the International Organization of the Flavor Industry, Brussels, Belgium.

Zimmermann FK et al. (1985a). Genetic change may be caused by interference with protein–protein interactions. *Mutation Research*, 150:203–210.

Zimmermann FK et al. (1985b). Acetone, methyl ethyl ketone, ethyl acetate, acetonitrile and other polar aprotic solvents are strong inducers of aneuploidy in *Saccharomyces cerevisiae*. *Mutation Research*, 149:339–351.

Zimmermann FK et al. (1985c). Induction of mitotic aneuploidy in yeast with aprotic polar solvents. In: Zimmermann FK, Taylor-Mayer RE, eds. *Mutagenicity testing in environmental pollution control*. New York, NY, USA, Halsted Press, pp. 166–179.

Zimmermann FK et al. (1986). Genetic and anti-tubulin effects induced by pyridine derivatives. *Mutation Research*, 163:23–31.

SULFUR-CONTAINING HETEROCYCLIC COMPOUNDS (addendum)

First draft prepared by

B.A. Fields[1], M. DiNovi[2], A.G. Renwick[3] and P. Sinhaseni[4]

[1] Food Standards Australia New Zealand, Canberra, ACT, Australia
[2] Center for Food Safety and Applied Nutrition, Food and Drug Administration, College Park, Maryland, United States of America (USA)
[3] School of Medicine, University of Southampton, Southampton, England
[4] Community Risk Analysis Research and Development Center, Pathumwan, Bangkok, Thailand

1. Evaluation 277
   1.1 Introduction 277
   1.2 Assessment of dietary exposure 278
   1.3 Absorption, distribution, metabolism and elimination 278
   1.4 Application of the Procedure for the Safety Evaluation of
       Flavouring Agents 288
   1.5 Consideration of combined intakes from use as flavouring
       agents 289
   1.6 Consideration of secondary components 289
   1.7 Conclusion 290
2. Relevant background information 290
   2.1 Explanation 290
   2.2 Additional considerations on dietary exposure 290
   2.3 Biological data 290
       2.3.1 Biochemical data: absorption, distribution, metabolism
             and elimination 290
             (a) Structurally related compound: 2-acetylthiophene 290
             (b) Structurally related compound: 2-phenylthiophene 291
             (c) 4-Amino-5,6-dimethylthieno[2,3-d]pyrimidin-2(1H)-
                 one hydrochloride (No. 2117) 291
       2.3.2 Toxicological studies 291
             (a) Acute toxicity 291
             (b) Short-term studies of toxicity 291
             (c) Genotoxicity 296
3. References 298

1. EVALUATION

1.1 Introduction

The Committee evaluated an additional 12 flavouring agents belonging to the group of sulfur-containing heterocyclic compounds. The additional flavouring agents comprised five thiophenes (Nos 2106, 2107 and 2110–2112), four thiazoles (Nos 2108, 2109, 2113 and 2114), one thiazoline (No. 2115), one dithiazine (No. 2116) and one thiophene-pyrimidine derivative (No. 2117). The evaluations were conducted using the Procedure for the Safety Evaluation of Flavouring Agents (see Figure 1, Introduction) (Annex 1, reference *131*). None of these flavouring agents have been previously evaluated by the Committee. One of the flavouring agents in this group—namely, 4-amino-5,6-dimethylthieno[2,3-d]pyrimidin-2(1H)-one hydrochloride (No. 2117)—is reported to be a flavour modifier.

The Committee evaluated 47 other members of this group of flavouring agents at its fifty-ninth and sixty-eighth meetings and concluded that none would give rise to safety concerns based on estimated dietary exposures (Annex 1, references *160* and *187*).

Eight of the 12 flavouring agents (Nos 2106–2111, 2113 and 2116) in this group have been reported to occur naturally in coffee, cocoa, pig liver, sheep liver, chicken, beef, lamb, shrimp, squid, trassi, beans, soya bean, onion, shallot, leek, asparagus, oats, peanut butter, potato, coriander seed and sweet corn (Nijssen, van Ingen-Visscher & Donders, 2011).

1.2 Assessment of dietary exposure

The total annual volumes of production of the 12 sulfur-containing heterocyclic compounds are approximately 1000 kg in the USA, 1 kg in Europe and 1.5 kg in Japan (European Flavour and Fragrance Association, 2004; Japan Flavor and Fragrance Materials Association, 2005; Gavin, Williams & Hallagan, 2008; International Organization of the Flavor Industry, 2011). Approximately 99% of the total annual volume of production in the USA is accounted for by one flavouring agent in this group—namely, 4-amino-5,6-dimethylthieno[2,3-d]pyrimidin-2(1H)-one hydrochloride (No. 2117).

Dietary exposures were estimated using the maximized survey-derived intake (MSDI) method and the single portion exposure technique (SPET), with the highest estimates reported in Table 1. The highest estimated daily dietary exposure is for 4-amino-5,6-dimethylthieno[2,3-d]pyrimidin-2(1H)-one hydrochloride (No. 2117) (4500 μg, the SPET value from sugar substitutes). For the other flavouring agents, the estimated daily dietary exposures range from 0.05 to 90 μg, with the SPET yielding the highest estimates.

Annual volumes of production of this group of flavouring agents as well as the daily dietary exposures calculated using both the MSDI method and the SPET are summarized in Table 2.

1.3 Absorption, distribution, metabolism and elimination

The metabolism of sulfur-containing heterocyclic compounds was described in the reports of the fifty-ninth and sixty-eighth meetings of the Committee (Annex 1, references *160* and *187*).

Thiophene derivatives are metabolized primarily by *S*-oxidation, followed by conjugation with glutathione (GSH). Thiazole and its derivatives are metabolized primarily by side-chain oxidation or oxidation of the ring sulfur or nitrogen atoms. Dithiazine and thiazoline derivatives, being cyclic sulfides, are metabolized primarily by *S*-oxidation to yield the corresponding sulfoxides and sulfones. Other routes of

SULFUR-CONTAINING HETEROCYCLIC COMPOUNDS (addendum)

Table 1. Summary of the results of the safety evaluations of sulfur-containing heterocyclic compounds used as flavouring agents[a,b,c]

Flavouring agent	No.	CAS No. and structure	Step B3[d] Does estimated dietary exposure exceed the threshold of concern?	Follow-on from step B3[e] Are additional data available for flavouring agent with an estimated dietary exposure exceeding the threshold of concern? Step B4[e] Adequate margin of exposure for the flavouring agent or a related substance?	Comments on predicted metabolism	Related structure name (No.) and structure (if applicable)	Conclusion based on current estimated dietary exposure
Structural class II							
2-Pentylthiophene	2106	4861-58-9	No, SPET: 40	B4: Yes. The NOEL of 3 mg/kg bw per day in a 28-day study in rats (Marr & Watson, 2007) is 4500 times the estimated daily dietary exposure to No. 2106 when used as a flavouring agent.	Note 1		No safety concern
2-Acetyl-5-methylthiophene	2107	13679-74-8	No, SPET: 0.3	B4: Yes. The NOEL of 3 mg/kg bw per day in a 28-day study in rats for the structurally related 2-pentylthiophene (No. 2106) (Marr & Watson, 2007) is 600 000 times the estimated daily dietary exposure to No. 2107 when used as a flavouring agent.	Note 1	2-Pentylthiophene (No. 2106)	No safety concern

continued

Table 1 (continued)

Flavouring agent	No.	CAS No. and structure	Step B3[a] Does estimated dietary exposure exceed the threshold of concern?	Follow-on from step B3[e] Are additional data available for flavouring agent with an estimated dietary exposure exceeding the threshold of concern? Step B4[e] Adequate margin of exposure for the flavouring agent or a related substance?	Comments on predicted metabolism	Related structure name (No.) and structure (if applicable)	Conclusion based on current estimated dietary exposure
2-Pentylthiazole	2108	37645-62-8	No, SPET: 0.5	B4: Yes. The NOEL of 0.92 mg/kg bw per day in a 90-day study in rats for the structurally related 2,4-dimethyl-5-vinylthiazole (No. 1039) (Posternak, Linder & Vodoz, 1969) is 110 000 times the estimated daily dietary exposure to No. 2108 when used as a flavouring agent.	Note 2	2,4-Dimethyl-5-vinylthiazole (No. 1039)	No safety concern
4,5-Dimethyl-2-isobutylthiazole	2109	53498-32-1	No, SPET: 0.3	B4: Yes. The NOEL of 0.92 mg/kg bw per day in a 90-day study in rats for the structurally related 2,4-dimethyl-5-vinylthiazole (No. 1039) (Posternak, Linder & Vodoz, 1969) is 180 000 times the estimated daily dietary exposure to No. 2109 when used as a flavouring agent.	Note 2	2,4-Dimethyl-5-vinylthiazole (No. 1039)	No safety concern

Structural class III

3,4-Dimethylthiophene	2110	632-15-5	No, SPET: 2	B4: Yes. The NOEL of 3 mg/kg bw per day in a 28-day study in rats for the structurally related 2-pentylthiophene (No. 2106) (Marr & Watson, 2007) is 90 000 times the estimated daily dietary exposure to No. 2110 when used as a flavouring agent.	Note 1	2-Pentylthiophene (No. 2106)	No safety concern
2-Thienylmethanol	2111	636-72-6	No, SPET: 0.3	B4: Yes. The NOEL of 3 mg/kg bw per day in a 28-day study in rats for the structurally related 2-pentylthiophene (No. 2106) (Marr & Watson, 2007) is 600 000 times the estimated daily dietary exposure to No. 2111 when used as a flavouring agent.	Note 1	2-Pentylthiophene (No. 2106)	No safety concern
1-(2-Thienyl)-ethanethiol	2112	94089-02-8	No, SPET: 0.05	B4: Yes. The NOEL of 3 mg/kg bw per day in a 28-day study in rats for the structurally related 2-pentylthiophene (No. 2106) (Marr & Watson, 2007) is 3.6 million times the estimated daily dietary exposure to No. 2112 when used as a flavouring agent.	Notes 1 and 3	2-Pentylthiophene (No. 2106)	No safety concern

continued

Table 1 (continued)

Flavouring agent	No.	CAS No. and structure	Step B3[a] Does estimated dietary exposure exceed the threshold of concern?	Follow-on from step B3[a] Are additional data available for flavouring agent with an estimated dietary exposure exceeding the threshold of concern? Step B4[e] Adequate margin of exposure for the flavouring agent or a related substance?	Comments on predicted metabolism	Related structure name (No.) and structure (if applicable)	Conclusion based on current estimated dietary exposure
5-Ethyl-2-methylthiazole	2113	19961-52-5	No, SPET: 2	B4: Yes. The NOEL of 0.92 mg/kg bw per day in a 90-day study in rats for the structurally related 2,4-dimethyl-5-vinylthiazole (No. 1039) (Posternak, Linder & Vodoz, 1969) is 28 000 times the estimated daily dietary exposure to No. 2113 when used as a flavouring agent.	Note 2	2,4-Dimethyl-5-vinylthiazole (No. 1039)	No safety concern
2-Ethyl-2,5-dihydro-4-methylthiazole	2114	41803-21-8	No, SPET: 90	B4: Yes. The NOEL of 1.2 mg/kg bw per day in a 90-day study in rats for the structurally related 2-(2-butyl)-4,5-dimethyl-3-thiazoline (No. 1059) (Babish, 1978) is 800 times the estimated daily dietary exposure to No. 2114 when used as a flavouring agent. The margin of exposure calculated using the MSDI of 0.01 μg/day is 7.2 million.	Note 4	2-(2-Butyl)-4,5-dimethyl-3-thiazoline (No. 1059)	No safety concern

SULFUR-CONTAINING HETEROCYCLIC COMPOUNDS (addendum)

4-Methyl-3-thiazoline	2115	52558-99-3	No, SPET: 4	B4: Yes. The NOEL of 1.2 mg/kg bw per day in a 90-day study in rats for the structurally related 2-(2-butyl)-4,5-dimethyl-3-thiazoline (No. 1059) (Babish, 1978) is 18 000 times the estimated daily dietary exposure to No. 2115 when used as a flavouring agent.	Note 4	2-(2-Butyl)-4,5-dimethyl-3-thiazoline (No. 1059)	No safety concern
2-Ethyl-4,6-dimethyldihydro-1,3,5-dithiazine	2116	54717-14-5	No, SPET: 9	B4: Yes. The NOEL of 11 mg/kg bw per day in a 14-day study in rats for the structurally related mixture of 2-isopropyl-4,6-dimethyldihydro-1,3,5-dithiazine and 4-isopropyl-2,6-dimethyldihydro-1,3,5-dithiazine (No. 1047) (Rush, 1989) is 73 000 times the estimated daily dietary exposure to No. 2116 when used as a flavouring agent.	Note 4	A mixture of of 2-isopropyl-4,6-dimethyldihydro-1,3,5-dithiazine and 4-isopropyl-2,6-dimethyldihydro-1,3,5-dithiazine (No. 1047) 44% 27%	No safety concern

continued

Table 1 (continued)

Flavouring agent	No.	CAS No. and structure	Step B3[b] Does estimated dietary exposure exceed the threshold of concern?	Follow-on from step B3[b] Are additional data available for flavouring agent with an estimated dietary exposure exceeding the threshold of concern? Step B4[e] Adequate margin of exposure for the flavouring agent or a related substance?	Comments on predicted metabolism	Related structure name (No.) and structure (if applicable)	Conclusion based on current estimated dietary exposure
4-Amino-5,6-dimethylthieno[2,3-d]pyrimidin-2(1H)-one hydrochloride	2117	1033366-59-4	Yes, SPET: 4500	Yes. The NOAEL of 60 mg/kg bw per day in a 90-day rat study (Ross, 2008b) provides margins of exposure of 800 in relation to the dietary exposure calculated using the SPET value (4500 μg/day) and 33 000 using the MSDI (110 μg/day) when No. 2117 is used as a flavouring agent.	Note 1		No safety concern

bw, body weight; CAS, Chemical Abstracts Service; NOAEL, no-observed-adverse-effect level; NOEL, no-observed-effect level

[a] Forty-seven flavouring agents in this group were previously evaluated by the Committee (Annex 1, references *160* and *187*).
[b] *Step 1*: Four flavouring agents (Nos 2106–2109) are in structural class II, and eight flavouring agents (Nos 2110–2117) are in structural class III.
[c] *Step 2*: None of the flavouring agents in this group can be predicted to be metabolized to innocuous products.
[d] The thresholds of human dietary exposure for structural classes II and III are 540 and 90 μg/person per day, respectively. All dietary exposure values are expressed in μg/day. The dietary exposure value listed represents the highest estimated dietary exposure calculated using either the SPET or the MSDI method. The highest estimates were all derived using the SPET.
[e] The margins of exposure were calculated based on the estimated dietary exposures calculated using the SPET. In cases where the resulting margin of exposure was relatively low, a comparison with the MSDI was also made.

Notes:
1. Thiophene derivatives are metabolized primarily by S-oxidation, followed by conjugation with GSH.
2. Thiazole and its derivatives are metabolized primarily by side-chain oxidation or oxidation of the ring sulfur or nitrogen atoms.
3. The thiol group can undergo S-oxidation followed by conjugation with GSH or may form mixed disulfides.
4. Dithiazine and thiazoline derivatives, being cyclic sulfides, are metabolized primarily by S-oxidation to yield the corresponding sulfoxides and sulfones.

Table 2. Annual volumes of production and daily dietary exposures for sulfur-containing heterocyclic compounds used as flavouring agents in Europe, the USA and Japan

Flavouring agent (No.)	Most recent annual volume of production (kg)[a]	Dietary exposure					Annual volume of consumption via natural occurrence in foods (kg)[d]	Consumption ratio[e]
		MSDI[b]			SPET[c]			
		µg/day	µg/kg bw per day		µg/day	µg/kg bw per day		
2-Pentylthiophene (2106)								
Europe	1	0.11	0.002		40	0.7	7	70
USA	0.1	0.01	0.0002					
Japan	0.1	0.03	0.0005					
2-Acetyl-5-methylthiophene (2107)								
Europe	ND	ND	ND		0.3	0.005	1165	NA
USA	ND	ND	ND					
Japan	0.1	0.03	0.0005					
2-Pentylthiazole (2108)								
Europe	ND	ND	ND		0.5	0.01	+	NA
USA	ND	ND	ND					
Japan	0.2	0.06	0.001					
4,5-Dimethyl-2-isobutylthiazole (2109)								
Europe	ND	ND	ND		0.3	0.005	+	NA
USA	ND	ND	ND					
Japan	0.1	0.03	0.0005					
3,4-Dimethylthiophene (2110)								
Europe	ND	ND	ND		2	0.03	+	NA
USA	ND	ND	ND					
Japan	0.2	0.06	0.001					

continued

Table 2 (continued)

Flavouring agent (No.)	Most recent annual volume of production (kg)[a]	Dietary exposure				Annual volume of consumption via natural occurrence in foods (kg)[d]	Consumption ratio[e]
		MSD[b]		SPET[c]			
		µg/day	µg/kg bw per day	µg/day	µg/kg bw per day		
2-Thienylmethanol (2111)							
Europe	ND	ND	ND			2736	NA
USA	ND	ND	ND	0.3	0.005		
Japan	0.6	0.2	0.003				
1-(2-Thienyl)ethanethiol (2112)							
Europe	ND	ND	ND			–	NA
USA	ND	ND	ND	0.05	0.001		
Japan	0.1	0.03	0.0005				
5-Ethyl-2-methylthiazole (2113)							
Europe	ND	ND	ND			+	NA
USA	0.1	0.01	0.0002	2	0.03		
Japan	ND	ND	ND				
2-Ethyl-2,5-dihydro-4-methylthiazole (2114)							
Europe	ND	ND	ND			–	NA
USA	0.1	0.01	0.0002	90	2		
Japan	ND	ND	ND				
4-Methyl-3-thiazoline (2115)							
Europe	ND	ND	ND			–	NA
USA	ND	ND	ND	4	0.07		
Japan	0.1	0.03	0.0005				
2-Ethyl-4,6-dimethyldihydro-1,3,5-dithiazine (2116)							
Europe	ND	ND	ND	9	0.2	+	NA

USA	0.1	0.01	0.0002	—	NA
Japan	ND	ND	ND		
4-Amino-5,6-dimethylthieno[2,3-d]pyrimidin-2(1H)-one hydrochloride (2117)					
Europe	ND	ND	ND		
USA	1000	110	2	4500	75
Japan	ND	ND	ND		
Total					
Europe	1				
USA	1000				
Japan	1.5				

NA, not applicable; ND, no data reported; +, reported to occur naturally in foods, but no quantitative data; −, not reported to occur naturally in foods

[a] From European Flavour and Fragrance Association (2004), Japan Flavor and Fragrance Materials Association (2005), Gavin, Williams & Hallagan (2008) and International Organization of the Flavor Industry (2011). Values greater than 0 kg but less than 0.1 kg were reported as 0.1 kg.

[b] MSDI (μg/person per day) calculated as follows:
(annual volume, kg) × (1 × 10^9 μg/kg)/(population × survey correction factor × 365 days), where population (10%, "eaters only") = 32 × 10^6 for Europe, 31 × 10^6 for the USA and 13 × 10^6 for Japan; and where survey correction factor = 0.8 for the surveys in the USA, Europe and Japan, representing the assumption that only 80% of the annual flavour volume was reported in the poundage surveys (European Flavour and Fragrance Association, 2004; Japan Flavor and Fragrance Materials Association, 2005; Gavin, Williams & Hallagan, 2008; International Organization of the Flavor Industry, 2011). MSDI (μg/kg bw per day) calculated as follows:
(μg/person per day)/body weight, where body weight = 60 kg. Slight variations may occur from rounding.

[c] SPET (μg/person per day) calculated as follows:
(standard food portion, g/day) × (average use level) (International Organization of the Flavor Industry, 2011). The dietary exposure from the single food category leading to the highest dietary exposure from one portion is taken as the SPET estimate.
SPET (μg/kg bw per day) calculated as follows:
(μg/person per day)/body weight, where body weight = 60 kg. Slight variations may occur from rounding.

[d] Quantitative data for the USA reported by Stofberg & Grundschober (1987).

[e] The consumption ratio is calculated as follows:
(annual consumption via natural occurrence in foods in the USA, kg)/(most recent reported annual volume of production as a flavouring agent in the USA, kg).

metabolism for sulfur-containing heterocyclic compounds, including ring oxidation and cleavage, are also possible.

1.4 Application of the Procedure for the Safety Evaluation of Flavouring Agents

Step 1. In applying the Procedure for the Safety Evaluation of Flavouring Agents to the additional flavouring agents in this group, the Committee assigned four flavouring agents (Nos 2106–2109) to structural class II. The remaining eight flavouring agents (Nos 2110–2117) were assigned to structural class III (Cramer, Ford & Hall, 1978).

Step 2. None of the flavouring agents in this group are predicted to be metabolized to innocuous products. Therefore, the evaluation of these flavouring agents proceeded via the B-side of the Procedure.

Step B3. The highest dietary exposure for one flavouring agent (No. 2117) is above the threshold of concern (i.e. 90 µg/person per day for class III). Accordingly, for this flavouring agent, data are required on the flavouring agent or a closely related substance in order to perform a safety evaluation. For the remaining 11 flavouring agents, the dietary exposures are below the thresholds of concern (i.e. 540 µg/person per day for class II and 90 µg/person per day for class III). Accordingly, evaluation of these flavouring agents proceeded to step B4.

Step B4. For 2-pentylthiophene (No. 2106), available data give a no-observed-effect level (NOEL) of 3 mg/kg of body weight (bw) per day from a 28-day study in rats (Marr & Watson, 2007). This provides a margin of exposure of 4500 in relation to the dietary exposure to No. 2106 (SPET = 40 µg/day) when used as a flavouring agent.

For 2-acetyl-5-methylthiophene (No. 2107), 3,4-dimethylthiophene (No. 2110), 2-thienylmethanol (No. 2111) and 1-(2-thienyl)ethanethiol (No. 2112), the NOEL of 3 mg/kg bw per day for the structurally related 2-pentylthiophene (No. 2106) provides respective margins of exposure of 600 000, 90 000, 600 000 and 3.6 million in relation to the dietary exposures to No. 2107 (SPET = 0.3 µg/day), No. 2110 (SPET = 2 µg/day), No. 2111 (SPET = 0.3 µg/day) and No. 2112 (SPET = 0.05 µg/day) when used as flavouring agents.

For 2-pentylthiazole (No. 2108), 4,5-dimethyl-2-isobutylthiazole (No. 2109) and 5-ethyl-2-methylthiazole (No. 2113), available data on the structurally related 2,4-dimethyl-5-vinylthiazole (No. 1039) give a NOEL of 0.92 mg/kg bw per day from a 90-day study in rats (Posternak, Linder & Vodoz, 1969). This provides respective margins of exposure of 110 000, 180 000 and 28 000 in relation to the dietary exposures to No. 2108 (SPET = 0.5 µg/day), No. 2109 (SPET = 0.3 µg/day) and No. 2113 (SPET = 2 µg/day) when used as flavouring agents.

For 2-ethyl-2,5-dihydro-4-methylthiazoline (No. 2114) and 4-methyl-3-thiazoline (No. 2115), available data on the structurally related 2-(2-butyl)-4,5-dimethyl-3-thiazoline (No. 1059) give a NOEL of 1.2 mg/kg bw per day from a 90-day study in rats (Babish, 1978). This provides respective margins of exposure of 800 and 18 000 in relation to the dietary exposures to No. 2114 (SPET = 90 µg/day) and No. 2115 (SPET = 4 µg/day) when used as flavouring agents. For No. 2114, the margin of exposure calculated using the MSDI of 0.01 µg/day is 7.2 million.

For 2-ethyl-4,6-dimethyldihydro-1,3,5-dithiazine (No. 2116)[1], available data on the structurally related 2-isopropyl-4,6-dimethyldihydro-1,3,5-dithiazine and 4-isopropyl-2,6-dimethyldihydro-1,3,5-dithiazine (mixture of isomers; No. 1047) give a NOEL of 11 mg/kg bw per day from a 14-day study in rats (Rush, 1989). This provides a margin of exposure of 73 000 in relation to the dietary exposure to No. 2116 (SPET = 9 µg/day) when used as a flavouring agent.

Consideration of the flavouring agent with high exposure evaluated via the B-side of the decision-tree:

Short-term toxicity data are available on 4-amino-5,6-dimethylthieno[2,3-d]-pyrimidin-2(1H)-one hydrochloride (No. 2117). The no-observed-adverse-effect level (NOAEL) of 60 mg/kg bw per day in a 90-day rat study (Ross, 2008b) provides margins of exposure of 800 in relation to the dietary exposure calculated using the SPET (4500 µg/day) and 33 000 in relation to the MSDI (110 µg/day) when No. 2117 is used as a flavouring agent. Therefore, No. 2117 is not considered to pose a safety concern at current estimated dietary exposure.

The Committee therefore concluded that none of the 12 additional flavouring agents (Nos 2106–2117) belonging to the group of sulfur-containing heterocyclic compounds would pose a safety concern at current estimated dietary exposures. Table 1 summarizes the evaluations of these additional flavouring agents.

1.5 Consideration of combined intakes from use as flavouring agents

The highest MSDI for members of the current group of flavouring agents is 110 µg/day (No. 2117). No. 2117 does not share a close structural relationship with any other members of the current group or with those members evaluated previously. The MSDI values of the remaining members of the current group are negligible (≤0.2 µg/day). Combined intakes are therefore not a safety concern.

1.6 Consideration of secondary components

Two flavouring agents in this group (Nos 2114 and 2116) have minimum assay values of less than 95% (see Annex 5). The secondary components of 2-ethyl-2,5-dihydro-4-methylthiazole (No. 2114) are 2-ethyl-4-methyl-4,5-dihydrothiazole-4-ol (2–3%), 3,4-dimethylthiophene (2–3%) and 2-ethyl-4-methylthiazole (2–3%). 3,4-Dimethylthiophene (No. 2110) is a member of the current group. 2-Ethyl-4-methylthiazole (No. 1044) was evaluated at the fifty-ninth meeting (Annex 1, reference *160*) and was concluded to be of no safety concern at estimated dietary exposure when used as a flavouring agent. 2-Ethyl-4-methyl-4,5-dihydrothiazole-4-ol is anticipated to undergo further oxidative metabolism and/or conjugate formation with subsequent elimination in urine. It does not present a safety concern at current estimated dietary exposure.

---

[1] The Committee noted that No. 2116 was originally submitted as a mixture of two isomers, but that only one of the two isomers is in commercial use.

The secondary components of 2-ethyl-4,6-dimethyldihydro-1,3,5-dithiazine (No. 2116)—namely, 2,4,6-trimethyldihydro-4H-1,3,5-dithiazine (No. 1049) and 3,5-diethyl-1,2,4-trithiolane (No. 1686)—were evaluated at the fifty-ninth and sixty-eighth meetings of the Committee, respectively (Annex 1, references *160* and *187*), and concluded to be of no safety concern at estimated dietary exposures.

1.7 Conclusion

In the previous evaluations of members of this group, studies of acute toxicity, short-term studies of toxicity, and studies of genotoxicity and reproductive and developmental toxicity were available. The toxicity data available for this evaluation supported the previous evaluations.

The Committee concluded that none of the 12 flavouring agents evaluated at the present meeting, which are additions to the group of sulfur-containing heterocyclic compounds evaluated previously, raise any safety concerns at current estimated dietary exposures.

2. RELEVANT BACKGROUND INFORMATION

2.1 Explanation

This monograph addendum summarizes key aspects relevant to the safety evaluation of 12 sulfur-containing heterocyclic compounds, which are additions to a group of 47 flavouring agents evaluated previously by the Committee at its fifty-ninth and sixty-eighth meetings (Annex 1, references *160* and *187*).

2.2 Additional considerations on dietary exposure

Annual volumes of production and estimated dietary exposures calculated using both the MSDI method and the SPET for each flavouring agent are reported in Table 2. There is no additional information on estimated dietary exposures.

2.3 Biological data

2.3.1 Biochemical data: absorption, distribution, metabolism and elimination

General information on the metabolism of sulfur-containing heterocyclic compounds was previously provided in the reports of the fifty-ninth and sixty-eighth meetings of the Committee (Annex 1, references *160* and *187*). Information on two compounds structurally related to members of the current group is provided below. Specific information on flavouring agent No. 2117 is also provided below.

(a) Structurally related compound: 2-acetylthiophene

2-Acetylthiophene (10 µmol/l) was incubated with human liver microsomal proteins in potassium phosphate buffer (50 mmol/l, pH 7.4) supplemented with GSH and isotopically labelled GSH (^{15}N, ^{13}C) premixed at an equal molar ratio. After incubation at 37° C for 5 minutes, reactions were initiated by the addition of reduced nicotinamide adenine dinucleotide phosphate (NADPH)-generating solution. After 60 minutes, the presence of the GSH conjugate at the C5 position of 2-acetylthiophene was identified by liquid chromatography–tandem mass spectrometry (Yan et al., 2005).

(b)   Structurally related compound: 2-phenylthiophene

2-Phenylthiophene was incubated with rat liver microsomes in the presence of NADPH and GSH. Three classes of metabolites were identified by ^{1}H nuclear magnetic resonance and mass spectrometry: 1) dimers of 2-phenylthiophene-S-oxide formed via Diels-Alder reactions; 2) GSH conjugates of 2-phenylthiophene-S-oxide formed by 1,4-Michael-type addition; and 3) a mixture of GSH adducts resulting from nucleophilic attack of GSH to an intermediary metabolite predicted to be the 4,5-epoxide of 2-phenylthiophene. The same metabolites were formed when 2-phenylthiophene was incubated with human cytochrome P450 1A1 in the presence of NADPH and GSH (Dansette, Bertho & Mansuy, 2005).

(c)   4-Amino-5,6-dimethylthieno[2,3-d]pyrimidin-2(1H)-one hydrochloride (No. 2117)

In a pharmacokinetic study, 4-amino-5,6-dimethylthieno[2,3-d]pyrimidin-2(1H)-one and its hydrochloride salt were separately administered to groups of Sprague-Dawley rats (six of each sex) as single oral gavage doses of 100 mg/kg bw. Blood samples were collected at eight time points for up to 48 hours post-dosing (pre-dosing, 0.5, 1, 2, 4, 8, 24 and 48 hours). Both compounds appear as 4-amino-5,6-dimethylthieno[2,3-d]pyrimidin-2(1H)-one in plasma. The hydrochloride salt form was rapidly absorbed following oral administration, with a time to peak plasma concentration (C_{max}), or T_{max}, of 2.0 hours. Slower absorption was observed with the free base form (T_{max} of 4.0 hours). Maximum plasma concentrations (C_{max}) were 28.6 µg/ml (salt form) and 25.1 µg/ml (free base form). Elimination of 4-amino-5,6-dimethylthieno[2,3-d]pyrimidin-2(1H)-one from plasma followed first-order kinetics for both the salt and free base forms. Areas under the plasma concentration versus time curves over 48 hours ($AUC_{0-48\,h}$) were 240 µg·h/ml for the salt form and 201 µg·h/ml for the free base form (Taylor, 2008).

2.3.2   Toxicological studies

(a)   Acute toxicity

An oral acute toxicity study has been conducted on 4-amino-5,6-dimethylthieno[2,3-d]pyramidin-2(1H)-one (i.e. the free base form of No. 2117). No deaths occurred following gavage administration to rats at doses up to 50 mg/kg bw, the highest dose tested (Arulnesan, 2007).

(b)   Short-term studies of toxicity

Short-term studies of the toxicity of 2-pentylthiophene (No. 2106), 4-amino-5,6-dimethylthieno[2,3-d]pyrimidin-2(1H)-one and its hydrochloride salt (No. 2117) are described below and summarized in Table 3.

Table 3. Results of short-term studies of the toxicity of sulfur-containing heterocyclic compounds used as flavouring agents

No.	Flavouring agent	Species; sex	No. of test groups[a] / no. per group[b]	Route	Duration (days)	NOEL/ NOAEL (mg/kg bw per day)	Reference
2106	2-Pentylthiophene	Rat; M, F	3/10	Gavage	28	NE	Dhinsa, Watson & Brooks (2006)
2106	2-Pentylthiophene	Rat; M, F	1/10	Gavage	28	3[c]	Marr & Watson (2007)
2117	4-Amino-5,6-dimethylthieno[2,3-d]pyrimidin-2(1H)-one hydrochloride[d]	Rat; M, F	3/10	Diet	21	100[e]	Ross (2008a)
2117	4-Amino-5,6-dimethylthieno[2,3-d]pyrimidin-2(1H)-one hydrochloride	Rat; M, F	3/40	Diet	90	60[e]	Ross (2008b)

F, female; M, male; NE, not established (treatment-related adverse effects were observed at each dose level)

[a] Total number of test groups does not include control animals.
[b] Total number per test group includes both male and female animals.
[c] NOEL (the only dose tested).
[d] Test article was 4-amino-5,6-dimethylthieno[2,3-d]pyrimidin-2(1H)-one (i.e. the free base form of No. 2117).
[e] NOAEL (the highest dose tested).

(i)   2-Pentylthiophene (No. 2106)

In a good laboratory practice (GLP)-compliant study conducted according to Organisation for Economic Co-operation and Development (OECD) Test Guideline 407, 2-pentylthiophene (No. 2106) was administered by oral gavage to Sprague-Dawley rats (five of each sex per group) for 28 days at dose levels of 15, 150 or 500 mg/kg bw per day. The control group (five of each sex) was dosed with vehicle alone (arachis oil). Clinical signs, motor activity, grip strength, sensory reactivity, body weight and feed and water consumption were monitored during the study. Haematology and clinical chemistry were evaluated for all animals at the end of the study. All animals were subjected to gross necropsy examination, and histopathological evaluation of selected tissues was performed.

There were no treatment-related deaths. Males and females receiving 500 mg/kg bw per day displayed increased salivation soon after dosing from day 3 until the end of the treatment period. Coloured staining (yellow/orange/brown) was also noted on the cage tray liners from this dose group from day 8. Similar effects were also noted for animals of both sexes treated with 150 mg/kg bw per day, although to a much lesser extent. No treatment-related clinical signs were evident at 15 mg/kg bw per day. There were no treatment-related changes in motor activity, grip strength or sensory reactivity.

No adverse effects on body weight or feed consumption were detected for treated animals in comparison with controls. An increase in water consumption was observed in males and females treated with 500 mg/kg bw per day. Haemolytic anaemia was evident in males and females treated with 500 and 150 mg/kg bw per day, evidenced by reductions in haemoglobin, haematocrit and erythrocyte count, and reticulocytosis was also observed. Increased alanine aminotransferase activity was observed in males and females treated with 500 mg/kg bw per day; however, the increase was statistically significant only in males ($P < 0.01$). Dose-dependent increases in bilirubin level were observed in males and females at 15, 150 and 500 mg/kg bw per day. The increases ranged from 1.4- to 4.8-fold in males and from 2.6-fold to 12-fold in females. In males, these increases were statistically significant at 150 mg/kg bw per day ($P < 0.01$) and 500 mg/kg bw per day ($P < 0.01$), whereas in females, the increases were statistically significant at all dose levels ($P < 0.05$ at 15 mg/kg bw per day and $P < 0.01$ at 150 and 500 mg/kg bw per day). Males and females treated with 500 mg/kg bw per day showed increases in absolute and relative liver, kidney and spleen weights, with effects extending into the 150 mg/kg bw per day dose group. At necropsy, males and females receiving 500 mg/kg bw per day displayed dark and enlarged spleens, with effects extending into the 150 mg/kg bw per day dose group.

Treatment-related histopathological effects at 150 and 500 mg/kg bw per day were observed for the liver, spleen, kidney, bladder and thyroid gland. Centrilobular hepatocyte enlargement in males and females was observed at 500 and 150 mg/kg bw per day. Haemosiderin pigment deposits were also observed for most animals treated with 500 mg/kg bw per day and for one female treated with 150 mg/kg bw per day. For the spleen, greater incidences or higher grades of severity of extramedullary haematopoiesis and of haemosiderin accumulation were observed in males and females at 500 and 150 mg/kg bw per day. Associated hyperaemia was observed in animals treated with 500 and 150 mg/kg bw per day. For the kidney, haemosiderin pigment deposits were seen in the tubular epithelium of males and females treated with 500 mg/kg bw per day. Tubular hypertrophy, tubular dilatation and tubular basophilia were also observed at this dose level. Pyelitis with associated hyperplasia of the renal papillary/pelvic epithelium was seen in two females at 500 mg/kg bw per day. Globular accumulations of eosinophilic material were observed in the renal tubules of two males treated with 500 mg/kg bw per day and for three males treated with 150 mg/kg bw per day. For the bladder, epithelial hyperplasia and associated epithelial and subepithelial inflammatory cell infiltrates were observed in two females treated with 500 mg/kg bw per day. For the thyroid, hypertrophy of follicle lining cells was observed in males and females at 500 mg/kg bw per day and in males at 150 mg/kg bw per day.

A NOAEL cannot be assigned in this study because of the dose-dependent increases in bilirubin level with onset at the low dose of 15 mg/kg bw per day (Dhinsa, Watson & Brooks, 2006).

In a follow-up GLP-compliant study, 2-pentylthiophene (No. 2106) was administered by oral gavage at a dose level of 3 mg/kg bw per day to one group of five male and five female Sprague-Dawley rats for 28 days. A control group of five males and five females was given vehicle alone (arachis oil). Clinical signs, body weight and feed and water consumption monitored during the study revealed no treatment-related differences between test and control animals. Haematology and clinical chemistry were evaluated for all animals at the end of the study and showed no treatment-related changes. No macroscopic abnormalities were detected at necropsy. There were no treatment-related changes in spleen weights for treated animals in comparison with controls. Weights of other organs or tissues were not evaluated. Histopathology evaluations were not conducted. The NOEL was considered to be 3 mg/kg bw per day, the only dose tested (Marr & Watson, 2007).

(ii) *4-Amino-5,6-dimethylthieno[2,3-d]pyrimidin-2(1H)-one and the salt form 4-amino-5,6-dimethylthieno[2,3-d]pyrimidin-2(1H)-one hydrochloride (No. 2117)*

In a GLP-compliant study, groups of Sprague-Dawley rats (five of each sex per group) received 4-amino-5,6-dimethylthieno[2,3-d]pyrimidin-2(1H)-one via the diet over a period of 21 days. Target dose levels were 10, 30 and 100 mg/kg bw per day. Controls (five of each sex) received untreated diet throughout the treatment period. Survival, clinical observations, body weight, feed consumption, organ weights and macroscopic evaluations of all animals were assessed. Additionally, the livers of all control and high-dose animals were subjected to histopathological examination. Overall achieved doses were 10.3, 29.4 and 101 mg/kg bw per day for males and 10.9, 31.1 and 103 mg/kg bw per day for females.

There were no deaths, no clinical signs attributable to treatment and no treatment-related effects on body weight or feed consumption. Body weight–relative liver weights were increased in males (by 11%) and females (by 12%) receiving 100 mg/kg bw per day; however, the increase was statistically significant only in females ($P < 0.05$), and there were no macroscopic or microscopic changes in the liver.

The NOAEL was considered to be 100 mg/kg bw per day, the highest dose tested (Ross, 2008a).

In a GLP-compliant study designed to meet OECD and United States Food and Drug Administration Redbook guidelines, groups of Sprague-Dawley rats (20 of each sex per group) received 4-amino-5,6-dimethylthieno[2,3-d]pyrimidin-2(1H)-one hydrochloride (No. 2117) via the diet over a period of 13 weeks. Target dose levels were 10, 30 and 60 mg/kg bw per day. Controls (20 of each sex per group) received untreated diet throughout the treatment period. During the study, clinical condition, detailed physical and arena observations, sensory reactivity, grip strength, motor activity, body weight, feed consumption, ophthalmic examination, haematology, clinical chemistry, urine analysis, organ weight, gross pathology and histopathology investigations were undertaken. Blood samples for haematology and clinical chemistry investigations were taken on day 14 and during weeks 6 and 13 of treatment.

Overall achieved doses were 10.3, 30.8 and 61.9 mg/kg bw per day for males and 10.3, 30.6 and 60.3 mg/kg bw per day for females. The appearance and behaviour of the animals were unaffected by treatment. There was no effect of treatment on sensory reactivity, grip strength or motor activity. Two incidental deaths occurred, one low-dose female in week 1 and one high-dose male due to damage sustained accidentally during the blood sampling procedure. Body weight gain and feed consumption were not affected by treatment. There were no treatment-related effects on ophthalmology, haematology or urine analysis.

There were changes in several clinical chemistry parameters that were considered attributable to treatment, as shown in Table 4: increased creatinine in females receiving 10, 30 or 60 mg/kg bw per day; increased total cholesterol in females receiving 10, 30 or 60 mg/kg bw per day and in males receiving 60 mg/kg bw per day; increased total triglycerides in males and females receiving 60 mg/kg bw per day; increased glucose in females receiving 10, 30 or 60 mg/kg bw per day and in males at 60 mg/kg bw per day; and increased total protein in females receiving 60 mg/kg bw per day and in males receiving 30 or 60 mg/kg bw per day.

The effect on total protein in week 13 in the high-dose males was attributed to an increase in globulin, leading to a reduction of the albumin to globulin ratio.

After 13 weeks of treatment, body weight–relative liver weights were increased in a dose-related manner (8–17% compared with controls) in animals given 30 or 60 mg/kg bw per day ($P < 0.01$). Body weight–relative kidney weights were slightly increased (8%; $P < 0.05$) in males given 60 mg/kg bw per day, and body weight–relative thyroid plus parathyroid weights were increased by 15% ($P < 0.05$) in females given 60 mg/kg bw per day.

There were no treatment-related macroscopic changes after 13 weeks of treatment. Histopathological changes related to treatment were confined to minimal centrilobular hypertrophy in the livers of one male and three females given 60 mg/kg bw per day.

Changes in several clinical chemistry parameters (high glucose, creatinine, cholesterol, triglyceride and protein levels) can be attributed to an alteration of hepatic metabolism as a consequence of an adaptive response to treatment. The effect on cholesterol involved all treated groups of females, and the cause was not established. There was no evidence for an increase in hepatocyte vacuolation. Although the effects on cholesterol were consistent, there was no adverse response in any animal, and these findings are therefore considered unlikely to be of any toxicological significance. The small increase in body weight–relative thyroid weight in high-dose females was not associated with any histopathological finding. For the kidney, as there was no histopathological change in the organ and no findings in the blood or urine that suggest renal toxicity, the increase in body weight–relative kidney weight was considered an adaptive response to the excretion of the test article and/or its metabolites.

It is concluded that treatment with 4-amino-5,6-dimethylthieno[2,3-d]-pyrimidin-2(1H)-one hydrochloride produced adaptive changes in the liver and kidneys and a possible secondary effect on the thyroid. These changes are not

Table 4. Clinical chemistry parameters in rats administered 4-amino-5,6-dimethylthieno[2,3-d]pyrimidin-2(1H)-one hydrochloride in the diet for 13 weeks

Parameter	Time	Dose (mg/kg bw per day)			
		0	10	30	60
Males					
Glucose (mmol/l)	Week 13	8.09	8.41	7.29	10.17*
Total cholesterol (mmol/l)	Day 14	1.98	2.07	2.33	2.48*
Triglycerides (mmol/l)	Day 14	0.68	0.75	0.77	1.08**
Total protein (g/l)	Week 6	65	66	67*	68**
	Week 13	70	71	75	76*
Albumin (g/l)	Week 6	36	35	36	37**
Globulin (g/l)	Week 13	35	37	39	40*
Albumin/globulin	Week 13	1.00	0.94	0.94	0.89*
Females					
Creatinine (μmol/l)	Week 6	34	38*	38*	40**
	Week 13	40	47	44	53**
Glucose (mmol/l)	Day 14	5.94	5.76	6.34	7.15**
	Week 13	6.55	7.71*	7.69*	7.54*
Total cholesterol (mmol/l)	Day 14	1.87	2.42*	2.30*	2.73**
	Week 6	1.90	2.27	2.43*	2.74**
	Week 13	1.69	2.11*	2.26**	2.51**
Triglycerides (mmol/l)	Week 13	0.70	0.95	1.00	1.05*
Total protein (g/l)	Week 6	66	68	70	70*
	Week 13	72	72	72	77**
Albumin (g/l)	Week 13	38	38	38	40*

* $P < 0.05$; ** $P < 0.01$

considered to be toxicologically important. The NOAEL was therefore considered to be 60 mg/kg bw per day, the highest dose tested (Ross, 2008b).

(c) Genotoxicity

In vitro and in vivo genotoxicity studies have been conducted on one flavouring agent in this group (No. 2117), with uniformly negative results. Details of these studies are presented in Table 5.

Table 5. Results of genotoxicity studies with sulfur-containing heterocyclic compounds used as flavouring agents

No.	Flavouring agent	End-point	Test object	Concentration/dose	Results	Reference
In vitro						
2117	4-Amino-5,6-dimethylthieno[2,3-d]-pyrimidin-2(1H)-one hydrochloride[a]	Reverse mutation	*Salmonella typhimurium* TA98 and TA100	7–5000 µg/plate	Negative[b,c,d]	Zhang (2007)
2117	4-Amino-5,6-dimethylthieno[2,3-d]-pyrimidin-2(1H)-one hydrochloride	Reverse mutation	*S. typhimurium* TA98, TA100, TA1535, TA1537 and TA1538 and *Escherichia coli* WP2uvrA	21–5000 µg/plate	Negative[b,d,e]	Zhang (2008a)
2117	4-Amino-5,6-dimethylthieno[2,3-d]-pyrimidin-2(1H)-one hydrochloride	Chromosomal aberration	Chinese hamster ovary cells	21–5000 µg/ml	Negative[f,g]	Zhang (2008b)
2117	4-Amino-5,6-dimethylthieno[2,3-d]-pyrimidin-2(1H)-one hydrochloride	Chromosomal aberration	Chinese hamster ovary cells	190–1670 µg/ml	Negative[h]	Zhang (2008b)
In vivo						
2117	4-Amino-5,6-dimethylthieno[2,3-d]-pyrimidin-2(1H)-one hydrochloride	Micronucleus induction	CD-1 mouse bone marrow cells	500, 1000 and 2000 mg/kg bw[i]	Negative[i,j]	Zhang (2008c)

[a] Test article was 4-amino-5,6-dimethylthieno[2,3-d]pyrimidin-2(1H)-one (i.e. the free base form of No. 2117).
[b] With and without metabolic activation.
[c] Plate incorporation method.
[d] Precipitate observed at 5000 µg/plate.
[e] Separate assays using plate incorporation and preincubation methods.
[f] Two separate exposure conditions were tested: 1) 3-hour exposure in the absence of metabolic activation; and 2) 3-hour exposure in the presence of metabolic activation.
[g] Severe cytotoxicity was observed at 5000 µg/ml.
[h] Eighteen-hour exposure in the absence of metabolic activation.
[i] Administered by gavage.
[j] Bone marrow was harvested at 24 and 48 hours.

3. REFERENCES

Arulnesan N (2007). Acute oral toxicity study with S19752383 in rats. Unpublished report no. 188232 from Nucro-Technics, Toronto, Ontario, Canada. Submitted to WHO by the International Organization of the Flavor Industry, Brussels, Belgium.

Babish JG (1978). 90-day feeding study of IFF Code No. 16516 in Sprague Dawley rats. Unpublished interim report no. 5664b from Food and Drug Research Laboratories, Maspeth, NY, USA. Submitted to WHO by the International Organization of the Flavor Industry, Brussels, Belgium.

Cramer GM, Ford RA, Hall RL (1978). Estimation of toxic hazard—a decision tree approach. *Food and Cosmetics Toxicology*, 16:255–276.

Dansette PM, Bertho G, Mansuy D (2005). First evidence that cytochrome P450 may catalyze both S-oxidation and epoxidation of thiophene derivatives. *Biochemical and Biophysical Research Communications*, 338(1):450–455.

Dhinsa NK, Watson P, Brooks PN (2006). Flavor substance 0106: twenty-eight day repeated dose oral (gavage) toxicity study in the rat. Unpublished report no. 1834-0006 from SafePharm Laboratories, Derbyshire, England. Submitted to WHO by the International Organization of the Flavor Industry, Brussels, Belgium.

European Flavour and Fragrance Association (2004). European inquiry on volume use. Private communication to the Flavor and Extract Manufacturers Association, Washington, DC, USA. Submitted to WHO by the International Organization of the Flavor Industry, Brussels, Belgium.

Gavin CL, Williams MC, Hallagan JB (2008). *Flavor and Extract Manufacturers Association of the United States 2005 poundage and technical effects update survey.* Washington, DC, USA, Flavor and Extract Manufacturers Association of the United States.

International Organization of the Flavor Industry (2011). Interim inquiry on volume use and added use levels for flavoring agents to be presented at the 76th JECFA meeting. Private communication to the Flavor and Extract Manufacturers Association, Washington, DC, USA. Submitted to WHO by the International Organization of the Flavor Industry, Brussels, Belgium.

Japan Flavor and Fragrance Materials Association (2005). Japanese inquiry on volume use. Private communication to the Flavor and Extract Manufacturers Association, Washington, DC, USA. Submitted to WHO by the International Organization of the Flavor Industry, Brussels, Belgium.

Marr A, Watson P (2007). Flavor substance 0106: limited twenty-eight day repeated oral (gavage) toxicity study in the rat. Unpublished report no. 1834-0007 from SafePharm Laboratories, Derbyshire, England. Submitted to WHO by the International Organization of the Flavor Industry, Brussels, Belgium.

Nijssen LM, van Ingen-Visscher CA, Donders JJH (2011). Volatile Compounds in Food (VCF) database, version 13.1. Zeist, the Netherlands, TNO Triskelion (http://www.vcf-online.nl/VcfHome.cfm).

Posternak JM, Linder A, Vodoz CA (1969). Summaries of toxicological data: toxicological tests on flavoring matters. *Food and Cosmetics Toxicology*, 7:405–407.

Ross J (2008a). S2383: preliminary study by dietary administration to CD rats for 21 days. Unpublished report no. TEI0001 from Huntingdon Life Sciences, Cambridgeshire, England. Submitted to WHO by the International Organization of the Flavor Industry, Brussels, Belgium.

Ross J (2008b). S2383 HCl salt: toxicity study by dietary administration to CD rats for 13 weeks. Unpublished report no. TEI0003 from Huntingdon Life Sciences, Cambridgeshire, England. Submitted to WHO by the International Organization of the Flavor Industry, Brussels, Belgium.

Rush RE (1989). 14-day dietary toxicity study in rats with isopropyldimethyldihydrodithiazin 690 996. Unpublished report no. 3141.3B from Springborn Laboratories, Inc., Spencerville,

OH, USA. Submitted to WHO by the International Organization of the Flavor Industry, Brussels, Belgium.

Stofberg J, Grundschober F (1987). Consumption ratio and food predominance of flavoring materials. *Perfumer & Flavorist*, 12:27–68.

Taylor S (2008). Comparative bioavailability study of S2383 and S8475 by oral administration in Sprague-Dawley rats. Unpublished report no. 199097 from Nucro-Technics, Toronto, Ontario, Canada. Submitted to WHO by the International Organization of the Flavor Industry, Brussels, Belgium.

Yan Z et al. (2005). Rapid detection and characterization of minor reactive metabolites using stable-isotope trapping in combination with tandem mass spectrometry. *Rapid Communications in Mass Spectrometry*, 19(22):3322–3330.

Zhang B (2007). TA98 and TA100 reverse mutation test of S2383. Unpublished report no. 189965 from Nucro-Technics, Toronto, Ontario, Canada. Submitted to WHO by the International Organization of the Flavor Industry, Brussels, Belgium.

Zhang B (2008a). Bacterial reverse mutation test of S8475 (S2383 HCl salt). Unpublished report no. 194267 from Nucro-Technics, Toronto, Ontario, Canada. Submitted to WHO by the International Organization of the Flavor Industry, Brussels, Belgium.

Zhang B (2008b). Chromosome aberration test of S8475 (S2383 HCl salt) in cultured Chinese hamster ovary cells. Unpublished report no. 194279 from Nucro-Technics, Toronto, Ontario, Canada. Submitted to WHO by the International Organization of the Flavor Industry, Brussels, Belgium.

Zhang B (2008c). Mouse micronucleus test, in vivo, of S8475 (S2383 HCl salt). Unpublished report no. 194280 from Nucro-Technics, Toronto, Ontario, Canada. Submitted to WHO by the International Organization of the Flavor Industry, Brussels, Belgium.

ANNEXES

ANNEXES

ANNEX 1

REPORTS AND OTHER DOCUMENTS RESULTING FROM PREVIOUS MEETINGS OF THE JOINT FAO/WHO EXPERT COMMITTEE ON FOOD ADDITIVES

1. *General principles governing the use of food additives* (First report of the Joint FAO/WHO Expert Committee on Food Additives). FAO Nutrition Meetings Report Series, No. 15, 1957; WHO Technical Report Series, No. 129, 1957 (out of print).
2. *Procedures for the testing of intentional food additives to establish their safety for use* (Second report of the Joint FAO/WHO Expert Committee on Food Additives). FAO Nutrition Meetings Report Series, No. 17, 1958; WHO Technical Report Series, No. 144, 1958 (out of print).
3. *Specifications for identity and purity of food additives (antimicrobial preservatives and antioxidants)* (Third report of the Joint FAO/WHO Expert Committee on Food Additives). These specifications were subsequently revised and published as *Specifications for identity and purity of food additives, Vol. I. Antimicrobial preservatives and antioxidants*, Rome, Food and Agriculture Organization of the United Nations, 1962 (out of print).
4. *Specifications for identity and purity of food additives (food colours)* (Fourth report of the Joint FAO/WHO Expert Committee on Food Additives). These specifications were subsequently revised and published as *Specifications for identity and purity of food additives, Vol. II. Food colours*, Rome, Food and Agriculture Organization of the United Nations, 1963 (out of print).
5. *Evaluation of the carcinogenic hazards of food additives* (Fifth report of the Joint FAO/WHO Expert Committee on Food Additives). FAO Nutrition Meetings Report Series, No. 29, 1961; WHO Technical Report Series, No. 220, 1961 (out of print).
6. *Evaluation of the toxicity of a number of antimicrobials and antioxidants* (Sixth report of the Joint FAO/WHO Expert Committee on Food Additives). FAO Nutrition Meetings Report Series, No. 31, 1962; WHO Technical Report Series, No. 228, 1962 (out of print).
7. *Specifications for the identity and purity of food additives and their toxicological evaluation: emulsifiers, stabilizers, bleaching and maturing agents* (Seventh report of the Joint FAO/WHO Expert Committee on Food Additives). FAO Nutrition Meetings Series, No. 35, 1964; WHO Technical Report Series, No. 281, 1964 (out of print).
8. *Specifications for the identity and purity of food additives and their toxicological evaluation: food colours and some antimicrobials and antioxidants* (Eighth report of the Joint FAO/WHO Expert Committee on Food Additives). FAO Nutrition Meetings Series, No. 38, 1965; WHO Technical Report Series, No. 309, 1965 (out of print).
9. *Specifications for identity and purity and toxicological evaluation of some antimicrobials and antioxidants.* FAO Nutrition Meetings Report Series, No. 38A, 1965; WHO/Food Add/24.65 (out of print).
10. *Specifications for identity and purity and toxicological evaluation of food colours.* FAO Nutrition Meetings Report Series, No. 38B, 1966; WHO/Food Add/66.25.

11. *Specifications for the identity and purity of food additives and their toxicological evaluation: some antimicrobials, antioxidants, emulsifiers, stabilizers, flour treatment agents, acids, and bases* (Ninth report of the Joint FAO/WHO Expert Committee on Food Additives). FAO Nutrition Meetings Series, No. 40, 1966; WHO Technical Report Series, No. 339, 1966 (out of print).
12. *Toxicological evaluation of some antimicrobials, antioxidants, emulsifiers, stabilizers, flour treatment agents, acids, and bases.* FAO Nutrition Meetings Report Series, No. 40A, B, C; WHO/Food Add/67.29.
13. *Specifications for the identity and purity of food additives and their toxicological evaluation: some emulsifiers and stabilizers and certain other substances* (Tenth report of the Joint FAO/WHO Expert Committee on Food Additives). FAO Nutrition Meetings Series, No. 43, 1967; WHO Technical Report Series, No. 373, 1967.
14. *Specifications for the identity and purity of food additives and their toxicological evaluation: some flavouring substances and non nutritive sweetening agents* (Eleventh report of the Joint FAO/WHO Expert Committee on Food Additives). FAO Nutrition Meetings Series, No. 44, 1968; WHO Technical Report Series, No. 383, 1968.
15. *Toxicological evaluation of some flavouring substances and non nutritive sweetening agents.* FAO Nutrition Meetings Report Series, No. 44A, 1968; WHO/Food Add/68.33.
16. *Specifications and criteria for identity and purity of some flavouring substances and non-nutritive sweetening agents.* FAO Nutrition Meetings Report Series, No. 44B, 1969; WHO/Food Add/69.31.
17. *Specifications for the identity and purity of food additives and their toxicological evaluation: some antibiotics* (Twelfth report of the Joint FAO/WHO Expert Committee on Food Additives). FAO Nutrition Meetings Series, No. 45, 1969; WHO Technical Report Series, No. 430, 1969.
18. *Specifications for the identity and purity of some antibiotics.* FAO Nutrition Meetings Series, No. 45A, 1969; WHO/Food Add/69.34.
19. *Specifications for the identity and purity of food additives and their toxicological evaluation: some food colours, emulsifiers, stabilizers, anticaking agents, and certain other substances* (Thirteenth report of the Joint FAO/WHO Expert Committee on Food Additives). FAO Nutrition Meetings Series, No. 46, 1970; WHO Technical Report Series, No. 445, 1970.
20. *Toxicological evaluation of some food colours, emulsifiers, stabilizers, anticaking agents, and certain other substances.* FAO Nutrition Meetings Report Series, No. 46A, 1970; WHO/Food Add/70.36.
21. *Specifications for the identity and purity of some food colours, emulsifiers, stabilizers, anticaking agents, and certain other food additives.* FAO Nutrition Meetings Report Series, No. 46B, 1970; WHO/Food Add/70.37.
22. *Evaluation of food additives: specifications for the identity and purity of food additives and their toxicological evaluation: some extraction solvents and certain other substances; and a review of the technological efficacy of some antimicrobial agents* (Fourteenth report of the Joint FAO/WHO Expert Committee on Food Additives). FAO Nutrition Meetings Series, No. 48, 1971; WHO Technical Report Series, No. 462, 1971.
23. *Toxicological evaluation of some extraction solvents and certain other substances.* FAO Nutrition Meetings Report Series, No. 48A, 1971; WHO/Food Add/70.39.

24. *Specifications for the identity and purity of some extraction solvents and certain other substances.* FAO Nutrition Meetings Report Series, No. 48B, 1971; WHO/Food Add/70.40.
25. *A review of the technological efficacy of some antimicrobial agents.* FAO Nutrition Meetings Report Series, No. 48C, 1971; WHO/Food Add/70.41.
26. *Evaluation of food additives: some enzymes, modified starches, and certain other substances: Toxicological evaluations and specifications and a review of the technological efficacy of some antioxidants* (Fifteenth report of the Joint FAO/WHO Expert Committee on Food Additives). FAO Nutrition Meetings Series, No. 50, 1972; WHO Technical Report Series, No. 488, 1972.
27. *Toxicological evaluation of some enzymes, modified starches, and certain other substances.* FAO Nutrition Meetings Report Series, No. 50A, 1972; WHO Food Additives Series, No. 1, 1972.
28. *Specifications for the identity and purity of some enzymes and certain other substances.* FAO Nutrition Meetings Report Series, No. 50B, 1972; WHO Food Additives Series, No. 2, 1972.
29. *A review of the technological efficacy of some antioxidants and synergists.* FAO Nutrition Meetings Report Series, No. 50C, 1972; WHO Food Additives Series, No. 3, 1972.
30. *Evaluation of certain food additives and the contaminants mercury, lead, and cadmium* (Sixteenth report of the Joint FAO/WHO Expert Committee on Food Additives). FAO Nutrition Meetings Series, No. 51, 1972; WHO Technical Report Series, No. 505, 1972, and corrigendum.
31. *Evaluation of mercury, lead, cadmium and the food additives amaranth, diethylpyrocarbamate, and octyl gallate.* FAO Nutrition Meetings Report Series, No. 51A, 1972; WHO Food Additives Series, No. 4, 1972.
32. *Toxicological evaluation of certain food additives with a review of general principles and of specifications* (Seventeenth report of the Joint FAO/WHO Expert Committee on Food Additives). FAO Nutrition Meetings Series, No. 53, 1974; WHO Technical Report Series, No. 539, 1974, and corrigendum (out of print).
33. *Toxicological evaluation of some food additives including anticaking agents, antimicrobials, antioxidants, emulsifiers, and thickening agents.* FAO Nutrition Meetings Report Series, No. 53A, 1974; WHO Food Additives Series, No. 5, 1974.
34. *Specifications for identity and purity of thickening agents, anticaking agents, antimicrobials, antioxidants and emulsifiers.* FAO Food and Nutrition Paper, No. 4, 1978.
35. *Evaluation of certain food additives* (Eighteenth report of the Joint FAO/WHO Expert Committee on Food Additives). FAO Nutrition Meetings Series, No. 54, 1974; WHO Technical Report Series, No. 557, 1974, and corrigendum.
36. *Toxicological evaluation of some food colours, enzymes, flavour enhancers, thickening agents, and certain other food additives.* FAO Nutrition Meetings Report Series, No. 54A, 1975; WHO Food Additives Series, No. 6, 1975.
37. *Specifications for the identity and purity of some food colours, enhancers, thickening agents, and certain food additives.* FAO Nutrition Meetings Report Series, No. 54B, 1975; WHO Food Additives Series, No. 7, 1975.
38. *Evaluation of certain food additives: some food colours, thickening agents, smoke condensates, and certain other substances.* (Nineteenth report of the

Joint FAO/WHO Expert Committee on Food Additives). FAO Nutrition Meetings Series, No. 55, 1975; WHO Technical Report Series, No. 576, 1975.
39. *Toxicological evaluation of some food colours, thickening agents, and certain other substances.* FAO Nutrition Meetings Report Series, No. 55A, 1975; WHO Food Additives Series, No. 8, 1975.
40. *Specifications for the identity and purity of certain food additives.* FAO Nutrition Meetings Report Series, No. 55B, 1976; WHO Food Additives Series, No. 9, 1976.
41. *Evaluation of certain food additives* (Twentieth report of the Joint FAO/WHO Expert Committee on Food Additives). FAO Food and Nutrition Meetings Series, No. 1, 1976; WHO Technical Report Series, No. 599, 1976.
42. *Toxicological evaluation of certain food additives.* WHO Food Additives Series, No. 10, 1976.
43. *Specifications for the identity and purity of some food additives.* FAO Food and Nutrition Series, No. 1B, 1977; WHO Food Additives Series, No. 11, 1977.
44. *Evaluation of certain food additives* (Twenty-first report of the Joint FAO/WHO Expert Committee on Food Additives). WHO Technical Report Series, No. 617, 1978.
45. *Summary of toxicological data of certain food additives.* WHO Food Additives Series, No. 12, 1977.
46. *Specifications for identity and purity of some food additives, including antioxidant, food colours, thickeners, and others.* FAO Nutrition Meetings Report Series, No. 57, 1977.
47. *Evaluation of certain food additives and contaminants* (Twenty-second report of the Joint FAO/WHO Expert Committee on Food Additives). WHO Technical Report Series, No. 631, 1978.
48. *Summary of toxicological data of certain food additives and contaminants.* WHO Food Additives Series, No. 13, 1978.
49. *Specifications for the identity and purity of certain food additives.* FAO Food and Nutrition Paper, No. 7, 1978.
50. *Evaluation of certain food additives* (Twenty-third report of the Joint FAO/WHO Expert Committee on Food Additives). WHO Technical Report Series, No. 648, 1980, and corrigenda.
51. *Toxicological evaluation of certain food additives.* WHO Food Additives Series, No. 14, 1980.
52. *Specifications for identity and purity of food colours, flavouring agents, and other food additives.* FAO Food and Nutrition Paper, No. 12, 1979.
53. *Evaluation of certain food additives* (Twenty-fourth report of the Joint FAO/WHO Expert Committee on Food Additives). WHO Technical Report Series, No. 653, 1980.
54. *Toxicological evaluation of certain food additives.* WHO Food Additives Series, No. 15, 1980.
55. *Specifications for identity and purity of food additives (sweetening agents, emulsifying agents, and other food additives).* FAO Food and Nutrition Paper, No. 17, 1980.
56. *Evaluation of certain food additives* (Twenty-fifth report of the Joint FAO/WHO Expert Committee on Food Additives). WHO Technical Report Series, No. 669, 1981.

57. *Toxicological evaluation of certain food additives.* WHO Food Additives Series, No. 16, 1981.
58. *Specifications for identity and purity of food additives (carrier solvents, emulsifiers and stabilizers, enzyme preparations, flavouring agents, food colours, sweetening agents, and other food additives).* FAO Food and Nutrition Paper, No. 19, 1981.
59. *Evaluation of certain food additives and contaminants* (Twenty-sixth report of the Joint FAO/WHO Expert Committee on Food Additives). WHO Technical Report Series, No. 683, 1982.
60. *Toxicological evaluation of certain food additives.* WHO Food Additives Series, No. 17, 1982.
61. *Specifications for the identity and purity of certain food additives.* FAO Food and Nutrition Paper, No. 25, 1982.
62. *Evaluation of certain food additives and contaminants* (Twenty-seventh report of the Joint FAO/WHO Expert Committee on Food Additives). WHO Technical Report Series, No. 696, 1983, and corrigenda.
63. *Toxicological evaluation of certain food additives and contaminants.* WHO Food Additives Series, No. 18, 1983.
64. *Specifications for the identity and purity of certain food additives.* FAO Food and Nutrition Paper, No. 28, 1983.
65. *Guide to specifications—General notices, general methods, identification tests, test solutions, and other reference materials.* FAO Food and Nutrition Paper, No. 5, Rev. 1, 1983.
66. *Evaluation of certain food additives and contaminants* (Twenty-eighth report of the Joint FAO/WHO Expert Committee on Food Additives). WHO Technical Report Series, No. 710, 1984, and corrigendum.
67. *Toxicological evaluation of certain food additives and contaminants.* WHO Food Additives Series, No. 19, 1984.
68. *Specifications for the identity and purity of food colours.* FAO Food and Nutrition Paper, No. 31/1, 1984.
69. *Specifications for the identity and purity of food additives.* FAO Food and Nutrition Paper, No. 31/2, 1984.
70. *Evaluation of certain food additives and contaminants* (Twenty-ninth report of the Joint FAO/WHO Expert Committee on Food Additives). WHO Technical Report Series, No. 733, 1986, and corrigendum.
71. *Specifications for the identity and purity of certain food additives.* FAO Food and Nutrition Paper, No. 34, 1986.
72. *Toxicological evaluation of certain food additives and contaminants.* WHO Food Additives Series, No. 20. Cambridge University Press, 1987.
73. *Evaluation of certain food additives and contaminants* (Thirtieth report of the Joint FAO/WHO Expert Committee on Food Additives). WHO Technical Report Series, No. 751, 1987.
74. *Toxicological evaluation of certain food additives and contaminants.* WHO Food Additives Series, No. 21. Cambridge University Press, 1987.
75. *Specifications for the identity and purity of certain food additives.* FAO Food and Nutrition Paper, No. 37, 1986.
76. *Principles for the safety assessment of food additives and contaminants in food.* WHO Environmental Health Criteria, No. 70. Geneva, World Health

Organization, 1987 (out of print). The full text is available electronically at www.who.int/pcs.
77. *Evaluation of certain food additives and contaminants* (Thirty-first report of the Joint FAO/WHO Expert Committee on Food Additives). WHO Technical Report Series, No. 759, 1987, and corrigendum.
78. *Toxicological evaluation of certain food additives.* WHO Food Additives Series, No. 22. Cambridge University Press, 1988.
79. *Specifications for the identity and purity of certain food additives.* FAO Food and Nutrition Paper, No. 38, 1988.
80. *Evaluation of certain veterinary drug residues in food* (Thirty-second report of the Joint FAO/WHO Expert Committee on Food Additives). WHO Technical Report Series, No. 763, 1988.
81. *Toxicological evaluation of certain veterinary drug residues in food.* WHO Food Additives Series, No. 23. Cambridge University Press, 1988.
82. *Residues of some veterinary drugs in animals and foods.* FAO Food and Nutrition Paper, No. 41, 1988.
83. *Evaluation of certain food additives and contaminants* (Thirty-third report of the Joint FAO/WHO Expert Committee on Food Additives). WHO Technical Report Series, No. 776, 1989.
84. *Toxicological evaluation of certain food additives and contaminants.* WHO Food Additives Series, No. 24. Cambridge University Press, 1989.
85. *Evaluation of certain veterinary drug residues in food* (Thirty-fourth report of the Joint FAO/WHO Expert Committee on Food Additives). WHO Technical Report Series, No. 788, 1989.
86. *Toxicological evaluation of certain veterinary drug residues in food.* WHO Food Additives Series, No. 25, 1990.
87. *Residues of some veterinary drugs in animals and foods.* FAO Food and Nutrition Paper, No. 41/2, 1990.
88. *Evaluation of certain food additives and contaminants* (Thirty-fifth report of the Joint FAO/WHO Expert Committee on Food Additives). WHO Technical Report Series, No. 789, 1990, and corrigenda.
89. *Toxicological evaluation of certain food additives and contaminants.* WHO Food Additives Series, No. 26, 1990.
90. *Specifications for identity and purity of certain food additives.* FAO Food and Nutrition Paper, No. 49, 1990.
91. *Evaluation of certain veterinary drug residues in food* (Thirty-sixth report of the Joint FAO/WHO Expert Committee on Food Additives). WHO Technical Report Series, No. 799, 1990.
92. *Toxicological evaluation of certain veterinary drug residues in food.* WHO Food Additives Series, No. 27, 1991.
93. *Residues of some veterinary drugs in animals and foods.* FAO Food and Nutrition Paper, No. 41/3, 1991.
94. *Evaluation of certain food additives and contaminants* (Thirty-seventh report of the Joint FAO/WHO Expert Committee on Food Additives). WHO Technical Report Series, No. 806, 1991, and corrigenda.
95. *Toxicological evaluation of certain food additives and contaminants.* WHO Food Additives Series, No. 28, 1991.
96. *Compendium of food additive specifications (Joint FAO/WHO Expert Committee on Food Additives (JECFA)). Combined specifications from 1st through the*

37th meetings, 1956–1990. Rome, Food and Agriculture Organization of the United Nations, 1992 (2 volumes).
97. *Evaluation of certain veterinary drug residues in food* (Thirty-eighth report of the Joint FAO/WHO Expert Committee on Food Additives). WHO Technical Report Series, No. 815, 1991.
98. *Toxicological evaluation of certain veterinary residues in food.* WHO Food Additives Series, No. 29, 1991.
99. *Residues of some veterinary drugs in animals and foods.* FAO Food and Nutrition Paper, No. 41/4, 1991.
100. *Guide to specifications—General notices, general analytical techniques, identification tests, test solutions, and other reference materials.* FAO Food and Nutrition Paper, No. 5, Ref. 2, 1991.
101. *Evaluation of certain food additives and naturally occurring toxicants* (Thirty-ninth report of the Joint FAO/WHO Expert Committee on Food Additives). WHO Technical Report Series No. 828, 1992.
102. *Toxicological evaluation of certain food additives and naturally occurring toxicants.* WHO Food Additives Series, No. 30, 1993.
103. *Compendium of food additive specifications: addendum 1.* FAO Food and Nutrition Paper, No. 52, 1992.
104. *Evaluation of certain veterinary drug residues in food* (Fortieth report of the Joint FAO/WHO Expert Committee on Food Additives). WHO Technical Report Series, No. 832, 1993.
105. *Toxicological evaluation of certain veterinary drug residues in food.* WHO Food Additives Series, No. 31, 1993.
106. *Residues of some veterinary drugs in animals and food.* FAO Food and Nutrition Paper, No. 41/5, 1993.
107. *Evaluation of certain food additives and contaminants* (Forty-first report of the Joint FAO/WHO Expert Committee on Food Additives). WHO Technical Report Series, No. 837, 1993.
108. *Toxicological evaluation of certain food additives and contaminants.* WHO Food Additives Series, No. 32, 1993.
109. *Compendium of food additive specifications: addendum 2.* FAO Food and Nutrition Paper, No. 52, Add. 2, 1993.
110. *Evaluation of certain veterinary drug residues in food* (Forty-second report of the Joint FAO/WHO Expert Committee on Food Additives). WHO Technical Report Series, No. 851, 1995.
111. *Toxicological evaluation of certain veterinary drug residues in food.* WHO Food Additives Series, No. 33, 1994.
112. *Residues of some veterinary drugs in animals and foods.* FAO Food and Nutrition Paper, No. 41/6, 1994.
113. *Evaluation of certain veterinary drug residues in food* (Forty-third report of the Joint FAO/WHO Expert Committee on Food Additives). WHO Technical Report Series, No. 855, 1995, and corrigendum.
114. *Toxicological evaluation of certain veterinary drug residues in food.* WHO Food Additives Series, No. 34, 1995.
115. *Residues of some veterinary drugs in animals and foods.* FAO Food and Nutrition Paper, No. 41/7, 1995.

116. *Evaluation of certain food additives and contaminants* (Forty-fourth report of the Joint FAO/WHO Expert Committee on Food Additives). WHO Technical Report Series, No. 859, 1995.
117. *Toxicological evaluation of certain food additives and contaminants.* WHO Food Additives Series, No. 35, 1996.
118. *Compendium of food additive specifications: addendum 3.* FAO Food and Nutrition Paper, No. 52, Add. 3, 1995.
119. *Evaluation of certain veterinary drug residues in food* (Forty-fifth report of the Joint FAO/WHO Expert Committee on Food Additives). WHO Technical Report Series, No. 864, 1996.
120. *Toxicological evaluation of certain veterinary drug residues in food.* WHO Food Additives Series, No. 36, 1996.
121. *Residues of some veterinary drugs in animals and foods.* FAO Food and Nutrition Paper, No. 41/8, 1996.
122. *Evaluation of certain food additives and contaminants* (Forty-sixth report of the Joint FAO/WHO Expert Committee on Food Additives). WHO Technical Report Series, No. 868, 1997.
123. *Toxicological evaluation of certain food additives.* WHO Food Additives Series, No. 37, 1996.
124. *Compendium of food additive specifications, addendum 4.* FAO Food and Nutrition Paper, No. 52, Add. 4, 1996.
125. *Evaluation of certain veterinary drug residues in food* (Forty-seventh report of the Joint FAO/WHO Expert Committee on Food Additives). WHO Technical Report Series, No. 876, 1998.
126. *Toxicological evaluation of certain veterinary drug residues in food.* WHO Food Additives Series, No. 38, 1996.
127. *Residues of some veterinary drugs in animals and foods.* FAO Food and Nutrition Paper, No. 41/9, 1997.
128. *Evaluation of certain veterinary drug residues in food* (Forty-eighth report of the Joint FAO/WHO Expert Committee on Food Additives). WHO Technical Report Series, No. 879, 1998.
129. *Toxicological evaluation of certain veterinary drug residues in food.* WHO Food Additives Series, No. 39, 1997.
130. *Residues of some veterinary drugs in animals and foods.* FAO Food and Nutrition Paper, No. 41/10, 1998.
131. *Evaluation of certain food additives and contaminants* (Forty-ninth report of the Joint FAO/WHO Expert Committee on Food Additives). WHO Technical Report Series, No. 884, 1999.
132. *Safety evaluation of certain food additives and contaminants.* WHO Food Additives Series, No. 40, 1998.
133. *Compendium of food additive specifications: addendum 5.* FAO Food and Nutrition Paper, No. 52, Add. 5, 1997.
134. *Evaluation of certain veterinary drug residues in food* (Fiftieth report of the Joint FAO/WHO Expert Committee on Food Additives). WHO Technical Report Series, No. 888, 1999.
135. *Toxicological evaluation of certain veterinary drug residues in food.* WHO Food Additives Series, No. 41, 1998.
136. *Residues of some veterinary drugs in animals and foods.* FAO Food and Nutrition Paper, No. 41/11, 1999.

137. *Evaluation of certain food additives* (Fifty-first report of the Joint FAO/WHO Expert Committee on Food Additives). WHO Technical Report Series, No. 891, 2000.
138. *Safety evaluation of certain food additives.* WHO Food Additives Series, No. 42, 1999.
139. *Compendium of food additive specifications, addendum 6.* FAO Food and Nutrition Paper, No. 52, Add. 6, 1998.
140. *Evaluation of certain veterinary drug residues in food* (Fifty-second report of the Joint FAO/WHO Expert Committee on Food Additives). WHO Technical Report Series, No. 893, 2000.
141. *Toxicological evaluation of certain veterinary drug residues in food.* WHO Food Additives Series, No. 43, 2000.
142. *Residues of some veterinary drugs in animals and foods.* FAO Food and Nutrition Paper, No. 41/12, 2000.
143. *Evaluation of certain food additives and contaminants* (Fifty-third report of the Joint FAO/WHO Expert Committee on Food Additives). WHO Technical Report Series, No. 896, 2000.
144. *Safety evaluation of certain food additives and contaminants.* WHO Food Additives Series, No. 44, 2000.
145. *Compendium of food additive specifications, addendum 7.* FAO Food and Nutrition Paper, No. 52, Add. 7, 1999.
146. *Evaluation of certain veterinary drug residues in food* (Fifty-fourth report of the Joint FAO/WHO Expert Committee on Food Additives). WHO Technical Report Series, No. 900, 2001.
147. *Toxicological evaluation of certain veterinary drug residues in food.* WHO Food Additives Series, No. 45, 2000.
148. *Residues of some veterinary drugs in animals and foods.* FAO Food and Nutrition Paper, No. 41/13, 2000.
149. *Evaluation of certain food additives and contaminants* (Fifty-fifth report of the Joint FAO/WHO Expert Committee on Food Additives). WHO Technical Report Series, No. 901, 2001.
150. *Safety evaluation of certain food additives and contaminants.* WHO Food Additives Series, No. 46, 2001.
151. *Compendium of food additive specifications: addendum 8.* FAO Food and Nutrition Paper, No. 52, Add. 8, 2000.
152. *Evaluation of certain mycotoxins in food* (Fifty-sixth report of the Joint FAO/WHO Expert Committee on Food Additives). WHO Technical Report Series, No. 906, 2002.
153. *Safety evaluation of certain mycotoxins in food.* WHO Food Additives Series, No. 47/FAO Food and Nutrition Paper 74, 2001.
154. *Evaluation of certain food additives and contaminants* (Fifty-seventh report of the Joint FAO/WHO Expert Committee on Food Additives). WHO Technical Report Series, No. 909, 2002.
155. *Safety evaluation of certain food additives and contaminants.* WHO Food Additives Series, No. 48, 2002.
156. *Compendium of food additive specifications: addendum 9.* FAO Food and Nutrition Paper, No. 52, Add. 9, 2001.

157. *Evaluation of certain veterinary drug residues in food* (Fifty-eighth report of the Joint FAO/WHO Expert Committee on Food Additives). WHO Technical Report Series, No. 911, 2002.
158. *Toxicological evaluation of certain veterinary drug residues in food.* WHO Food Additives Series, No. 49, 2002.
159. *Residues of some veterinary drugs in animals and foods.* FAO Food and Nutrition Paper, No. 41/14, 2002.
160. *Evaluation of certain food additives and contaminants* (Fifty-ninth report of the Joint FAO/WHO Expert Committee on Food Additives). WHO Technical Report Series, No. 913, 2002.
161. *Safety evaluation of certain food additives and contaminants.* WHO Food Additives Series, No. 50, 2003.
162. *Compendium of food additive specifications: addendum 10.* FAO Food and Nutrition Paper, No. 52, Add. 10, 2002.
163. *Evaluation of certain veterinary drug residues in food* (Sixtieth report of the Joint FAO/WHO Expert Committee on Food Additives). WHO Technical Report Series, No. 918, 2003.
164. *Toxicological evaluation of certain veterinary drug residues in food.* WHO Food Additives Series, No. 51, 2003.
165. *Residues of some veterinary drugs in animals and foods.* FAO Food and Nutrition Paper, No. 41/15, 2003.
166. *Evaluation of certain food additives and contaminants* (Sixty-first report of the Joint FAO/WHO Expert Committee on Food Additives). WHO Technical Report Series, No. 922, 2004.
167. *Safety evaluation of certain food additives and contaminants.* WHO Food Additives Series, No. 52, 2004.
168. *Compendium of food additive specifications: addendum 11.* FAO Food and Nutrition Paper, No. 52, Add. 11, 2003.
169. *Evaluation of certain veterinary drug residues in food* (Sixty-second report of the Joint FAO/WHO Expert Committee on Food Additives). WHO Technical Report Series, No. 925, 2004.
170. *Residues of some veterinary drugs in animals and foods.* FAO Food and Nutrition Paper, No. 41/16, 2004.
171. *Toxicological evaluation of certain veterinary drug residues in food.* WHO Food Additives Series, No. 53, 2005.
172. *Compendium of food additive specifications: addendum 12.* FAO Food and Nutrition Paper, No. 52, Add. 12, 2004.
173. *Evaluation of certain food additives* (Sixty-third report of the Joint FAO/WHO Expert Committee on Food Additives). WHO Technical Report Series, No. 928, 2005.
174. *Safety evaluation of certain food additives.* WHO Food Additives Series, No 54, 2005.
175. *Compendium of food additive specifications: addendum 13.* FAO Food and Nutrition Paper, No. 52, Add. 13 (with Errata), 2005.
176. *Evaluation of certain food contaminants* (Sixty-fourth report of the Joint FAO/WHO Expert Committee on Food Additives). WHO Technical Report Series, No. 930, 2005.
177. *Safety evaluation of certain contaminants in food.* WHO Food Additives Series, No. 55/FAO Food and Nutrition Paper, No. 82, 2006.

178. *Evaluation of certain food additives* (Sixty-fifth report of the Joint FAO/WHO Expert Committee on Food Additives). WHO Technical Report Series, No. 934, 2006.
179. *Safety evaluation of certain food additives.* WHO Food Additives Series, No. 56, 2006.
180. *Combined compendium of food additive specifications.* FAO JECFA Monographs 1, Volumes 1–4, 2005, 2006.
181. *Evaluation of certain veterinary drug residues in food* (Sixty-sixth report of the Joint FAO/WHO Expert Committee on Food Additives). WHO Technical Report Series, No. 939, 2006.
182. *Residue evaluation of certain veterinary drugs.* FAO JECFA Monographs 2, 2006.
183. *Toxicological evaluation of certain veterinary drug residues in food.* WHO Food Additives Series, No. 57, 2006.
184. *Evaluation of certain food additives and contaminants* (Sixty-seventh report of the Joint FAO/WHO Expert Committee on Food Additives). WHO Technical Report Series, No. 940, 2007.
185. *Compendium of food additive specifications.* FAO JECFA Monographs 3, 2006.
186. *Safety evaluation of certain food additives and contaminants.* WHO Food Additives Series, No. 58, 2007.
187. *Evaluation of certain food additives and contaminants* (Sixty-eighth report of the Joint FAO/WHO Expert Committee on Food Additives). WHO Technical Report Series, No. 947, 2007.
188. *Safety evaluation of certain food additives and contaminants.* WHO Food Additives Series, No. 59, 2008.
189. *Compendium of food additive specifications.* FAO JECFA Monographs 4, 2007.
190. *Evaluation of certain food additives* (Sixty-ninth report of the Joint FAO/WHO Expert Committee on Food Additives). WHO Technical Report Series, No. 952, 2009.
191. *Safety evaluation of certain food additives.* WHO Food Additives Series, No. 60, 2009.
192. *Compendium of food additive specifications.* FAO JECFA Monographs 5, 2009.
193. *Evaluation of certain veterinary drug residues in food* (Seventieth report of the Joint FAO/WHO Expert Committee on Food Additives). WHO Technical Report Series, No. 954, 2009.
194. *Toxicological evaluation of certain veterinary drug residues in food.* WHO Food Additives Series, No. 61, 2009.
195. *Residue evaluation of certain veterinary drugs.* FAO JECFA Monographs 6, 2009.
196. *Evaluation of certain food additives* (Seventy-first report of the Joint FAO/WHO Expert Committee on Food Additives). WHO Technical Report Series, No. 956, 2010.
197. *Safety evaluation of certain food additives.* WHO Food Additives Series, No. 62, 2010.
198. *Compendium of food additive specifications.* FAO JECFA Monographs 7, 2009.

199. *Evaluation of certain contaminants in food* (Seventy-second report of the Joint FAO/WHO Expert Committee on Food Additives). WHO Technical Report Series, No. 959, 2011.
200. *Safety evaluation of certain contaminants in food.* WHO Food Additives Series, No. 63/FAO JECFA Monographs 8, 2011.
201. *Residue evaluation of certain veterinary drugs.* FAO JECFA Monographs 9, 2010.
202. *Evaluation of certain food additives and contaminants* (Seventy-third report of the Joint FAO/WHO Expert Committee on Food Additives). WHO Technical Report Series, No. 960, 2011.
203. *Safety evaluation of certain food additives and contaminants.* WHO Food Additives Series, No. 64, 2011.
204. *Compendium of food additive specifications.* FAO JECFA Monographs 10, 2010.
205. *Evaluation of certain food additives and contaminants* (Seventy-fourth report of the Joint FAO/WHO Expert Committee on Food Additives). WHO Technical Report Series, No. 966, 2011.
206. *Safety evaluation of certain food additives and contaminants.* WHO Food Additives Series, No. 65, 2011.
207. *Compendium of food additive specifications.* FAO JECFA Monographs 11, 2011.
208. *Evaluation of certain veterinary drug residues in food* (Seventy-fifth report of the Joint FAO/WHO Expert Committee on Food Additives). WHO Technical Report Series, No. 969, 2012.
209. *Toxicological evaluation of certain veterinary drug residues in food.* WHO Food Additives Series, No. 66, 2012.
210. *Residue evaluation of certain veterinary drugs.* FAO JECFA Monographs 12, 2012.
211. *Evaluation of certain food additives* (Seventy-sixth report of the Joint FAO/WHO Expert Committee on Food Additives). WHO Technical Report Series, No. 974, 2012.
212. *Safety evaluation of certain food additives.* WHO Food Additives Series, No. 67, 2012.
213. *Compendium of food additive specifications.* FAO JECFA Monographs 13, 2012.

ANNEX 2

ABBREVIATIONS USED IN THE MONOGRAPHS

ADI	acceptable daily intake
ALT	alanine aminotransferase
APTT	activated partial thromboplastin time
AST	aspartate aminotransferase
AUC	area under the plasma concentration–time curve
BrdU	bromodeoxyuridine
bw	body weight
CL	clearance
C_{max}	peak concentration in plasma
DAPI	4′,6′-diamidino-2-phenylindole hydrochloride
DNA	deoxyribonucleic acid
dUTP	2′-deoxyuridine 5′-triphosphate
EFSA	European Food Safety Authority
EU	European Union
F	female; bioavailability
FAO	Food and Agriculture Organization of the United Nations
FTU	phytase unit
GLP	good laboratory practice
GSFA	Codex General Standard for Food Additives
GSH	glutathione
GST	glutathione S-transferase
INS	International Numbering System
IUIS	International Union of Immunological Societies
JECFA	Joint FAO/WHO Committee on Food Additives
KMTU	Kilo Microbial Trypsin Unit
LD_{50}	median lethal dose
M	male
MCHC	mean corpuscular haemoglobin concentration
MSDI	maximized survey-derived intake
MTDI	maximum tolerable daily intake
N^2-MFdG	N^2-[(furan-2-yl)methyl]-2′-deoxyguanosine
N^6-MFdA	N^6-[(furan-2-yl)methyl]-2′-deoxyadenosine
NA	not applicable
NADPH	nicotinamide adenine dinucleotide phosphate (reduced)
ND	not determined
NOAEL	no-observed-adverse-effect level
NOEL	no-observed-effect level
OECD	Organisation for Economic Co-operation and Development
PCE	polychromatic erythrocyte
PROT	protease unit
S9	9000 × g supernatant fraction of rat liver homogenate
SCE	sister chromatid exchange
SPET	single portion exposure technique
SULT	sulfotransferase

$t_{1/2}$	half-life
T_{max}	time to C_{max}
TOS	total organic solids
TUNEL	terminal deoxynucleotidyl transferase–mediated 2′-deoxyuridine 5′-triphosphate nick-end labelling
UDS	unscheduled DNA synthesis
USA	United States of America
USFDA	United States Food and Drug Administration
V_{ss}	volume of distribution at steady state
WHO	World Health Organization
w/w	weight per weight

ANNEX 3

JOINT FAO/WHO EXPERT COMMITTEE ON FOOD ADDITIVES

Geneva, 5–14 June 2012

MEMBERS

Professor J. Alexander, Norwegian Institute of Public Health, Oslo, Norway

Dr M. DiNovi, Center for Food Safety and Applied Nutrition, Food and Drug Administration, College Park, MD, USA

Dr Y. Kawamura, Division of Food Additives, National Institute of Health Sciences, Tokyo, Japan

Dr Madduri Veerabhadra Rao, Department of the President's Affairs, Al Ain, United Arab Emirates

Dr A. Mattia, Center for Food Safety and Applied Nutrition, Food and Drug Administration, College Park, MD, USA (*Chairperson*)

Mrs I. Meyland, Birkerød, Denmark (*Vice-Chairperson*)

Dr U. Mueller, Food Standards Australia New Zealand, Barton, ACT, Australia

Professor A. Renwick, Emeritus Professor, Faculty of Medicine, University of Southampton, Southampton, England (*Joint Rapporteur*)

Dr J. Schlatter, Nutritional and Toxicological Risks Section, Swiss Federal Office of Public Health, Zurich, Switzerland

Dr P. Sinhaseni, Community Risk Analysis Research and Development Center, Bangkok, Thailand

Mrs H. Wallin, Finnish Food Safety Authority (Evira), Helsinki, Finland (*Joint Rapporteur*)

SECRETARIAT

Dr D. Benford, Food Standards Agency, London, England (*WHO Expert*)

Dr A. Bruno, Joint FAO/WHO Food Standards Programme, Food and Agriculture Organization of the United Nations, Rome, Italy (*FAO Codex Secretariat*)

Dr S. Choudhuri, Center for Food Safety and Applied Nutrition, Food and Drug Administration, College Park, MD, USA (*WHO Expert*)

Professor L.R.M. dos Santos, Faculdade de Engenharia de Alimentos, Universidade Estadual de Campinas, São Paulo, Brazil (*FAO Expert*)

Dr R. Ellis, Nutrition and Consumer Protection Division, Food and Agriculture Organization of the United Nations, Rome, Italy (*FAO Joint Secretary*)

Dr B. Fields, Food Standards Australia New Zealand, Canberra, ACT, Australia (*WHO Expert*)

Dr D. Folmer, Center for Food Safety and Applied Nutrition, Food and Drug Administration, College Park, MD, USA (*FAO Expert*)

Dr S.M.F. Jeurissen, Centre for Substances and Integrated Risk Assessment, National Institute for Public Health and the Environment, Bilthoven, the Netherlands (*WHO Expert*)

Dr F. Kayama, School of Medicine, Jichi Medical University, Tochigi, Japan (*WHO Expert*)

Dr J.-C. Leblanc, Agence nationale de sécurité sanitaire de l'alimentation, de l'environnement et du travail (ANSES), Maisons-Alfort, France (*FAO Expert*)

Dr S.Y. Lee, Department of Food Safety and Zoonoses, World Health Organization, Geneva, Switzerland (*WHO Staff Member*)

Professor S.M. Mahungu, Department of Dairy, Food Science and Technology, Egerton University, Egerton, Kenya (*FAO Expert*)

Professor S. Rath, Department of Analytical Chemistry, University of Campinas, Campinas, São Paulo, Brazil (*FAO Expert*)

Ms M. Sheffer, Ottawa, Canada (*WHO Editor*)

Professor I.G. Sipes, College of Medicine, University of Arizona, Tucson, AZ, USA (*WHO Expert*)

Dr J.R. Srinivasan, Center for Food Safety and Applied Nutrition, Food and Drug Administration, College Park, MD, USA (*FAO Expert*)

Dr A. Tritscher, Department of Food Safety and Zoonoses, World Health Organization, Geneva, Switzerland (*WHO Joint Secretary*)

Dr T. Umemura, Biological Safety Research Center, National Institute of Health Sciences, Tokyo, Japan (*WHO Expert*)

Professor G.M. Williams, Department of Pathology, New York Medical College, Valhalla, NY, USA (*WHO Expert*)

ANNEX 4

TOXICOLOGICAL INFORMATION AND INFORMATION ON SPECIFICATIONS

Food additives considered for specifications only

Food additive	Specifications[a]
Ethyl cellulose	R
Mineral oil (medium viscosity)	N[b]
Modified starches	R
Titanium dioxide	R

[a] N, new specifications; R, existing specifications revised.
[b] The existing specifications for mineral oil (medium and low viscosity) were withdrawn (see below).

Food additives evaluated toxicologically and assessed for dietary exposure

Food additive	Specifications[a]	Acceptable or tolerable daily intakes and other toxicological recommendations
Magnesium dihydrogen diphosphate	N	Although an acceptable daily intake (ADI) "not specified"[b] has been established for a number of magnesium salts used as food additives, the estimated chronic dietary exposures to magnesium (960 mg/day for a 60 kg adult at the 95th percentile) from the proposed uses of magnesium dihydrogen diphosphate are up to twice the background exposures from food previously noted by the Committee (180–480 mg/day) and in the region of the minimum laxative effective dose of approximately 1000 mg of magnesium when taken as a single dose. The estimates of dietary exposure to phosphorus from the proposed uses of magnesium dihydrogen diphosphate were in the region of, or slightly exceeded, the maximum tolerable daily intake (MTDI) of 70 mg/kg body weight (bw) for phosphate salts, expressed as phosphorus, from this source alone. Thus, the MTDI is further exceeded when other sources of phosphate in the diet are taken into account. **The Committee therefore concluded that the proposed use levels and food categories result in an estimated dietary exposure to magnesium dihydrogen diphosphate that is of potential concern.** The Committee emphasized that in evaluating individual phosphate-containing food additives, there is a need for assessment of total dietary exposure to phosphorus. The Committee recommended that total dietary exposure to magnesium from food additives and other sources in the diet should be assessed.

Food additive	Specifications[a]	Acceptable or tolerable daily intakes and other toxicological recommendations
		The information submitted to the Committee and in the scientific literature did not indicate that the MTDI of 70 mg/kg bw for phosphate salts, expressed as phosphorus, is insufficiently health protective. On the contrary, because the basis for its derivation might not be relevant to humans, it could be overly conservative. Therefore, **the Committee recommended that the toxicological basis of the MTDI for phosphate salts expressed as phosphorus be reviewed.**
Mineral oil (medium and low viscosity) classes II and III	W	**The Committee concluded that the newly submitted data did not adequately address its previous requests for information** on the relevance to humans of the response of F344 and Sprague-Dawley rats to mineral oil (medium and low viscosity) classes II and III. The studies were conducted with a single administration, and it was not possible to predict the concentration in the target organ (liver) at steady state, or the potential for accumulation, in humans. Information requested at the forty-fourth meeting on compositional factors of mineral oils that influence absorption and toxicity had not been provided for materials meeting the criteria of mineral oil (medium and low viscosity) classes II and III.

The Committee noted that hydrocarbon deposits with carbon numbers consistent with mineral oils, including those of classes II and III, and associated lesions have been reported in human tissues, demonstrating the potential relevance to humans of the effects in the F344 rat. Because all blood levels were below the limit of detection in the single-dose human toxicokinetic study, it was not possible to reach conclusions on the rate of elimination of mineral oils in humans or on the concentration in the liver at steady state following prolonged exposure. Therefore, the new data did not provide information that would allow an ADI to be established based on internal exposure.

Similarly, it was not possible to establish an ADI based on external dose in the absence of information on the relative accumulation potential of classes II and III mineral oils in humans compared with rats.

The Committee noted that the temporary group ADI for mineral oil (medium and low viscosity) classes II and III had been established in 1995 and extended on a number of occasions. As data supporting the establishment of a full ADI had not been made available, **the previously established temporary group ADI was withdrawn.** |

Food additive	Specifications[a]	Acceptable or tolerable daily intakes and other toxicological recommendations
3-Phytase from *Aspergillus niger* expressed in *Aspergillus niger*	N	Comparing the conservative exposure estimate with the no-observed-adverse-effect level (NOAEL) from the 13-week study of oral toxicity in rats, the margin of exposure is approximately 250. **The Committee allocated an ADI "not specified"[b] for 3-phytase enzyme preparation from *A. niger* expressed in *A. niger*, used in the applications specified and in accordance with good manufacturing practice.**
Serine protease (chymotrypsin) from *Nocardiopsis prasina* expressed in *Bacillus licheniformis*	N	Comparing the exposure estimate with the NOAEL from the 13-week study of oral toxicity in rats, the margin of exposure is approximately 350. **The Committee allocated an ADI "not specified"[b] for serine protease (chymotrypsin) enzyme preparation from *N. prasina* expressed in the production strain *B. licheniformis*, used in the applications specified and in accordance with good manufacturing practice.**
Serine protease (trypsin) from *Fusarium oxysporum* expressed in *Fusarium venenatum*	N	Comparing the dietary exposure estimate with the NOAEL from the 13-week study of oral toxicity in rats, the margin of exposure is approximately 1200. **The Committee allocated an ADI "not specified"[b] for serine protease (trypsin) enzyme preparation from *F. oxysporum* expressed in the production strain *F. venenatum*, used in the applications specified and in accordance with good manufacturing practice.**

[a] N, new specifications; W, existing specifications withdrawn.
[b] ADI "not specified" is used to refer to a food substance of very low toxicity that, on the basis of the available data (chemical, biochemical, toxicological and other) and the total dietary exposure to the substance arising from its use at the levels necessary to achieve the desired effects and from its acceptable background levels in food, does not, in the opinion of the Committee, represent a hazard to health. For that reason, and for the reasons stated in the individual evaluations, the establishment of an ADI expressed in numerical form is not deemed necessary. An additive meeting this criterion must be used within the bounds of good manufacturing practice—i.e. it should be technologically efficacious and should be used at the lowest level necessary to achieve this effect, it should not conceal food of inferior quality or adulterated food, and it should not create a nutritional imbalance.

Flavouring agents evaluated by the Procedure for the Safety Evaluation of Flavouring Agents[1]

A. Aliphatic and aromatic amines and amides

Flavouring agent	No.	Specifications[a]	Conclusion based on current estimated dietary exposure
Structural class I			
2-Aminoacetophenone	2043	N	No safety concern
Structural class III			
(2E,6E/Z,8E)-N-(2-Methylpropyl)-2,6,8-decatrienamide	2077	N	No safety concern
(2S,5R)-N-[4-(2-Amino-2-oxoethyl)-phenyl]-5-methyl-2-(propan-2-yl)-cyclohexanecarboxamide	2078	N	No safety concern
(1R,2S,5R)-N-(4-Methoxyphenyl)-5-methyl-2-(1-methylethyl)cyclohexanecarboxamide	2079	N	No safety concern
N-Cyclopropyl-5-methyl-2-isopropylcyclohexanecarboxamide	2080	N	No safety concern
N-(2-Methylcyclohexyl)-2,3,4,5,6-pentafluorobenzamide	2081	N	No safety concern
3[(4-Amino-2,2-dioxido-1H-2,1,3-benzothiadiazin-5-yl)oxy]-2,2-dimethyl-N-propylpropanamide	2082	N	No safety concern

[a] N, new specifications.

---

[1] The flavouring agent **2-phenyl-2-methyl-2-hexenal (No. 2069)** was submitted for evaluation in the group of aliphatic linear α,β-unsaturated aldehydes, acids and related alcohols, acetals and esters; the Committee considered that it did not belong to this group of flavouring agents, and therefore it was not further considered. The safety of the submitted substance **(3R)-4-[[(1S)-1-benzyl-2-methoxy-2-oxo-ethyl]amino]-3-[3-(3-hydroxy-4-methoxyphenyl)propylamino]-4-oxo-butanoic acid hydrate (Advantame, No. 2124)** in the group of amino acids and related substances was not assessed; the Committee decided that it would not be appropriate to evaluate this substance as a flavouring agent, because it is a low-calorie intense sweetener. The safety of the two submitted substances **rebaudioside C (No. 2168)** and **rebaudioside A (No. 2169)** in the group of phenol and phenol derivatives was not assessed; the Committee decided that it would not be appropriate to evaluate these substances as flavouring agents, as they had already been evaluated as food additives (sweeteners).

B. Aliphatic and aromatic ethers

Flavouring agent	No.	Specifications[a]	Conclusion based on current estimated dietary exposure
Structural class II			
3,6-Dimethyl-2,3,3a,4,5,7a-hexahydrobenzofuran	2133	N	No safety concern
Ethyl linalyl ether	2134	N	No safety concern
Linalool oxide pyranoid	2135	N	No safety concern
Nerolidol oxide	2137	N	**Additional data required to complete evaluation**
Methyl hexyl ether	2138	N	No safety concern
Myrcenyl methyl ether	2139	N	No safety concern
Digeranyl ether	2142	N	No safety concern
Structural class III			
Isoamyl phenethyl ether	2136	N	No safety concern
5-Isopropyl-2,6-diethyl-2-methyltetrahydro-2H-pyran	2140	N	No safety concern
Butyl β-naphthyl ether	2141	N	No safety concern

[a] N, new specifications.

C. Aliphatic hydrocarbons, alcohols, aldehydes, ketones, carboxylic acids and related esters, sulfides, disulfides and ethers containing furan substitution

The Committee concluded that the Procedure could not be applied to this group because of unresolved toxicological concerns. Studies that could assist in the safety evaluation include investigations of the influence of the nature and position of furan ring substitutions on metabolism and covalent binding to macromolecules, demonstration of the ring opening and reactivity of the resulting products. Depending on the findings, additional genotoxicity or other studies might be needed.

Flavouring agent	No.	Specifications[a]
2-Pentylfuran	1491	M
2-Heptylfuran	1492	M
2-Decylfuran	1493	M
3-Methyl-2-(3-methylbut-2-enyl)-furan	1494	M
3-(2-Furyl)acrolein	1497	M
3-(5-Methyl-2-furyl)prop-2-enal	1499	M
2-Furyl methyl ketone	1503	M
2-Acetyl-5-methylfuran	1504	M

Flavouring agent	No.	Specifications[a]
2-Acetyl-3,5-dimethylfuran	1505	M
2-Butyrylfuran	1507	M
(2-Furyl)-2-propanone	1508	M
2-Pentanoylfuran	1509	M
1-(2-Furyl)butan-3-one	1510	M
4-(2-Furyl)-3-buten-2-one	1511	M
Ethyl 3-(2-furyl)propanoate	1513	M
Isobutyl 3-(2-furan)propionate	1514	M
Isoamyl 3-(2-furan)propionate	1515	M
Isoamyl 4-(2-furan)butyrate	1516	M
Phenethyl 2-furoate	1517	M
Furfuryl methyl ether	1520	M
Ethyl furfuryl ether	1521	M
Difurfuryl ether	1522	M
2,5-Dimethyl-3-furanthiol acetate	1523	M
Furfuryl 2-methyl-3-furyl disulfide	1524	M
3-[(2-Methyl-3-furyl)thio]-2-butanone	1525	M
O-Ethyl S-(2-furylmethyl)thiocarbonate	1526	M
2,3-Dimethylbenzofuran	1495	M
2,4-Difurfurylfuran	1496	M
2-Methyl-3(2-furyl)acrolein	1498	M
3-(5-Methyl-2-furyl)-butanal	1500	M
2-Furfurylidene-butyraldehyde	1501	M
2-Phenyl-3-(2-furyl)prop-2-enal	1502	M
3-Acetyl-2,5-dimethylfuran	1506	M
Pentyl 2-furyl ketone	1512	M
Propyl 2-furanacrylate	1518	M
2,5-Dimethyl-3-oxo-(2H)-fur-4-yl butyrate	1519	M
(E)-Ethyl 3-(2-furyl)acrylate	2103	N
Di-2-furylmethane	2104	N
2-Methylbenzofuran	2105	N

[a] M, specifications maintained; N, new specifications.

D. Aliphatic linear α,β-unsaturated aldehydes, acids and related alcohols, acetals and esters

Flavouring agent	No.	Specifications[a]	Conclusion based on current estimated dietary exposure
Structural class I			
trans-2-Nonenyl acetate	2163	N	No safety concern
Propyl sorbate	2164	N	No safety concern
cis-2-Octenol	2165	N	No safety concern
trans-2-Tridecenol	2166	N	No safety concern
Ethyl 2-hexenoate (mixture of isomers)	2167	N	No safety concern

[a] N, new specifications.

E. Amino acids and related substances

Flavouring agent	No.	Specifications[a]	Conclusion based on current estimated dietary exposure
Structural class I			
L-Ornithine (as the monochlorohydrate)	2120	N	No safety concern
L-Alanyl-L-glutamine	2121	N	No safety concern
L-Methionylglycine	2122	N	No safety concern
Glutamyl-valyl-glycine	2123	N	No safety concern

[a] N, new specifications.

The Committee considered that the use of the Procedure for the Safety Evaluation of Flavouring Agents was inappropriate for two members of this group—namely, L-isoleucine (No. 2118) and L-threonine (No. 2119). In view of the fact that these substances are macronutrients and normal components of protein, the Committee concluded that the use of these substances as flavouring agents would not raise any safety concerns at current estimated dietary exposures.

Flavouring agent	No.	Specifications[a]
L-Isoleucine	2118	N
L-Threonine	2119	N

[a] N, new specifications.

F. Epoxides

Flavouring agent	No.	Specifications[a]	Conclusion based on current estimated dietary exposure
Structural class III			
Ethyl α-ethyl-β-methyl-β-phenylglycidate	2143	N	No safety concern
Methyl β-phenylglycidate	2144	N	No safety concern
d-8-p-Menthene-1,2-epoxide	2145	N	No safety concern
l-8-p-Menthene-1,2-epoxide	2146	N	No safety concern
2,3-Epoxyoctanal	2147	N	**Additional data required to complete evaluation**
2,3-Epoxyheptanal	2148	N	**Additional data required to complete evaluation**
2,3-Epoxydecanal	2149	N	Additional data required to complete evaluation

[a] N, new specifications.

G. Furfuryl alcohol and related substances

New in vitro and in vivo studies raise concerns regarding the potential genotoxicity of furfuryl alcohol and derivatives that can be metabolized to furfuryl alcohol (e.g. furfuryl esters). The Committee concluded that this group of flavouring agents could not be evaluated according to the Procedure because of the unresolved concerns regarding genotoxicity. In addition, the group ADI previously established by the Committee will need to be reconsidered at a future meeting.

Flavouring agent	No.	Specifications[a]
5-Methylfurfuryl alcohol	2099	N
Furfural propyleneglycol acetal	2100	N
Furfuryl formate	2101	N
Furfuryl decanoate	2102	N

[a] N, new specifications.

H. Linear and branched-chain aliphatic, unsaturated, unconjugated alcohols, aldehydes, acids and related esters

Flavouring agent	No.	Specifications[a]	Conclusion based on current estimated dietary exposure
Structural class I			
cis-3-Nonen-1-ol	2177	N	No safety concern
trans-3-Nonen-1-ol	2178	N	No safety concern
cis,cis-3,6-Nonadienyl acetate	2179	N	No safety concern
trans-3-Hexenyl acetate	2180	N	No safety concern
cis-3-Hexenoic acid	2181	N	No safety concern
cis-3-Nonenyl acetate	2182	N	No safety concern
cis-6-Nonenyl acetate	2183	N	No safety concern
(Z)-5-Octenyl acetate	2184	N	No safety concern
(E)-4-Undecenal	2185	N	No safety concern

[a] N, new specifications.

I. Miscellaneous nitrogen-containing compounds

Flavouring agent	No.	Specifications[a]	Conclusion based on current estimated dietary exposure
Structural class II			
3-(1-((3,5-Dimethylisoxazol-4-yl)methyl)-1H-pyrazol-4-yl)-1-(3-hydroxybenzyl)-imidazolidine-2,4-dione	2161	N	No safety concern
3-(1-((3,5-Dimethylisoxazol-4-yl)methyl)-1H-pyrazol-4-yl)-1-(3-hydroxybenzyl)-5,5-dimethylimidazolidine-2,4-dione	2162	N	No safety concern

[a] N, new specifications.

J. Phenol and phenol derivatives

Flavouring agent	No.	Specifications[a]	Conclusion based on current estimated dietary exposure
Structural class III			
3′,7-Dihydroxy-4′-methoxyflavan	2170	N	No safety concern
Trilobatin	2171	N	No safety concern
(±)-Eriodictyol	2172	N	No safety concern

[a] N, new specifications.

K. Pyrazine derivatives

Flavouring agent	No.	Specifications[a]	Conclusion based on current estimated dietary exposure
Structural class II			
Isopropenylpyrazine	2125	N	No safety concern
5-Ethyl-2,3-dimethylpyrazine	2126	N	No safety concern
2-Methyl-5-vinylpyrazine	2127	N	No safety concern
Mixture of 2,5-dimethyl-6,7-dihydro-5H-cyclopentapyrazine and 2,7-dimethyl-6,7-dihydro-5H-cyclopentapyrazine	2128	N	No safety concern
2-Ethoxy-3-isopropylpyrazine	2065	N	No safety concern
Structural class III			
Mixture of 3,5-dimethyl-2-isobutylpyrazine and 3,6-dimethyl-2-isobutylpyrazine	2130	N	No safety concern
2-Ethoxy-3-ethylpyrazine	2131	N	No safety concern
2-Ethyl-3-methylthiopyrazine	2132	N	No safety concern

[a] N, new specifications.

L. Pyridine, pyrrole and quinoline derivatives

Flavouring agent	No.	Specifications[a]	Conclusion based on current estimated dietary exposure
Structural class II			
1-Ethyl-2-pyrrolecarboxaldehyde	2150	N	Additional data required to complete evaluation
2,4-Dimethylpyridine	2151	N	No safety concern (temporary)[b]
1-Methyl-1H-pyrrole-2-carboxaldehyde	2152	N	Additional data required to complete evaluation
Structural class III			
2-Acetyl-4-isopropenylpyridine	2153	T	No safety concern
4-Acetyl-2-isopropenylpyridine	2154	T	No safety concern
2-Acetyl-4-isopropylpyridine	2155	N	No safety concern
2-Methoxypyridine	2156	N	Additional data required to complete evaluation
6-Methoxyquinoline	2157	N	No safety concern
1-(2-Hydroxyphenyl)-3-(pyridin-4-yl)-propan-1-one	2158	N	Additional data required to complete evaluation

Flavouring agent	No.	Specifications[a]	Conclusion based on current estimated dietary exposure
1-(2-Hydroxy-4-isobutoxyphenyl)-3-(pyridin-2-yl)propan-1-one	2159	N	Additional data required to complete evaluation
1-(2-Hydroxy-4-methoxyphenyl)-3-(pyridin-2-yl)propan-1-one	2160	N	Additional data required to complete evaluation

[a] N, new specifications; T, tentative specifications.
[b] The evaluation for No. 2151 is temporary pending receipt of additional toxicological data.

M. Saturated aliphatic acyclic branched-chain primary alcohols, aldehydes and acids

Flavouring agent	No.	Specifications[a]	Conclusion based on current estimated dietary exposure
Structural class I			
3-Methylhexanal	2173	N	No safety concern
6-Methylheptanal	2174	N	No safety concern
6-Methyloctanal	2175	N	No safety concern
3,7-Dimethyloctanal	2176	N	No safety concern

[a] N, new specifications.

N. Simple aliphatic and aromatic sulfides and thiols

Flavouring agent	No.	Specifications[a]	Conclusion based on current estimated dietary exposure
Subgroup ii: Acyclic sulfides with oxidized side-chains			
Structural class I			
1-(Methylthio)-3-octanone	2086	N	No safety concern
Subgroup iii: Cyclic sulfides			
Structural class III			
4-Methyl-2-propyl-1,3-oxathiane	2089	N	No safety concern
Subgroup iv: Simple thiols			
Structural class I			
3-Pentanethiol	2083	N	No safety concern

Flavouring agent	No.	Specifications[a]	Conclusion based on current estimated dietary exposure
Subgroup v: Thiols with oxidized side-chains			
Structural class I			
4-Mercapto-3-methyl-2-butanol	2084	N	No safety concern
Ethyl 2-mercapto-2-methylpropionate	2085	N	No safety concern
Subgroup vi: Dithiols			
Structural class III			
1,1-Propanedithiol	2087	N	No safety concern
Subgroup viii: Disulfides with oxidized side-chains			
Structural class III			
1-Methyldithio-2-propanone	2088	N	No safety concern

[a] N, new specifications.

O. Sulfur-containing heterocyclic compounds

Flavouring agent	No.	Specifications[a]	Conclusion based on current estimated dietary exposure
Structural class II			
2-Pentylthiophene	2106	N	No safety concern
2-Acetyl-5-methylthiophene	2107	N	No safety concern
2-Pentylthiazole	2108	N	No safety concern
4,5-Dimethyl-2-isobutylthiazole	2109	N	No safety concern
Structural class III			
3,4-Dimethylthiophene	2110	N	No safety concern
2-Thienylmethanol	2111	N	No safety concern
1-(2-Thienyl)ethanethiol	2112	N	No safety concern
5-Ethyl-2-methylthiazole	2113	N	No safety concern
2-Ethyl-2,5-dihydro-4-methylthiazole	2114	N	No safety concern
4-Methyl-3-thiazoline	2115	N	No safety concern
2-Ethyl-4,6-dimethyldihydro-1,3,5-dithiazine	2116	N	No safety concern
4-Amino-5,6-dimethylthieno[2,3-d]-pyrimidin-2(1H)-one hydrochloride	2117	N	No safety concern

[a] N, new specifications.

P. Sulfur-substituted furan derivatives

Flavouring agent	No.	Specifications[a]	Conclusion based on current estimated dietary exposure
Structural class III			
5-Methylfurfuryl mercaptan	2090	N	No safety concern
2-Methyl-3-furyl methylthiomethyl disulfide	2091	N	No safety concern
2-Methyl-3-furyl 2-methyl-3-tetrahydrofuryl disulfide	2092	N	No safety concern
2-Tetrahydrofurfuryl 2-mercaptopropionate	2093	N	**Additional data required to complete evaluation**
Methyl 3-(furfurylthio)propionate	2094	N	No safety concern
3-[(2-Methyl-3-furyl)thio]butanal	2095	N	No safety concern
1-(2-Furfurylthio)-propanone	2096	N	No safety concern
2-Methyl-4,5-dihydrofuran-3-thiol	2097	N	No safety concern
2-Methyltetrahydrofuran-3-thiol acetate	2098	N	No safety concern

[a] N, new specifications.

ANNEX 5

SUMMARY OF THE SAFETY EVALUATION OF SECONDARY COMPONENTS FOR FLAVOURING AGENTS WITH MINIMUM ASSAY VALUES OF LESS THAN 95%

JECFA No.	Flavouring agent	Minimum assay value	Secondary components	Comments on secondary components
Simple aliphatic and aromatic sulfides and thiols				
2088	1-Methyldithio-2-propanone	90	2–3% 1-mercapto-2-propanone; 2–3% 1,1′-disulfanediyldipropan-2-one; 1–3% 1,3-dimethyltrisulfane	1-Mercapto-2-propanone (No. 557) and 1,3-dimethyltrisulfane (No. 582) were evaluated by the Committee at its fifty-third meeting (Annex 1, reference *149*) and were concluded to be of no safety concern at estimated dietary exposures. 1,1′-Disulfanediyldipropan-2-one is expected to be hydrolysed to 1-mercapto-2-propanone (No. 557) and is therefore considered not to present a safety concern at current estimated dietary exposures to the flavouring agent.
Sulfur-substituted furan derivatives				
2097	2-Methyl-4,5-dihydrofuran-3-thiol	55	35–40% 2-methyl-3-tetrahydrofuranthiol; 5–7% 2-methyl-3-furanthiol	2-Methyl-3-tetrahydrofuranthiol (No. 1090) and 2-methyl-3-furanthiol (No. 1060) were evaluated by the Committee at the fifty-ninth meeting (Annex 1, reference *160*) and were concluded to be of no safety concern at estimated dietary exposures.
Sulfur-containing heterocyclic compounds				
2114	2-Ethyl-2,5-dihydro-4-methylthiazole	90	2–3% 2-ethyl-4-methyl-4,5-dihydrothiazole-4-ol; 2–3% 3,4-dimethylthiophene; 2–3% 2-ethyl-4-methylthiazole	2-Ethyl-4-methylthiazole (No. 1044) was evaluated by the Committee at the fifty-ninth meeting (Annex 1, reference *160*) and was concluded to be of no safety concern at estimated dietary exposures. 2-Ethyl-4-methyl-4,5-dihydrothiazole-4-ol is anticipated to undergo glucuronic acid conjugate formation and elimination in the urine. It does not present a safety concern at current estimated dietary exposures. 3,4-Dimethylthiophene (No. 2110) is a member of the current group and is expected to undergo side-chain oxidation and subsequent conjugation and elimination in the urine. It does not present a safety concern at current estimated dietary exposures to the flavouring agent.

No.	Name	%	Details
2116	2-Ethyl-4,6-dimethyldihydro-1,3,5-dithiazine	90	3–5% 3,5-diethyl-1,2,4-trithiolane and 2–3% 2,4,6-trimethyldihydro-4H-1,3,5-dithiazine	3,5-Diethyl-1,2,4-trithiolane (No. 1686) and 2,4,6-trimethyldihydro-4H-1,3,5-dithiazine (No. 1049) were evaluated at the sixty-eighth (Annex 1, reference *187*) and the fifty-ninth (Annex 1, reference *160*) meetings, respectively, and were concluded to be of no safety concern at estimated dietary exposures.

Aliphatic and aromatic ethers

| 2135 | Linalool oxide pyranoid | 92 | 3–5% linalool | Linalool (No. 356) was evaluated by the Committee at the fifty-first meeting (Annex 1, reference *137*) and was concluded to be of no safety concern at estimated dietary exposures. |

Epoxides

2144	Methyl β-phenylglycidate	85	10–12% ethyl β-phenylglycidate	Ethyl β-phenylglycidate (No. 1576) was evaluated by the Committee at the sixty-fifth meeting (Annex 1, reference *178*) and was concluded to be of no safety concern at estimated dietary exposures.
2148	2,3-Epoxyheptanal	94	2–3% *trans*-2-heptenal	*trans*-2-Heptenal (No. 1360) was evaluated by the Committee at the sixty-third meeting (Annex 1, reference *173*) and was concluded to be of no safety concern at estimated dietary exposures.
2149	2,3-Epoxydecanal	94	2–3% *trans*-2-decenal	*trans*-2-Decenal (No. 1349) was evaluated by the Committee at the sixty-third meeting (Annex 1, reference *173*) and was concluded to be of no safety concern at estimated dietary exposures.